# Advanced Physics Study Course

W. Bolton

* A comprehensive coverage of the A-Level Physics core
* Can be used for independent learning
* Over 1000 self-assessment questions, with answers and discussion

John Murray

First published 1986
by John Murray (Publishers) Ltd
50 Albemarle Street, London W1X 4BD



Typeset by Fakenham Photosetting Ltd
Fakenham, Norfolk
Printed and bound in Great Britain by
The Bath Press, Avon

British Library Cataloguing in Publication Data

Bolton, W. (William) *1933–*
Advanced physics study course.
1. Physics
I. Title
530 QC23

ISBN 0–7195–4252–9

# Preface

This book offers a graded course through Advanced Level Physics with two aims: success in the main examination and, almost as important, development of the techniques needed to study independently and successfully in Higher Education.

There are 60 *lessons* in eight chapters, representing about one lesson per week during a conventional two-year course, allowing time for revision. Each gives *aims*, *prerequisite knowledge*, *reading assignments* and graded *self-assessment questions*, followed by a summary of *key points* and a discussion of the *answers* to every self-assessment question. Together, the lessons cover the main syllabus requirements of all the Examining Boards. Although designed as a continuous course, the order of studies can be varied to suit institutional or individual needs.

## Grading of self-assessment questions

Self-assessment questions are presented at three levels.
Level One (unmarked): preparatory questions designed to develop understanding.
Level Two (marked*): questions of core A-Level standard that could appear as either part of or a whole examination question.
Level Three (marked**): either extensions of the subject beyond core A-Level, or of a greater difficulty than A-Level questions, aimed at those students preparing for Special or entrance papers.

## Acknowledgements

Thanks are due to Peter Knight for his advice, and to the artist Tony Langham.

# Contents

# References

*B* ST1, 2, 3, 4, 5, 6, 7 or 8 Bolton, W., *Study Topics in Physics*, 8 volumes (Butterworths, 1980).

*B* PP Bolton, W., *Patterns in Physics* (McGraw-Hill, 1974).

*D* Duncan, T., *Physics: A Textbook for Advanced Level Students* (John Murray, 1982).
References are to this combined version. If you are using the two-volume version, *Advanced Physics: Materials and Mechanics* and *Advanced Physics: Fields, Waves and Atoms*, ch. 1–10 refer to ch. 1–10 in *Materials and Mechanics*, and ch. 11–22 become ch. 1–12 in *Fields, Waves and Atoms*.

*N* Nelkon, M. and Parker, P., *Advanced Level Physics*, 5th ed. (Heinemann, 1982).

*W et al.* Wenham, E. J., Dorling, G. W., Snell, J. A. N. and Taylor, B., *Physics Concepts and Models* (Addison-Wesley).
References without brackets are to the 1st ed. (1972), those in brackets are to the 2nd ed. (1984).

*Wh & H* Whelan, P. M. and Hodgson, M. J., *Essential Principles of Physics* (John Murray, 1978).

# CHAPTER 1

# Mechanics

## Lesson 1 Equilibrium

### Aims

After completing this lesson you should be able to

- find the resultant of a number of coplanar forces acting at a point
- resolve a force into two components
- solve problems involving the equilibrium of coplanar forces.

## STUDY

**I Read** about determining the resultant of a number of coplanar forces and the resolution of a force into two components.
References: *B* ST1 ch. 3, *D* ch. 6, *N* ch. 4, *W et al.* ch. 3(2), *Wh&H* ch. 1.

**Self-assessment questions**

**1.1** Distinguish between *vector* and *scalar* quantities and give an example of each.

**1.2** Is force a vector or a scalar quantity?

**1.3** State the *parallelogram law* for finding the resultant of a pair of forces acting at a point.

**1.4** Forces of 10 N and 20 N act at the same point on an object. If the lines of action of the two forces are 60° apart, what is the magnitude and direction of the resultant?

**1.5** Determine the resultant force acting on an object if it is acted on by two forces of magnitudes 10 kN and 30 kN and the angle between them is 45°.

**1.6** If four forces in the same plane act at the same point on an object, how could you determine the resultant force?

**1.7** The terms *coplanar* and *concurrent* are sometimes used in problems involving forces. What do these terms mean?

**1.8** What is meant by 'resolving a force into its components'?

**1.9** A force of 40 N acts at an angle of 30° to the horizontal. What are the horizontal and vertical components of the force?

**1.10** A truck is on a hill where the ground is at an angle of 20° to the horizontal. If the gravitational force acting on the truck is 20 kN, what are the components of this force acting at right angles to the sloping ground and parallel to the ground (Figure 1.1)?

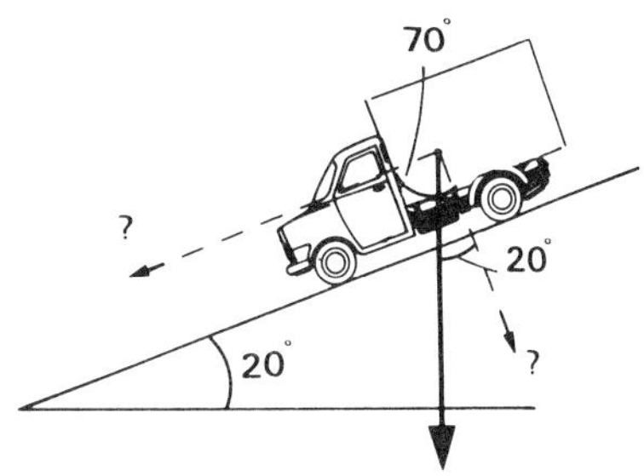


**Figure 1.1**

**II Read** about the moment of a force, torques and couples.
References: *B* ST2 ch. 2, *D* ch. 6, *N* ch. 4, *W et al.* (ch. 6), *Wh&H* ch. 7.

**Self-assessment questions**

**1.11** What is meant by (a) the moment of a force, (b) the torque of a force, (c) a couple and (d) the moment of a couple?

**1.12** A lever has a force of 200 N applied at a perpendicular distance of 0.6 m from the pivot. What is the moment of this force about the pivot?

**1.13** What is the torque produced on a nut by a force of 100 N being applied via a spanner, if the perpendicular distance between the line of action of the force and the axis of the nut is 0.4 m?

**1.14** A car's steering wheel has a diameter of 350 mm. A driver applies a force of 20 N with each hand to the wheel (Figure 1.2). If the hands hold the wheel at the opposite ends of the wheel diameter and are applying the forces tangentially to the wheel, what is the moment of the couple?

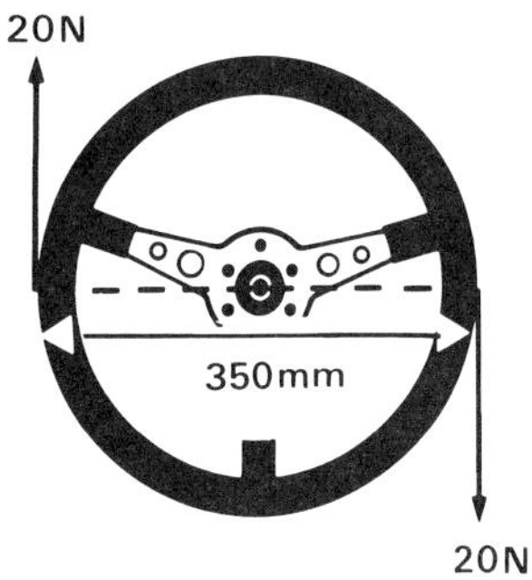


**Figure 1.2**

**III** **Read** about the equilibrium of coplanar forces, the conditions for such equilibrium and the solution of equilibrium problems.
References: *B* ST1 ch. 3, *D* ch. 6, *N* ch. 4, *W et al.* (ch. 6), *Wh&H* ch. 7.

**Self-assessment questions**

**1.15** What is the condition for there to be no rotation of a body?

**1.16** If three forces maintain a body in equilibrium, what can be said about their lines of action?

**1.17** State the *triangle of forces theorem.*

**1.18** If an object is in equilibrium what can be said about the sums of the resolved parts of the forces on the object in any two mutually perpendicular directions?

**1.19** An object is acted on by a force of 100 N pulling it in a northerly direction and a force of 50 N pulling it in an easterly direction. In what direction should a force be applied, and of what size, for the object to be in equilibrium?

**1.20** An object of mass 5.0 kg is suspended by two strings which are at angles of 30° and 70° to the vertical (Figure 1.3). What are the tensions in the strings? ($g = 9.8\ \mathrm{m\ s^{-2}}$)

**1.21** An object of mass 2 kg is hung from the 20 cm mark of a 100 cm long horizontal rod, and an object of mass 4 kg is hung from the 60 cm mark. Find the point at which a pivot should be placed so that the rod balances. Assume the rod to have insignificant mass.

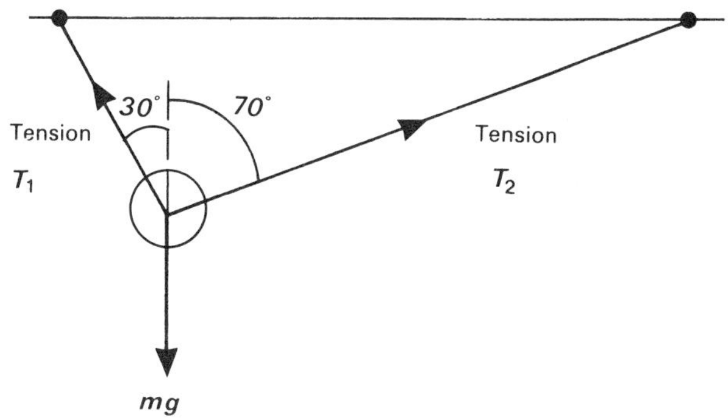


**Figure 1.3**

**1.22** What is meant by *centre of mass* (or *centre of gravity*)?

**1.23** If the rod in SAQ 1.21 is uniform and homogeneous, at what point will its centre of mass be? If the rod has a mass of 1 kg where will the balance point be, taking the mass of the rod into account?

***1.24** For the strut shown in Figure 1.4, what is the tension in the cable and the horizontal and vertical components of the force exerted by the wall on the strut? The strut has a mass of 10 kg and its centre of mass can be assumed to be at its centre. ($g = 9.8\ \mathrm{m\ s^{-2}}$)

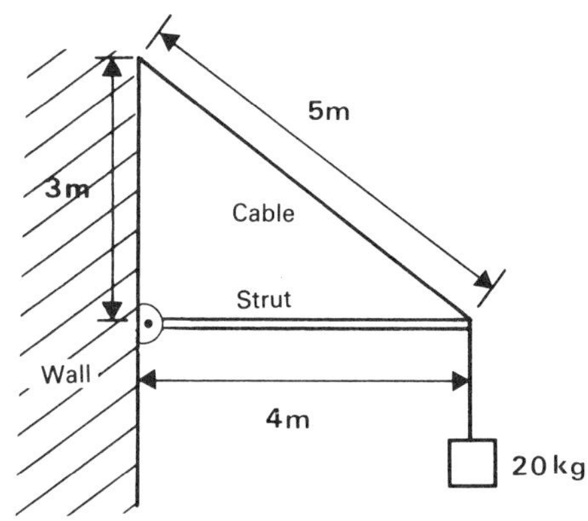


**Figure 1.4**

## SELF-ASSESSMENT

### Key points

**The parallelogram law:** if two forces acting at a point are represented in magnitude and direction by the adjacent sides of a parallelogram, then the resultant is represented by the diagonal of the parallelogram drawn from the junction of the two adjacent sides.

Any single force $F$ may be replaced by two forces mutually at right angles, these **components** being $F \sin \theta$ and $F \cos \theta$, where $\theta$ is the angle between the $F \cos \theta$ component and the force $F$.
The **moment**, or **torque**, of a force about an axis is the product of the force and the perpendicular distance from the line of action of the force to the axis.
A **couple** consists of two equal and oppositely directed parallel forces.
The **moment of a couple** is the product of one of the forces and the perpendicular distance between the forces.
If three forces are in equilibrium they must be **coplanar** and **concurrent**, and if the forces are represented in magnitude and direction by arrow-headed lines then these lines when taken in the sequence of the forces must form a triangle—the so-called **triangle of forces**. If more than three forces are involved then a polygon must be produced. The **general conditions for equilibrium** are: the sums of the resolved parts of the forces in any two perpendicular directions must be zero, and the sum of the clockwise moments about any axis must equal the sum of the anticlockwise moments about the same axis.

## Answers I

**1.1** With a scalar quantity only the magnitude has any relevance; with a vector quantity both the magnitude and the direction need to be stated for the effect of the quantity to be determined. Volume and mass are scalar quantities; velocity and force are vector quantities.

**1.2** vector

**1.3** See the key points.

**1.4** See Figure A1.1. The resultant is about 26.4 N at 41° to the 10 N force.

**1.5** See Figure A1.2. The resultant is about 37.7 kN at 10.8° to the 30 kN force.

**1.6** The resultant of a pair of them could be found by the use of the parallelogram law. Then this resultant could be combined with another force to give, using the parallelogram law, a new resultant. Finally this could be combined with the fourth force to give the final resultant.

**1.7** *Coplanar* means all in the same plane, *concurrent* means that all the lines of action of the forces pass through the same point.

**1.8** Replacing a single force by two forces mutually at right angles, the components of force $F$ being $F \cos \theta$ and $F \sin \theta$, where $\theta$ is the angle between the $F \cos \theta$ component and the force $F$.

**1.9** Horizontal component: $40 \cos 30° = 34.6$ N. Vertical component: $40 \sin 30° = 20.0$ N.

**1.10** At right angles to the slope: $20 \cos 20° = 18.8$ kN. Parallel to and down the slope: $20 \sin 20° = 6.8$ kN.

## Answers II

**1.11** See the key points.

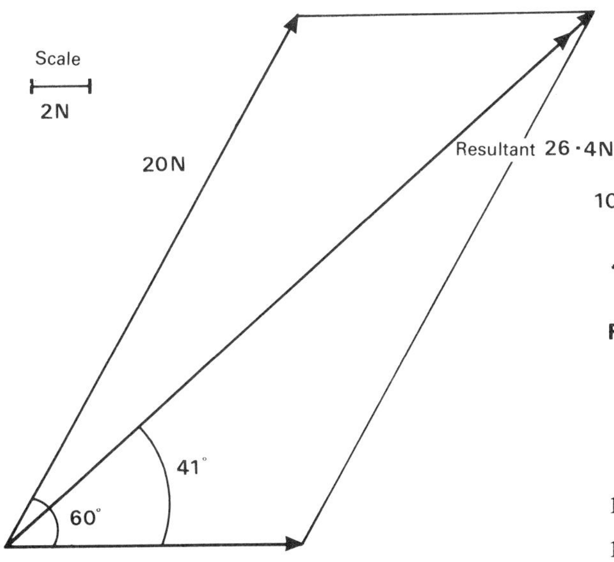

Figure A1.1

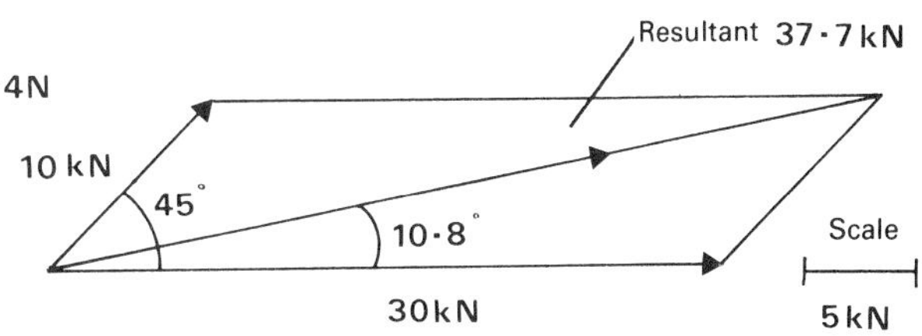

Figure A1.2

**1.12** moment $= 200 \times 0.6 = 120$ N m

**1.13** torque $= 100 \times 0.4 = 40$ N m

**1.14** moment of couple $= 20 \times 0.350 = 7.0$ N m

## Answers III

**1.15** The sum of the clockwise moments about any axis must equal the sum of the anticlockwise moments about the same axis.

**1.16** They must all pass through the same point, i.e. be concurrent.

**1.17** See the key points.

**1.18** The sums of the resolved parts of the forces in any two mutually perpendicular directions must be zero.

**1.19** See Figure A1.3. The required force is 111.8 N in a direction 63° south of west.

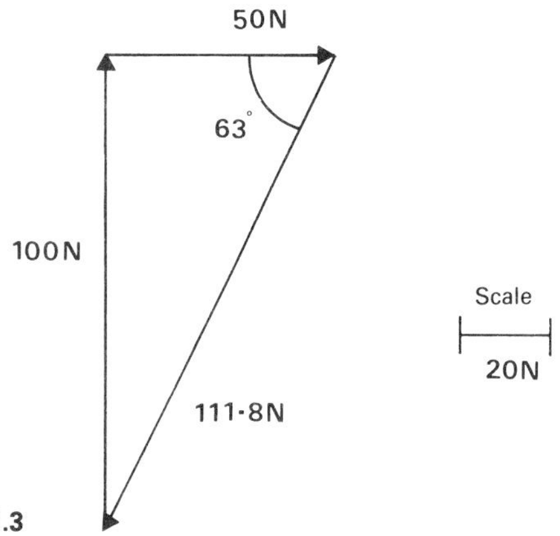


**Figure A1.3**

**1.20** The triangle of forces could be used or the forces resolved into the horizontal and vertical directions and the sum in each direction taken to be zero.
Figure A1.4 shows the triangle of forces method; a line of a length related to 5.0 × 9.8 N is drawn and then lines to represent $T_1$ and $T_2$ are drawn from each end of it at the required angles. $T_1$ and $T_2$ can then be obtained from measurement of the sides of the triangle.
Resolving horizontally: $T_1 \sin 30° = T_2 \sin 70°$.
Resolving vertically: $T_1 \cos 30° + T_2 \cos 70° = 5.0 \times 9.8$.
Hence, as $0.5T_1 = 0.94T_2$ we have $T_1 = 1.88T_2$.
Substituting this in the vertical component equation gives $T_2 = 24.9$ N. Hence $T_1 = 46.8$ N.

**1.21** The distance between the two objects is 40 cm, thus if the pivot is a distance $L$ from the 2 kg mass it will be $(40 - L)$ from the other mass. Taking moments about the pivot gives: $2gL = 4g(40 - L)$. Hence $L = 26.7$ cm.

**1.22** The *centre of mass* is the point at which all the mass of an object can be considered to act for the purpose of forces acting on the object. The *centre of gravity* is the single point through which the line of action of the weight passes, however the object is oriented. The centre of mass and the centre of gravity coincide for bodies on the surface of the earth.

**1.23** The centre of mass is at its centre point, i.e. the 50 cm mark. If the pivot is a distance $L$ from the 2 kg mass, taking moments about the pivot gives: $2gL + 1g(L - 30) = 4g(40 - L)$. Hence $L = 27.1$ cm.

**1.24** Taking moments about the strut at the wall and remembering that it is the distance to the line of action of the force that is required: $T \times (12/5) = 20g \times 4 + 10g \times 2$. Hence the tension in the cable $T = 408.3$ N. As both the vertical and the horizontal forces on the strut must total zero, the horizontal component of the force of the wall on the strut must be $408.3 \times (4/5) = 326.6$ N. For the vertical component, $F_v + 408.3 \times (3/5) = 20g + 10g$. Hence the vertical component is 49.0 N.

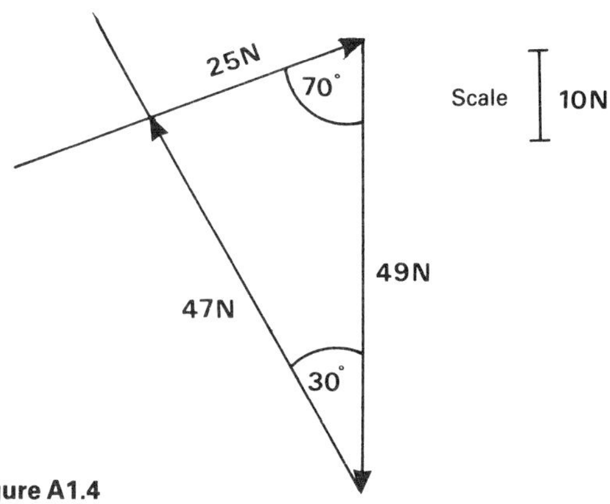


**Figure A1.4**

# Lesson 2 Linear motion

## Aims

After completing this lesson you should be able to

- define *average* and *instantaneous speed*, *velocity* and *acceleration*
- derive the equations for uniformly accelerated linear motion and solve problems involving them
- interpret distance–time and velocity–time graphs
- solve problems involving vertical motion under gravity.

# STUDY

**I Read** about definitions of average and instantaneous speed, velocity and acceleration, the equations for uniformly accelerated motion and distance–time and velocity–time graphs.
References: *B* PP ch. 1, *B* ST1 ch. 1, *D* ch. 6, *N* ch. 1, *Wh & H* ch. 2.

**Self-assessment questions**

**2.1** Define (a) average speed, (b) instantaneous speed, (c) average velocity, (d) instantaneous velocity, (e) average acceleration, (f) instantaneous acceleration.

**2.2** Give an example of where an object has a constant speed but not a constant velocity.

**2.3** When is (a) a velocity and (b) an acceleration uniform?

**2.4** An object moves in a straight line with a uniform acceleration. Derive the equation $v = u + at$ for this motion, starting from the definition of average acceleration, with $u$ as the initial velocity, $v$ the final velocity after a time $t$ when the uniform acceleration is $a$.

**2.5** Average velocity $= \frac{1}{2}(u + v)$ for motion in a straight line with uniform acceleration, $a$. Use this equation and the one derived in SAQ 2.4 to derive the equation $s = ut + \frac{1}{2}at^2$, where $s$ is the distance covered in time $t$.

**2.6** Combine the equations arrived at in SAQ 2.4 and SAQ 2.5 to obtain the relationship $v^2 = u^2 + 2as$.

**2.7** A car is accelerated with a uniform acceleration of $2.0\,m\,s^{-2}$, from a velocity of $10\,m\,s^{-1}$, for 4.0 s in a straight line. What is (a) the distance covered in that time and (b) the velocity at the end of that time?

**2.8** A car starts from rest and accelerates uniformly in a straight line in first gear until it attains a velocity of $10\,m\,s^{-1}$. If this takes 5.0 s calculate the acceleration. The motorist then changes to second gear and accelerates for a further 5.0 s until the velocity is $30\,m\,s^{-1}$. What is the acceleration when the car is in second gear? What is the total distance travelled by the car since it started from rest?

**2.9** A car travelling at $16\,m\,s^{-1}$ in a straight line is uniformly retarded until it comes to rest after a distance of 200 m. What was the retardation and the time taken? Note: the term *retardation* is used for a negative acceleration, i.e. an acceleration where the final velocity is less than the initial velocity.

**2.10** The term *distance* should be used when the distance referred to is measured along the path taken by the moving object, however the path twists and turns. The term *displacement* is the distance measured in a straight line between two points on the path of the object. What would therefore be given by the gradients of (a) a distance–time graph and (b) a displacement–time graph of the motion of an object?

**2.11** What is given by the gradient of a velocity–time graph?

**2.12** What is given by the area under a velocity–time graph between two particular time ordinates?

**2.13** Figure 2.1 shows a displacement–time graph. At which of the marked points is the velocity (a) zero, (b) a maximum?

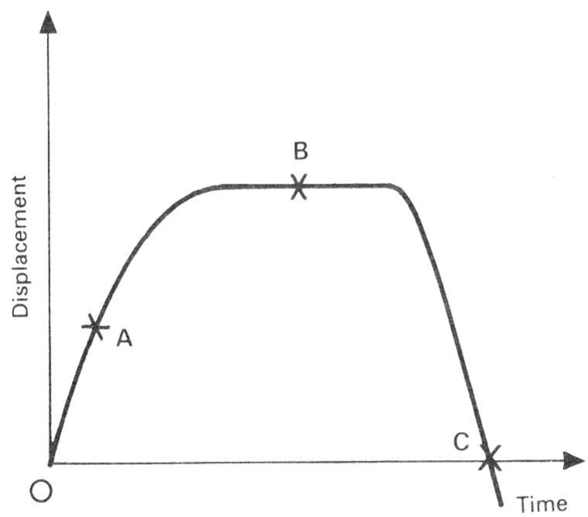


**Figure 2.1**

**2.14** Figure 2.2 shows a velocity–time graph. At which of the marked points is the acceleration (a) zero, (b) a maximum?

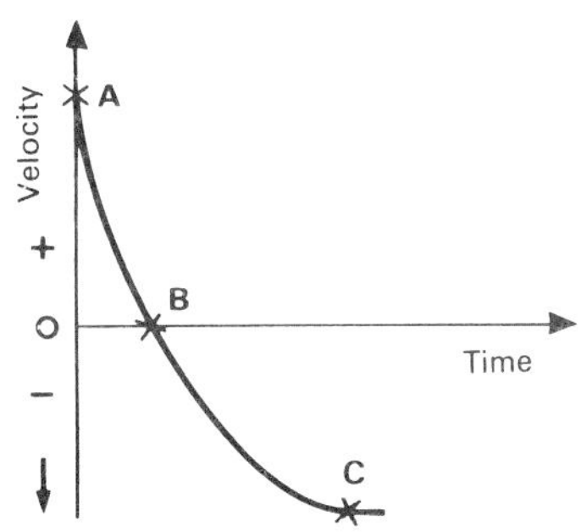


**Figure 2.2**

**2.15** Sketch the displacement–time graph and the velocity–time graph for an object moving with (a) constant velocity, (b) constant acceleration.

**2.16** Figure 2.3 shows a velocity–time graph for an object.
(a) Estimate the acceleration 0.5 s after the start.
(b) Estimate the displacement in this 0.5 s.
(c) What is the acceleration after 1.5 s?
(d) Estimate the total displacement after 1.5 s.

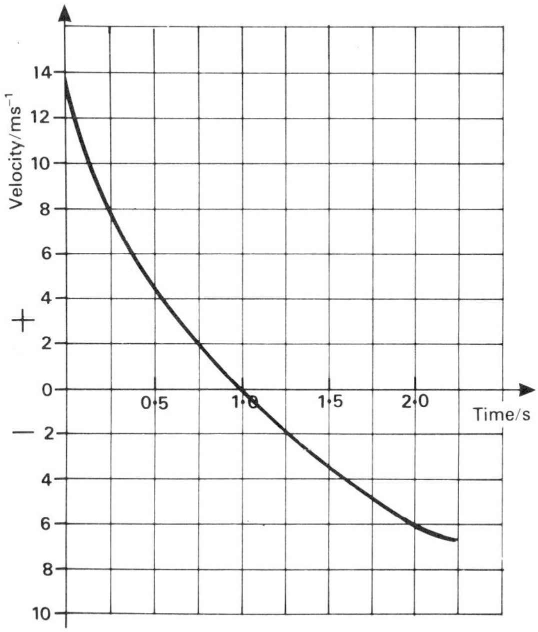


**Figure 2.3**

**II Read** about free fall under gravity and the solution of problems involving vertical motion, either upwards or downwards.
References: *B* PP ch. 1, *B* ST1 ch. 1, *D* ch. 6, *N* ch. 1.

**Self-assessment questions**

**2.17** An object falls freely from rest; what distance will it fall in (a) 1 s, (b) 2 s? ($g = 9.8\,\text{m s}^{-2}$)

**2.18** An object is thrown vertically upwards with an initial velocity of $2\,\text{m s}^{-1}$. What is the greatest height the object will attain and how long will it take to reach that height?

# SELF-ASSESSMENT

## Key points

**Average speed** = distance covered $d$/time taken $t$.
**Uniform speed** is when equal distances are covered in equal intervals of time, however small a time interval is considered.
**Instantaneous speed** $= \lim_{\delta t \to 0} (\delta d/\delta t) = \mathrm{d}d/\mathrm{d}t$.
**Distance** is measured along the path followed by the moving object; **displacement** is the distance measured in a straight line between two points on the path.
**Average velocity** = displacement $s$/time taken $t$.
**Uniform velocity** is when equal displacements in the same straight line are covered in equal intervals of time, however small a time interval is considered.
**Instantaneous velocity** $v = \lim_{\delta t \to 0} (\delta s/\delta t) = \mathrm{d}s/\mathrm{d}t$.
**Average acceleration** = change in velocity/time taken.
**Uniform acceleration** is when the velocity is changing by equal amounts in equal intervals of time, however small a time interval is considered.
**Instantaneous acceleration** $a = \lim_{\delta t \to 0} (\delta v/\delta t) = \mathrm{d}v/\mathrm{d}t$.
The **equations for uniformly accelerated motion** are

$$v = u + at, \quad s = ut + \tfrac{1}{2}at^2, \quad v^2 = u^2 + 2as$$

The gradient of a displacement–time graph is the velocity; the gradient of a velocity–time graph is the acceleration.
The area under a velocity–time graph between two time ordinates is the displacement between those times; the area under a speed–time graph between two time ordinates is the distance covered between those times.

## Answers I

**2.1** See the key points. The instantaneous values are the changes in distance, displacement or velocity in a vanishingly small time interval.

**2.2** Covering equal distances in equal intervals of time in motion round a circle would give constant speed, but because the direction is continually changing the velocity would not be constant.

**2.3** See the key points.

**2.4** $a = (v - u)/t$, hence $at = v - u$ and so $v = u + at$.

**2.5** Average velocity $= s/t = \tfrac{1}{2}(u + v)$. Thus $s = \tfrac{1}{2}ut + \tfrac{1}{2}vt$. But $v = u + at$, hence $s = ut + \tfrac{1}{2}at^2$.

**2.6** $v^2 = (u + at)^2 = u^2 + 2uat + a^2t^2 = u^2 + 2a(ut + \tfrac{1}{2}at^2) = u^2 + 2as$

**2.7** (a) $s = ut + \tfrac{1}{2}at^2 = 10 \times 4.0 + \tfrac{1}{2} \times 2.0 \times 4.0^2 = 56\,\text{m}$ (b) $v = u + at = 10 + 2.0 \times 4.0 = 18\,\text{m s}^{-1}$

**2.8** $v = u + at$, hence $10 = 0 + 5.0a$ and so $a = 2.0\,\text{m s}^{-2}$. In second gear, $v = u + at$ and so $30 = 10 + 5.0a$ and $a = 4.0\,\text{m s}^{-2}$. Distance travelled in first gear $s = ut + \tfrac{1}{2}at^2 = 0 + \tfrac{1}{2} \times 2.0 \times 5.0^2 = 25\,\text{m}$. Distance travelled in second gear $s = ut + \tfrac{1}{2}at^2 = 10 \times 5.0 + \tfrac{1}{2} \times 4.0 \times 5.0^2 = 100\,\text{m}$. Total distance = 125 m.

**2.9** $v^2 = u^2 + 2as$, hence $0 = 16^2 + 2 \times 200a$ and so $a = -0.64\ \mathrm{m\,s^{-2}}$. As $v = u + at$ then $0 = 16 - 0.64t$ and so $t = 25\ \mathrm{s}$.

**2.10** (a) speed (b) velocity

**2.11** acceleration

**2.12** The displacement in that time interval.

**2.13** (a) B, where the gradient is zero. (b) C, where the gradient is the greatest. It does not matter that the displacement at this point is zero: the object has the maximum (negative) velocity while it is travelling through this zero-displacement position.

**2.14** (a) C, where the gradient is zero. (b) A, where the gradient is greatest. When the clock was started for the time measurement the object was already accelerating.

**2.15** See Figure A2.1. With constant velocity the gradient of the displacement–time graph must be constant. With constant acceleration the gradient of the velocity–time graph must be constant, the displacement–time graph being given by the equation $s = \frac{1}{2}at^2$.

**2.16** (a) Estimate the gradient of the graph at 0.5 s: about $-11\ \mathrm{m\,s^{-2}}$.
(b) Estimate the area between zero time and the 0.5 s ordinates: about 4 m.
(c) About $6\ \mathrm{m\,s^{-2}}$. Note that this is a positive acceleration while in (a) the acceleration is negative.
(d) The area above the axis is positive, below it negative. Hence total displacement is about $5 - 1 = 4$ m.

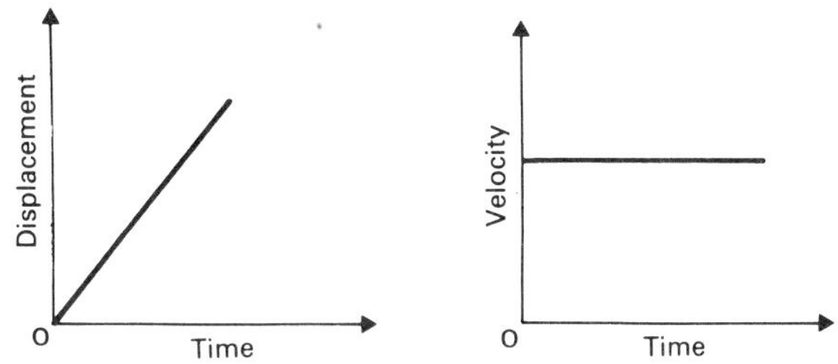


**(a)** Constant velocity

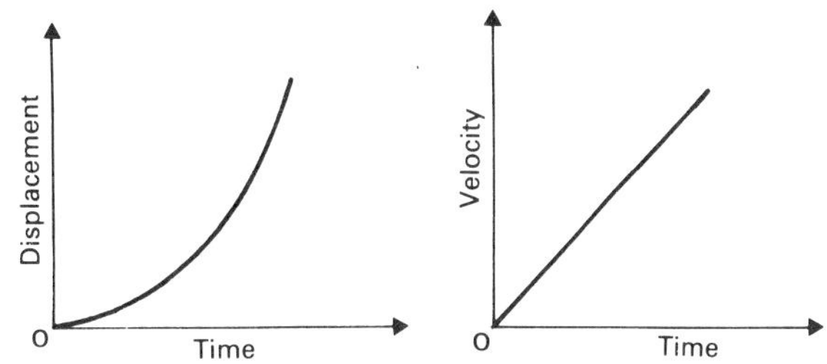


**(b)** Constant acceleration

**Figure A2.1**

## Answers II

**2.17** Using $s = \frac{1}{2}at^2$ we have (a) $s = \frac{1}{2} \times 9.8 \times 1^2 = 4.9$ m (b) $s = \frac{1}{2} \times 9.8 \times 2^2 = 19.6$ m.

**2.18** At the greatest height the object will have zero velocity. Using $v^2 = u^2 + 2as$, we have $0 = 2^2 - 2 \times 9.8\,s$ and so $s = 0.20$ m. Time can be obtained by using $v = u + at$, hence $0 = 2 - 9.8t$ and so $t = 0.20$ s. (With the upward motion the acceleration is negative because it is slowing the object down.)

# Lesson 3 Velocity as a vector

## Aims

After completing this lesson you should be able to

- determine relative velocities
- resolve velocities into components
- solve problems involving objects moving in two dimensions, e.g. projectiles.

## Prerequisites

Lesson 2 in this chapter, and Lesson 1 would also be useful.

## STUDY

**I Read** about relative velocity and its determination. References: *B* ST1 ch. 3; *N* ch.1; *W et al.* ch. 3 (7), *Wh & H* ch. 2.

### Self-assessment questions

**3.1** Object A is stationary and object B is moving away from it with a velocity of $2\ \mathrm{m\,s^{-1}}$. What is (a) the velocity of B relative to A, (b) the velocity of A relative to B?

**3.2** Two trains on parallel tracks approach each other with velocities of $80\ \text{km h}^{-1}$ and $60\ \text{km h}^{-1}$. What is the velocity of (a) the $80\ \text{km h}^{-1}$ train relative to the $60\ \text{km h}^{-1}$ train and (b) the $60\ \text{km h}^{-1}$ train relative to the $80\ \text{km h}^{-1}$ train?

**3.3** The procedure for finding the velocity of A relative to B is to subtract, using vector methods, the velocity of B from that of A. Explain how this method is used to determine the velocity of A relative to B if A is moving due north at $40\ \text{km h}^{-1}$ and B is moving due west at $30\ \text{km h}^{-1}$.

***3.4** When a train is stationary, rain is observed to be falling vertically. However, when the train is moving with a velocity of $40\ \text{km h}^{-1}$ due north the rain is seen to be making traces on the window which are inclined at 40° to the vertical. What is the velocity of the rain relative to the *stationary* train?

**3.5** An object is moving in a circular path with a constant speed of $4\ \text{m s}^{-1}$. What is the change in velocity of the object between two positions (a) at opposite ends of a diameter, (b) involving a movement through an arc of angle 90°?

**II Read** about the resolution of velocity into its components. This is the same as the resolution of any vector, so you could refer to Lesson 1 in this chapter. References: *B* ST1 ch. 3, *N* ch. 1.

**Self-assessment questions**

**3.6** A projectile is fired from a gun at an angle of 30° to the horizontal with a velocity of $100\ \text{m s}^{-1}$. What is the velocity component (a) parallel to the ground and (b) at right angles to the ground?

**3.7** An electron in an electron beam has at a particular instant a velocity of $2.0 \times 10^5\ \text{m s}^{-1}$ at an angle of 20° from the horizontal. What is the velocity component (a) horizontally and (b) vertically?

**III Read** about projectiles and motion in two dimensions in general.
References: *B* PP ch. 1, *B* ST1 ch. 2, *D* ch. 6, *N* ch. 1, *Wh&H* ch. 2.

**Self-assessment questions**

**3.8** (a) An object has a horizontal velocity of $4\ \text{m s}^{-1}$. How far has it moved after 1 s, after 2 s, after 3 s?
(b) An object is falling freely from rest. How far does it move in 1 s, in 2 s, in 3 s? ($g = 9.8\ \text{m s}^{-2}$)
(c) An object is projected from a cliff with a horizontal velocity of $4\ \text{m s}^{-1}$ and falls to the ground. What is its position relative to the edge of the cliff a time of 1 s, 2 s and 3 s after leaving the cliff edge?

**3.9** A projectile is fired from a gun with a velocity $v$ at an angle of $\theta$ to the horizontal.
(a) What are the vertical and horizontal components of this velocity?
(b) What are the vertical and horizontal components of the velocity when the projectile is at its greatest height?
(c) Considering only the vertical component of the velocity, derive a relationship for the time to the greatest height.
(d) Considering the horizontal component of the velocity, how far does the projectile move horizontally in the time taken for the object to reach the greatest height?
(e) The *range* of the projectile is the total horizontal distance covered by the projectile when it is back down at the same height as it started from. What is the time taken for this to occur?

***3.10** A gun fires a shell with a velocity of $80\ \text{m s}^{-1}$ at an angle of 30° to the horizontal. What is (a) the greatest height reached, (b) the range of the shell? ($g = 9.8\ \text{m s}^{-2}$)

****3.11** A ship's gun fires a shell with a muzzle velocity of $400\ \text{m s}^{-1}$ to hit an enemy ship 12 km away. Neglecting air resistance, calculate the angle to the horizontal at which the shell should be fired and its time of flight. ($g = 9.8\ \text{m s}^{-2}$)

# SELF-ASSESSMENT

## Key points

Velocity is a vector quantity.
To find the velocity of an object A relative to B, subtract by vector methods the velocity of B from that of A.
A velocity $v$ can be replaced by two components, $v \sin \theta$ and $v \cos \theta$, where $\theta$ is the angle between the $v \cos \theta$ component and $v$.
For a projectile the horizontal and vertical motions can be treated independently. The time to the greatest height is the time taken for the vertical component of the velocity to decrease to zero. During that time the projectile moves in a horizontal direction with a constant horizontal velocity component.

## Answers I

**3.1** (a) $2\ \text{m s}^{-1}$ (b) $-2\ \text{m s}^{-1}$, the minus sign indicating that the velocity is in the opposite direction.

**3.2** (a) $140\ \text{km h}^{-1}$ (b) $-140\ \text{km h}^{-1}$

**3.3** Subtract the velocity of B from that of A. Figure A3.1 shows the two velocities and the parallelogram of velocities. Subtraction means that a velocity opposite and equal to that of B has to be added to the velocity of A. The result is a relative velocity of 50 km $h^{-1}$ at 36.9° east of north.

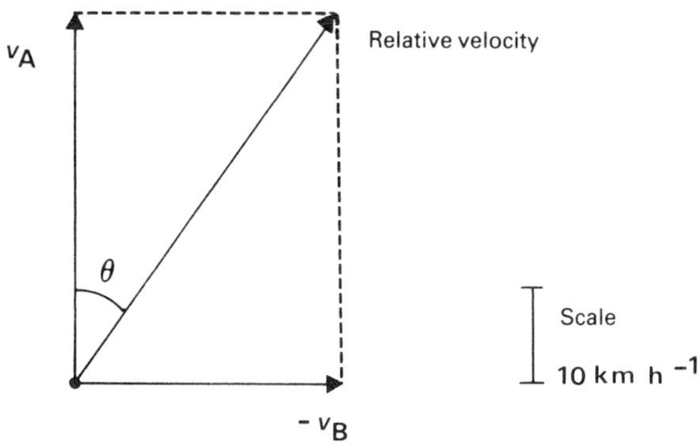


**Figure A3.1**

**3.4** Figure A3.2 shows the parallelogram of velocities; the angle of 40° to the vertical is the angle the rain's velocity makes relative to the moving train. The rain has a velocity of 47.7 km $h^{-1}$ vertically.

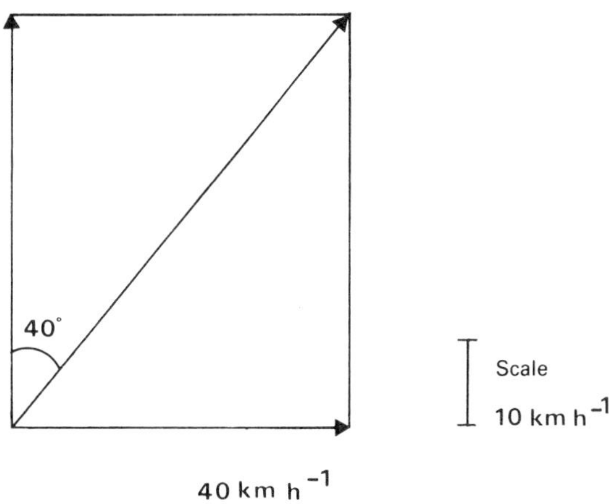


**Figure A3.2**

**3.5** (a) 8 m $s^{-1}$ (b) 5.7 m $s^{-1}$ at 135° to the original velocity. In this case use the parallelogram law to find the relative velocity.
This is a relative velocity problem, in that you want the final velocity relative to the initial velocity. This problem is the basis of the centripetal acceleration derivation in Lesson 7.

## Answers II

**3.6** (a) $100 \cos 30° = 86.6$ m $s^{-1}$ (b) $100 \sin 30° = 50.0$ m $s^{-1}$

**3.7** (a) $2.0 \times 10^5 \times \cos 20° = 1.9 \times 10^5$ m $s^{-1}$ (b) $2 \times 10^5 \times \sin 20° = 0.7 \times 10^5$ m $s^{-1}$

## Answers III

**3.8** (a) After 1 s, distance = 4 m; after 2 s, 8 m; after 3 s, 12 m.
(b) $s = \frac{1}{2}gt^2 = 4.9$ m after 1 s; after 2 s, 19.6 m; after 3 s, 44.1 m.
(c) After 1 s it is 4 m horizontally and 4.9 m vertically down, after 2 s it is 8 m horizontally and 19.6 m down, after 3 s it is 12 m horizontally and 44.1 m vertically down.

**3.9** (a) Vertical component $v \sin \theta$, horizontal component $v \cos \theta$.
(b) Vertical component zero, horizontal component $v \cos \theta$.
(c) Using $v = u + at$, we have $0 = v \sin \theta - gt$, hence $t = (v \sin \theta)/g$.
(d) Using $s = ut$ we have $s = (v \cos \theta) \times (v \sin \theta)/g$.
(e) Time = $2t$, where $t$ is the time to the greatest height, hence time = $2(v \sin \theta)/g$.

**3.10** (a) Vertical component = 80 sin 30°, hence using $v^2 = u^2 + 2as$ we have $0 = (80 \sin 30°)^2 - 2gh$ and $h = 81.6$ m.
(b) Using the answer to SAQ 3.9(e) we have time of flight = $(2 \times 80 \sin 30°)/9.8 = 8.16$ s. Using $s = ut$ we have $s = (80 \cos 30°) \times 8.16 = 565.3$ m.

**3.11** Using the equations derived in the answer to SAQ 3.9, range = $2\,(400^2 \cos \theta \sin \theta)/9.8 = 12 \times 10^3$. As $2 \sin \theta \cos \theta = \sin 2\theta$, then $\theta = 23.7°$. Time = $2(v \sin \theta)/g = 32.8$ s.

# Lesson 4 Newton's laws

## Aims

After completing this lesson you should be able to
- state and use Newton's laws
- use the principle of momentum conservation.

## Prerequisites

Lesson 2 (Linear motion) and the concept of force as a vector quantity.

# STUDY

**I Read** about Newton's laws and their use in solving problems involving force and motion.
References: *B* PP ch. 2, *B* ST1 ch. 2, *D* ch. 6, *N* ch. 1, *W et al.* ch. 6 (3, 4, 5), *Wh&H* ch. 3.

**Self-assessment questions**

**4.1** State Newton's first law of motion.

**4.2** In which of the following situations must there be a resultant force acting on an object?
(a) A freely falling object accelerates with the acceleration due to gravity.
(b) A car moves along a road with a constant velocity.
(c) An object moves in a circular path with a constant speed.

**4.3** Newton's second law of motion is often stated in terms of *momentum*.
(a) Define momentum and state its unit.
(b) State Newton's second law of motion in terms of momentum.

**4.4** An alternative way of stating Newton's second law is in terms of acceleration, i.e. $F = ma$, rather than as rate of change of momentum. Show that the two forms are equivalent.

**4.5** Define the unit of force.

**4.6** Distinguish between *weight* and *mass*.

**4.7** State Newton's third law of motion.

**4.8** A block of material has a length of (massless) string attached to it, the other end of which is attached to a second block. When the first block is pulled by a force $F$ and the whole system accelerates, what is the force $F$ and the force (tension) in the string in terms of the acceleration and the masses of the blocks, according to Newton's third law?

**4.9** A Saturn rocket develops an initial thrust of $7.4 \times 10^6$ N and has a mass at lift-off of $5.4 \times 10^5$ kg. What is the initial acceleration of the rocket at lift-off?

**4.10** An object of mass 2.0 kg is suspended from a spring balance in a lift. What will be the reading on the spring balance, in newtons, when the lift is (a) stationary, (b) accelerating upwards with an acceleration of $0.2\,m\,s^{-2}$, (c) moving upwards with a constant velocity of $1.0\,m\,s^{-1}$, (d) descending with an acceleration of $0.1\,m\,s^{-2}$? ($g = 9.8\,m\,s^{-2}$)

**4.11** A car of mass 600 kg pulls a trailer of mass 200 kg. If the two accelerate at $1.5\,m\,s^{-2}$ and frictional effects can be neglected, what is the tension in the tow-bar?

**II Read** about momentum and the conservation of momentum.
References: *B* PP ch. 2, *B* ST1 ch. 5, *D* ch. 6, *N* ch. 1, *W et al.* ch. 5 (4), *Wh&H* ch. 5.

**Self-assessment questions**

**4.12** State the *principle of conservation of momentum.*

**4.13** Define *impulse.*

**4.14** An object of mass 1.2 kg, moving with a velocity of $2\,m\,s^{-1}$, collides with a stationary object of mass 2.0 kg. If the two objects stick together with what velocity will they move off together?

**4.15** Two trolleys, one of mass 0.5 kg and the other 1.0 kg, are held at rest with a coiled spring between them. When the spring is released the trolleys spring apart. What is the relationship between their velocities?

**4.16** What force is needed to change the velocity of an object of mass 3 kg from $0.4\,m\,s^{-1}$ to $0.8\,m\,s^{-1}$ in 2.0 s?

**4.17** What impulse must be given to a tennis ball of mass 0.10 kg if it is to acquire a velocity of $10\,m\,s^{-1}$?

**4.18** Gases are ejected from the rear of a rocket with a velocity of $1200\,m\,s^{-1}$ and at the rate of $120\,kg\,s^{-1}$. What is the thrust experienced by the rocket?

**4.19** A jet of water has a velocity of $8\,m\,s^{-1}$ and a cross-sectional area of $4.0 \times 10^{-3}\,m^2$. When it strikes a wall at right angles it loses all its velocity.

(a) What is the mass of water striking the wall per second? (Density of water = 1000 kg m$^{-3}$)
(b) What is the change in momentum per second of the water hitting the wall?
(c) What is the force exerted on the wall?

***4.20** Rain falls vertically onto a flat roof of area 500 m$^2$. If the velocity of the rain is 8 m s$^{-1}$ and in 1 minute 0.5 mm falls, what is the force on the roof resulting from the rain? (Density of water = 1000 kg m$^{-3}$)

**4.21** Sand falls vertically at a rate of 0.15 kg s$^{-1}$ onto a horizontal conveyor belt moving with a velocity of 0.05 m s$^{-1}$. What force has to be applied to the conveyor belt for it to maintain its constant velocity?

***4.22** A helicopter hovers in still air well away from the ground. If the helicopter has a mass of 4000 kg and rotor blades of diameter 10 m, what is the velocity of air projected downwards by the blades? ($g$ = 9.8 m s$^{-2}$, density of air = 1.3 kg m$^{-3}$)

# SELF-ASSESSMENT

## Key points

**Newton's laws** can be expressed as:
*Law 1* Unless a body is acted on by a resultant force it remains at rest or in motion with a constant velocity.
*Law 2* The rate of change of momentum of a body is proportional to the resultant force acting on it.
*Law 3* If body A exerts a force on body B, then body B exerts an equal force on body A along the same line but in the opposite direction.
The **linear momentum** of a body is the product of its mass and its velocity.
Law 2 can be expressed as

$$F = \frac{\mathrm{d}}{\mathrm{d}t}(mv) = ma$$

The **impulse** of a force is the product of the force and the time for which it acts; it is also the change in momentum produced.
The **principle of conservation of momentum** is that the momentum of any closed system is conserved.

## Answers I

**4.1** See the key points.

**4.2** (a) yes (b) no (c) yes
Any time there is a change in velocity there must be a resultant force.

**4.3** (a) See the key points; unit kg m s$^{-1}$. (b) See the key points.

**4.4** $F = \mathrm{d}(mv)/\mathrm{d}t$, which for constant mass becomes $F = m\,\mathrm{d}v/\mathrm{d}t$. But the rate of change of velocity is the acceleration $a$, hence $F = ma$.

**4.5** The *newton* is that resultant force which causes a mass of 1 kg to accelerate at 1 m s$^{-2}$.

**4.6** The *mass* of a body is the quantitative measure of the *inertia* of the body in linear motion. The *weight* of the body can be defined as the force exerted by a body on its support. Another definition of weight is the gravitational force acting on the body.

**4.7** See the key points.

**4.8** See Figure A4.1. If the system has an acceleration $a$ and the first block has a mass $M$ and the second a mass $m$, then the force needed to accelerate the mass $m$ is $ma$. According to Newton's third law there is an opposite and equal force acting on $M$, hence the resultant force causing the acceleration of $M$ is $(F - ma) = Ma$, i.e. $F = (M + m)a$. The tension in the string is $ma$.

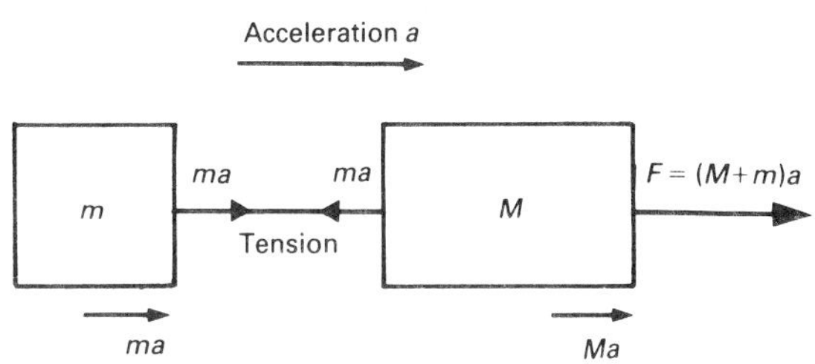


**Figure A4.1**

**4.9** $F = ma$, hence $a = (7.4 \times 10^6 - 5.4 \times 10^5 \times 9.8)/5.4 \times 10^5 = 3.9$ m s$^{-2}$.

**4.10** (a) $F = 2.0 \times 9.8 = 19.6$ N
(b) Resultant upward force on object = $F - mg$, where $F$ is the tension in the spring. Hence $F - mg = ma$ and so $F = m(a + g) = 20.0$ N
(c) $F = mg$ since $a = 0$, thus $F = 19.6$ N.
(d) Resultant downward force = $mg - F = ma$ and so $F = m(g - a) = 19.4$ N.
Note that when $a$ downwards equals $g$, $F = 0$. When using the definition of weight as the force exerted by a body on its support, we can refer to this condition as 'weightlessness'.

**4.11** See the answer to SAQ 4.8. Tension = 200 × 1.5 = 300 N. This assumes the tow-bar has negligible mass.

## Answers II

**4.12** See the key points.

**4.13** See the key points.

**4.14** Momentum before collision = $1.2 \times 2$ kg m s$^{-1}$. Momentum after collision = $3.2v$, hence, as momentum before equals momentum after, we have $v = 0.75$ m s$^{-1}$.

**4.15** Momentum before = 0, momentum after = $0.5\,v_1 + 1.0\,v_2$. Hence, as momentum before equals momentum after, $v_1 = -2v_2$.

**4.16** Rate of change of momentum = $3(0.8 - 0.4)/2.0 = 0.6$ N.

**4.17** Impulse = $0.10 \times 10 = 1$ kg m s$^{-1}$ (or N s).

**4.18** Mass ejected in 1 s = 120 kg. Change of momentum of this mass in 1 s = $120 \times 1200$ kg m s$^{-1}$. Hence force = $120 \times 1200$ N = $1.44 \times 10^5$ N.

**4.19** (a) All the water within 8 m of the wall will strike it in 1 s. Hence mass = $8 \times 4.0 \times 10^{-3} \times 1000 = 32$ kg. (b) $32 \times 8 = 256$ kg m s$^{-2}$ (c) 256 N

**4.20** Assuming that the rain hits the roof and does not bounce back up, then rate of change of momentum = mass hitting roof per second × velocity = $(0.5 \times 10^{-3}/60) \times 500 \times 1000 \times 8 = 33.3$ kg m s$^{-2}$. Force = 33.3 N.

**4.21** Change in mass per second of conveyor, assuming no sand is taken from it, = 0.15 kg. This extra mass has to be given a momentum of $0.15 \times 0.05 = 7.5 \times 10^{-3}$ kg m s$^{-1}$. Hence force = rate of change of momentum = $7.5 \times 10^{-3}$ N.

**4.22** Force required = $4000 \times 9.8$ N. This is provided by the rate of change of momentum of the air, hence $(\frac{1}{4}\pi \times 10^2 \times v \times 1.3)v = 4000 \times 9.8$, and so $v = 19.6$ m s$^{-1}$.

# Lesson 5 Work, energy and power

## Aims

After completing this lesson you should be able to

- define and use the terms *work*, *energy* and *power*
- define and use the terms *kinetic* and *potential* energy
- use the principle of energy conservation
- distinguish between elastic and inelastic collisions and solve problems involving such collisions.

## Prerequisites

Lessons 2 and 4, i.e. linear motion and Newton's laws, and in particular for the collisions the principle of conservation of momentum in Lesson 4.

## STUDY

**I Read** about the definitions of work, energy and power.
References: *B* PP ch. 3, *B* ST3 ch. 1, *D* ch. 6, *N* ch. 1, *W et al.* ch. 7(8), *Wh&H* ch. 4.

### Self-assessment questions

**5.1** Define *work* and state its unit.

**5.2** State the work done, or lack of work done, in the following cases:
(a) A constant force $F$ acts on an object during its movement through a distance $s$ in the direction of the force.
(b) A beam is balanced across a pivot at which there is a force $F$ acting to support the beam.
(c) A constant force $F$ acts on an object while it moves a distance $s$ in a direction at 30° to the direction of the force.
(d) An object moves in a circular path with a constant speed under the action of a force $F$ which always acts along a radius and so is at right angles to the motion.

**5.3** Is work a vector or a scalar quantity?

**5.4** Define *energy* and state its unit.

**5.5** Define *power* and state its unit.

**5.6** An object of mass $m$ is pushed by a force $F$ so that it moves with a constant velocity $v$ (other forces are also acting on the object, e.g. frictional forces if the object is pushed along the floor).
(a) How far does the object move in 1 s?

(b) If the object moves in the same direction as the force, what is the work done per second?
(c) What is the power of the force?

**5.7** A car travelling at 60 m s$^{-1}$ has resistive forces of 200 N opposing its motion. What must be the power generated by the engine to maintain the car at this velocity?

**II Read** about the definitions of potential and kinetic energy.
References: *B* PP ch. 3, *B* ST3 ch. 1, *D* ch. 6, *N* ch. 1, *W et al.* ch. 7(8), *Wh&H* ch. 4.

**Self-assessment questions**

**5.8** A constant force *F* acts on a body and causes it to accelerate in a straight line with an acceleration *a*. If the body had initially a velocity *u* and after a distance *s* this becomes *v*, derive a relationship for the work done on the body over this distance in terms of the initial and final velocities. Hence define *kinetic energy*.

**5.9** Define *potential energy* and show that the work done in lifting an object of mass *m* through a vertical height *h* is *mgh*. Hence define *gravitational* potential energy. Give an example of another form of potential energy.

**III Read** about the conservation of mechanical energy and energy conservation in general. Also read about elastic and inelastic collisions.
References: *B* PP ch. 3, *B* ST3 ch. 1, *D* ch. 6, *N* ch. 1, *W et al.* ch. 7(8), *Wh&H* ch. 4.

**Self-assessment questions**

**5.10** What is meant by the 'conservation of mechanical energy'?

**5.11** State the *principle of conservation of energy*.

**5.12** Distinguish between *elastic* and *inelastic* collisions and give an example of each type.

**5.13** A car of mass 1200 kg is travelling with a velocity of 60 km h$^{-1}$ along a horizontal straight road when the brakes are applied. These bring the car to rest in a distance of 45 m. What is (a) the kinetic energy lost by the car, (b) the work done in bringing the car to rest, (c) the average braking force, (d) the time taken for the car to come to rest?

***5.14** A car of mass 1000 kg travels up an incline of 1 in 25 at a constant velocity of 50 km h$^{-1}$. What power does the car engine have to develop if there is a resistive force of 300 N opposing the motion? ($g$ = 9.8 m s$^{-2}$)

***5.15** A billiard ball with a velocity of 0.50 m s$^{-1}$ collides head-on with another billiard ball of equal mass coming from the opposite direction with a velocity of 0.80 m s$^{-1}$. If the collision is elastic, what are the velocities of the two balls after they collide?

***5.16** A ball of mass 10 g is dropped from a height of 2.0 m to the horizontal ground. It then bounces back up to a height of 1.5 m.
(a) What is the loss of energy by the ball on contact with the ground? ($g$ = 9.8 m s$^{-2}$)
(b) If the ball loses the same percentage of its energy every time it hits the ground, what height will the ball attain after its third contact with the ground?

***5.17** In rocket propulsion hot gases are emitted from the rear of the rocket. Consider the momentum and kinetic energies of the gases and the rocket and state whether they are conserved.

# SELF-ASSESSMENT

## Key points

**Work** is done when a force moves its point of application so that some resolved part of the resulting displacement lies along the direction of the force. Work is defined by the equation

$$W = Fs \cos \theta$$

where *F* is the force, *s* the displacement and $\theta$ the angle between the displacement and the line of action of the force.
A body which can do work is said to possess **energy**.
**Power** is the rate at which energy is transferred or transformed from one form to another.
The **kinetic energy** of an object is the energy it possesses by virtue of its motion, being given by the equation

$$\text{kinetic energy} = \tfrac{1}{2}mv^2$$

**Potential energy** is the energy an object possesses by virtue of its position, being given in the case of gravitational potential energy by

$$\text{potential energy} = mgh$$

When **mechanical energy** is conserved the total amount of mechanical energy, i.e. potential and kinetic energy, remains constant.
The **principle of the conservation of energy** is that energy may be transformed from one form to another but it cannot be created or destroyed, i.e. the total energy must always remain constant.

In an **elastic collision** kinetic energy is conserved, this not being the case in an **inelastic** collision. In both types of collision momentum is conserved.

## Answers I

**5.1** See the key points; the unit is the joule.

**5.2** (a) $Fs$ (b) No work is done. (c) $Fs \cos 30°$ (d) No work is done.
For work to be done the point of application of the force must move and the movement must have a component along the line of action of the force.

**5.3** Scalar

**5.4** See the key points; the unit is the joule.

**5.5** See the key points; the unit is the watt. One watt is one joule per second.

**5.6** (a) $v$ (b) $W = Fv$ (c) power $= Fv$

**5.7** Power $= Fv = 200 \times 60 = 12\,\text{kW}$

## Answers II

**5.8** $v^2 = u^2 + 2as$, hence $W = Fs = F(v^2 - u^2)/2a$. But $F = ma$ and so $W = \frac{1}{2}mv^2 - \frac{1}{2}mu^2$. Kinetic energy $= \frac{1}{2}mv^2$.

**5.9** See the key points for definition. $W = Fh$, but $F = mg$ and so $W = mgh$, where $h$ is the height through which the object is lifted. Gravitational potential energy is thus $mgh$.

## Answers III

**5.10** See the key points.

**5.11** See the key points.

**5.12** See the key points. A collision between two electrons could be an elastic collision; a piece of Plasticine falling onto the ground is an inelastic collision.

**5.13** (a) Kinetic energy $= \frac{1}{2} \times 1200 \times (60 \times 10^3/3600)^2 = 1.67 \times 10^5\,\text{J}$.
(b) Work done is the same as kinetic energy lost, i.e. $1.67 \times 10^5\,\text{J}$.
(c) $Fs = 1.67 \times 10^5$ and so $F = 3.71 \times 10^3\,\text{N}$.
(d) Average deceleration $= F/m = 3.09\,\text{m s}^{-2}$. Time taken $= (60 \times 10^3/3600)/3.09 = 5.4\,\text{s}$.

**5.14** Distance travelled up the slope in 1 s is $50 \times 10^3/3600$ m. The resistive force acting on the car is 300 N down the slope plus the component of the weight acting down the slope, $mg \sin\theta$. Total resistive force $= 300 + 1000 \times 9.8 \times (1/25) = 692\,\text{N}$. Hence power = work done per second $= 692 \times 50 \times 10^3/3600 = 9.61\,\text{kW}$.

**5.15** $\frac{1}{2}mu_1^2 + \frac{1}{2}mu_2^2 = \frac{1}{2}mv_1^2 + \frac{1}{2}mv_2^2$. This can be simplified and rearranged to give $u_1^2 - v_1^2 = v_2^2 - u_2^2$ and so $(u_1 - v_1)(u_1 + v_1) = (v_2 - u_2)(v_2 + u_2)$. Also, from momentum conservation, we have $mu_1 + mu_2 = mv_1 + mv_2$, and so $u_1 - v_1 = v_2 - u_2$. Thus $u_1 + v_1 = v_2 + u_2$. This means $v_1 - v_2 = 0.80 - 0.50 = 0.30\,\text{m s}^{-1}$. But according to the momentum equation $v_1 + v_2 = 0.50 + 0.80 = 1.30\,\text{m s}^{-1}$. Hence adding these equations gives $2v_1 = 1.00$ and so $v_1 = 0.50\,\text{m s}^{-1}$. Therefore $v_2 = 0.80\,\text{m s}^{-1}$. The two objects have thus changed velocities.

**5.16** (a) The kinetic energy possessed by the ball just before contact with the ground equals the potential energy it had before it fell. Thus energy on hitting ground $= mgh = 0.01 \times 9.8 \times 2.0 = 0.196\,\text{J}$. The kinetic energy the ball has when it bounces back up from the ground is all converted into potential energy. Thus the energy is $mgh = 0.01 \times 9.8 \times 1.5 = 0.147\,\text{J}$. The loss in energy is thus 0.049 J.

(b) The percentage loss $= \dfrac{(2.0 - 1.5)}{2.0} \times 100 = 25\%$

at each contact with the ground.

$$\text{After 1st contact energy loss} = \frac{25}{100}(mg \times 2.0)$$

$$\text{and energy left} = (mg \times 2.0) - \frac{25}{100}(mg \times 2.0)$$

$$= \left(1 - \frac{25}{100}\right)(mg \times 2.0)$$

$$\text{After 2nd contact loss} = \frac{25}{100}\left(1 - \frac{25}{100}\right)(mg \times 2.0)$$

$$\text{and energy left} = \left(1 - \frac{25}{100}\right)^2(mg \times 2.0)$$

$$\text{After 3rd contact loss} = \frac{25}{100}\left(1 - \frac{25}{100}\right)^2(mg \times 2.0)$$

$$\text{and energy left} = \left(1 - \frac{25}{100}\right)^3(mg \times 2.0)$$

Hence height reached is given by

$$mgh_3 = \left(1 - \frac{25}{100}\right)^3(mg \times 2.0)$$

$$h_3 = \left(1 - \frac{25}{100}\right)^3 \times 2.0 = 0.84\,\text{m}$$

**5.17** The momentum gained by the rocket is opposite and equal in size to the momentum gained by the gases. The total momentum change is zero, momentum being conserved. Momentum is a vector quantity and hence the direction of the momentum has to be taken into account. Kinetic energy is a scalar quantity and hence direction is irrelevant. There is thus a gain in kinetic energy by the gases and by the rocket and so a net gain. This gain is supplied from the chemical energy in the fuel. Thus kinetic energy is not conserved.

# Lesson 6 Friction

## Aims

After completing this lesson you should be able to

- state the laws of friction
- solve both static and dynamic problems involving friction

## Prerequisites

Lessons 1 and 4 involving the consideration of static situations and equilibrium, and dynamics with Newton's laws.

# STUDY

**I Read** about the laws of friction.
References: *B* PP ch. 2, *B* ST1 ch. 2, *D* ch. 6, *N* ch. 7, *Wh&H* ch. 3.

### Self-assessment questions

**6.1** Figure 6.1 shows the forces acting on an object resting on a horizontal surface when it is subject to a horizontal force *P*.

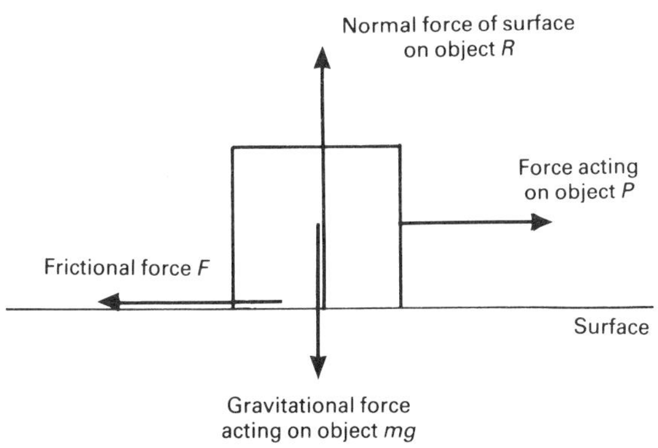


**Figure 6.1**

(a) What is the relationship between the normal force and the gravitational force acting on the object?
(b) When the force acting on the object is less than that necessary for the object to move, what can be said about the frictional force?
(c) When is the frictional force a maximum?
(d) What is the relationship between the maximum frictional force and the normal force?

**6.2** Define the coefficient of friction for both static and dynamic conditions.

**6.3** State the three laws of friction.

**6.4** An object of mass 4 kg rests on a horizontal surface. If the coefficient of friction is 0.6, what horizontal force has to act on the object to start it sliding? ($g = 9.8\,\text{m}\,\text{s}^{-2}$)

**6.5** An object rests on a horizontal surface. If the surface is slowly tilted the object begins to slide when the surface is at an angle of 20° to the horizontal. What is the coefficient of friction?

**6.6** An object is being slid along a horizontal surface where the coefficient of dynamic friction is 0.5. What is the force needed to keep the object, mass 2.0 kg, moving with a constant velocity? ($g = 9.8\,\text{m}\,\text{s}^{-2}$)

**II Revise** Lesson 1. The solution of statics problems involving friction follows directly from that lesson. The following reference gives a worked example.
Reference: *D* ch. 6.

### Self-assessment questions

****6.7** A uniform ladder of length 3.0 m rests with its upper end against a smooth vertical wall and its lower end on horizontal rough ground where the coefficient of friction is 0.30. What is the smallest angle the ladder can make with the horizontal ground before it begins to slide? ($g = 9.8\,\text{m}\,\text{s}^{-2}$)

**6.8 A uniform plank of length 5.0 m rests with its upper end against a smooth vertical wall and its lower end on rough horizontal ground. If the plank is on the point of slipping when the distance of its lower end from the wall is 2.0 m, what is the coefficient of friction?

**III Revise** Lesson 4. The solution of dynamics problems involving friction follows directly from that lesson.

**Self-assessment questions**

*6.9 An object of mass 3.0 kg is placed on a slope, inclined at 40° to the horizontal, and slides down it. How long will it take to slide 0.35 m down the slope, starting from rest, if the coefficient of dynamic friction is 0.20? ($g = 9.8\ \mathrm{m\,s^{-2}}$)

*6.10 A bullet of mass 10 g travelling horizontally at 400 m s$^{-1}$ hits and becomes embedded in a block of wood of mass 0.50 kg resting on the horizontal ground. After impact the block, together with the bullet, moves a distance of 18 m before coming to rest. What is the coefficient of friction between the block and the floor? ($g = 9.8\ \mathrm{m\,s^{-2}}$)

# SELF-ASSESSMENT

## Key points

**Laws of friction:**

1 The frictional force between two solid surfaces opposes their relative motion.

2 The frictional force is independent of the area of contact of the two surfaces.

3 The limiting (or maximum) frictional force $F$ is proportional to the normal force $R$:

$$F = \mu R$$

where $\mu$ is the *coefficient of friction*. The coefficient is called the *static* coefficient where $F$ is the frictional force needed to be overcome to start sliding. The coefficient is termed the *dynamic* or *kinetic* coefficient where $F$ is the force needed to overcome friction and maintain a constant velocity.

## Answers I

**6.1** (a) The normal force is opposite and equal to the gravitational force, i.e. $R = mg$.

(b) It is equal and opposite to the force acting on the object so that there is no resultant force acting on the object.

(c) The frictional force is a maximum when the object just begins to slide.

(d) $F_{max} = \mu R$

**6.2** The static coefficient of friction is the ratio of the limiting (or maximum) frictional force that has to be overcome to start motion and the normal force. The dynamic coefficient of friction is the ratio of the frictional force that occurs while an object is in motion with a constant velocity and the normal force.

**6.3** See the key points.

**6.4** When it starts to slide the applied force equals the limiting (or maximum) frictional force. Thus force = 0.6 × 4 × 9.8 = 23.5 N.

**6.5** See Figure A6.1. The components of the weight $mg$ are shown. $R = mg \cos \theta$ and just before the object starts to slide $F = mg \sin \theta$. But $F = \mu R$ and so $mg \sin \theta = \mu mg \cos \theta$ and so $\mu = \tan \theta$. Hence $\mu = \tan 20° = 0.36$.

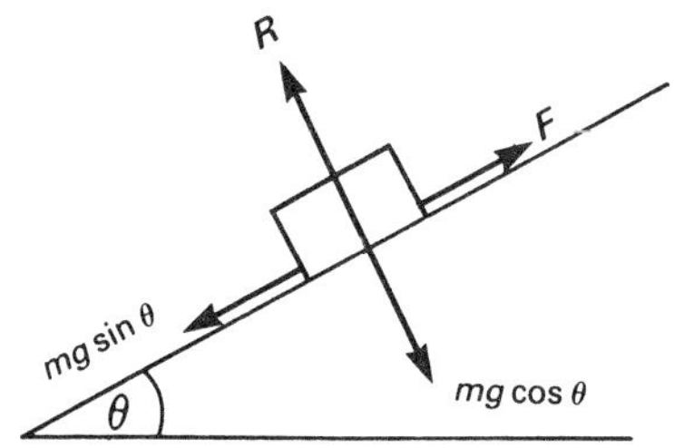


**Figure A6.1**

**6.6** $F = \mu R = 0.5 \times 2.0 \times 9.8 = 9.8$ N

## Answers II

**6.7** There are two ways in which this problem (Figure A6.2) can be tackled. One way is a general way that would apply if even more forces were involved, the other is applicable when there are just three forces.

For the general solution, equating the horizontal forces gives $F = S$ and equating the vertical forces gives $R = mg$. Hence $F = \mu R = \mu mg = S$. Taking moments about the upper contact point of the ladder with the wall gives $mg \times 1.5 \cos \theta + F \times 3.0 \sin \theta = R \times 3.0 \cos \theta$. Substituting for $F$ and $R$ gives $1.5 \cos \theta + 3.0\, \mu \sin \theta = 3.0 \cos \theta$. Thus $3.0\, \mu \sin \theta = 1.5 \cos \theta$ and so $\tan \theta = 1.5/(3.0 \times 0.30) = 1.67$ and $\theta = 59°$.

The other solution involves realizing that for three forces acting on an object to be in equilibrium their lines of action must pass through a single point. The two forces $R$ and $F$ both act at the same point and can be combined to give just one force $Z$ at that point. Figure A6.3 shows the forces involved. $Z$ must be at an angle $\phi$

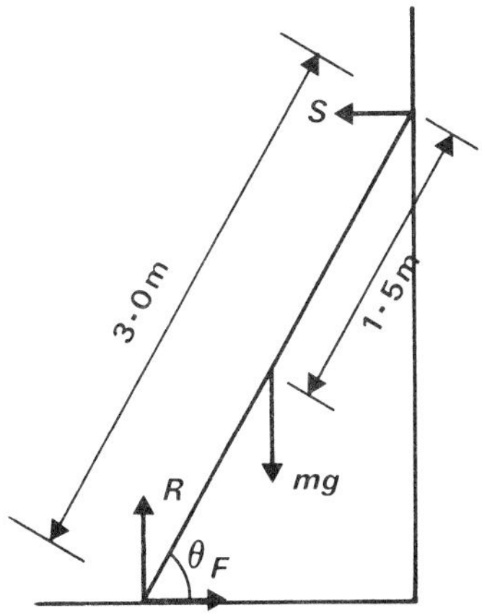


Figure A6.2

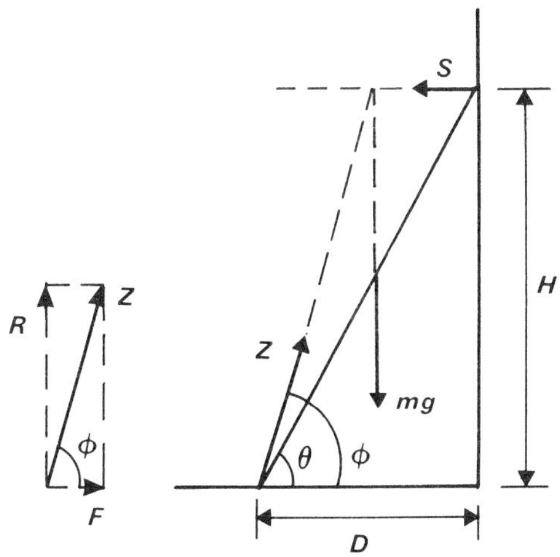


Figure A6.3

where tan $\phi = H/(D/2)$. But tan $\phi = R/F$. Thus, as $F = \mu R$ then tan $\phi = 1/\mu = H/(D/2)$. But tan $\theta = H/D$ and so $1/\mu = 2 \tan \theta$ and we have tan $\theta = 1/2\mu = 1.67$ and so $\theta = 59°$.

**6.8** Figure A6.4 shows the situation, $R$ and $S$ being the reaction forces. Two possible methods can be used, as in SAQ 6.7.

Using the general method, for the vertical forces to be in equilibrium $R = mg$ and for the horizontal forces $F = S$. As $F = \mu R$ then $F = \mu mg$. Taking moments about the upper contact point of the plank with the wall gives $mg \times 1.0 + F \times 4.6 = R \times 2.0$. Substituting for $F$ and $R$ gives $mg + 4.6\,\mu mg = 2.0\,mg$ and so $\mu = 1.0/4.6 = 0.22$.

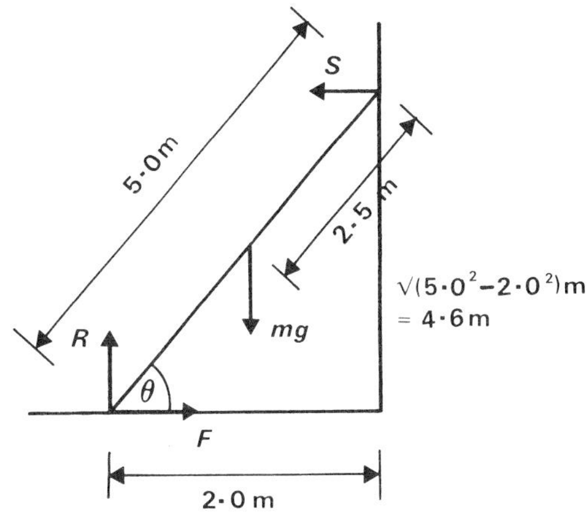


Figure A6.4

The solution involving the three concurrent forces is exactly the same as for SAQ 6.7, and so tan $\theta = 1/2\mu$. Hence, as tan $\theta = 4.6/2.0$, $\mu = 0.22$.

## Answers III

**6.9** Resultant force down the slope = component of weight down the slope − frictional force = $mg \sin \theta - \mu R$ and, as $R = mg \cos \theta$, resultant force = $mg \sin \theta - \mu mg \cos \theta$. As $F = ma$ the acceleration down the slope = $g \sin \theta - \mu g \cos \theta = 9.8 \sin 40° - 0.20 \times 9.8 \cos 40° = 4.8\ \mathrm{m\,s^{-2}}$. Using $s = ut + \frac{1}{2}at^2$, where $u = 0$, gives $0.35 = \frac{1}{2} \times 4.8t^2$ and so $t = 0.38$ s.

**6.10** Equating the momentum before the collision with the momentum after the collision gives $0.010 \times 400 = 0.51u$ and so the velocity $u$ after impact is $7.84\ \mathrm{m\,s^{-1}}$. Using $v^2 = u^2 + 2as$ with $v = 0$ when the block has come to rest, then $0 = 7.84^2 + 2 \times 18a$ and so $a = -1.71\ \mathrm{m\,s^{-2}}$. The force giving this retardation $F = ma = 0.51 \times 1.71 = 0.87$ N. $\mu = F/R$, but $R = mg = 0.51 \times 9.8$ N and so $\mu = 0.87/(0.51 \times 9.8) = 0.17$.

# Lesson 7 Motion in a circle

## Aims

After completing this lesson you should be able to
- explain how with circular motion at constant speed there is a *centripetal* acceleration
- derive and use the relationships for centripetal acceleration and force.

## Prerequisites

Lesson 3, where velocity is considered as a vector quantity, and Lesson 4 on Newton's laws.

## STUDY

**I Read** about motion in a circle and the derivation of the equations $a = v^2/r$ and $F = mv^2/r$.
References: *B* PP ch. 2, *B* ST1 ch. 4, *D* ch. 7, *N* ch. 2, *W et al.* ch. 8(7), *Wh&H* ch. 2.

### Self-assessment questions

**7.1** Consider an object moving in a circular path with a constant speed.
(a) What is the direction of the velocity at some instant and then quarter of a revolution later?
(b) Explain why, although the speed is constant, we say the velocity is changing.
(c) What would happen if there were no force acting on the object?

**7.2** Consider an object moving in a circular path with constant speed $v$, as in Figure 7.1(a).
(a) In time $\delta t$ the object moves from A to B round the arc. How is the distance covered related to the time $\delta t$?
(b) The arc length AB $= r\,\delta\theta$, hence using your answer from (a) arrive at the expression $\delta\theta = v\,\delta t/r$.
(c) The object changes velocity in going from A to B. Figure 7.1(b) shows the vector diagram to enable the change in velocity in going from A to B to be determined. Explain how this diagram is arrived at and why the change in velocity is represented by the line indicated.
(d) To a reasonable approximation, since we are considering a small angle $\delta\theta$, the length XZ can be considered to be an arc of a circle of radius $v$ centred at Y (the *size* of $v_A$ is the same as that of $v_B$, i.e. the speed $v$). Then, as arc length equals radius multiplied by angle subtended, we can write

$$\text{velocity at B relative to that at A} = v\,\delta\theta$$

But the velocity at B relative to that at A is the change in velocity occurring in the time $\delta t$. Hence we arrive at the relationship

$$a = \frac{v^2}{r}$$

(e) What is the direction of this acceleration?
(f) What is the centripetal force if the object has a mass $m$?

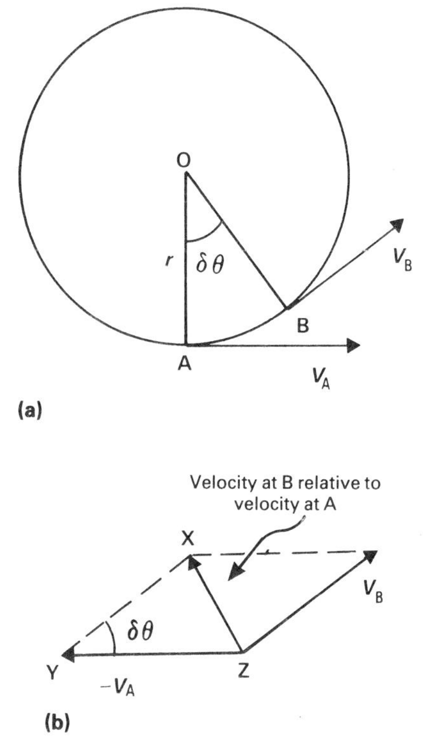


**Figure 7.1**

**7.3** What provides the centripetal force in the following situations?
(a) An object on the end of a string is whirled round in a horizontal circle.
(b) A car travels round a bend in the road.
(c) The moon orbits the earth.

**II Read** about the solution of problems involving circular motion, e.g. motion in horizontal and vertical circular paths, banking of roads.
References: *B* ST1 ch. 4, *D* ch. 7, *N* ch. 2, *Wh&H* ch. 3.

**Self-assessment questions**

**7.4** An object of mass 0.20 kg on the end of a string is whirled round in a horizontal circle of radius 0.80 m with a constant speed of 4.0 m $s^{-1}$. What is the tension in the string?

***7.5** An object of mass 0.50 kg on the end of a string is whirled round in a vertical circle of radius 1.0 m with a constant speed of 5.0 m $s^{-1}$. What are the maximum and minimum tensions in the string? ($g$ = 9.8 m $s^{-2}$)

***7.6** An object of mass 0.20 kg is whirled in a horizontal circle of radius 0.60 m by a string inclined at 30° to the vertical. Calculate (a) the tension in the string and (b) the speed, if the motion is one of constant speed. ($g$ = 9.8 m $s^{-2}$)

**7.7** Consider a car turning a corner of radius $r$ as a simple particle moving in a circular path. If the corner is banked at an angle $\theta$ to the horizontal and no frictional forces are involved, what is (a) the vertical upward force acting on the car, (b) the horizontal force responsible for the centripetal force, (c) the angle of banking in terms of the speed $v$, the radius $r$ and the acceleration due to gravity? Part (c) can be derived from parts (a) and (b).

***7.8** Calculate the angle of banking required on a corner of radius 80 m if cars are to travel round the corner at 80 km $h^{-1}$ without relying on frictional forces. ($g$ = 9.8 m $s^{-2}$)

***7.9** What is the (constant) speed of an aircraft moving in a horizontal circle of radius 300 m if the acceleration experienced by the pilot is 4$g$? ($g$ = 9.8 m $s^{-2}$)

# SELF-ASSESSMENT

## Key points

An object moving in a circular path with a constant speed has a **centripetal acceleration** given by $a = v^2/r$. This centripetal acceleration is provided by a **centripetal force** $F = mv^2/r$. In the absence of such a force an object will continue to move with a constant velocity, i.e. a constant speed in a constant straight line.

## Answers I

**7.1** (a) The velocity is tangential to the circular path and thus the two velocities are at right angles to each other.
(b) Speed can be constant if the direction is changing; velocity can only be constant if the direction does not change. Speed is a scalar and velocity a vector quantity. Thus in circular motion the velocity cannot be constant because the direction is continually changing.
(c) It would continue in linear motion with a constant velocity. A resultant force is needed to change the direction of the velocity.

**7.2** (a) arc AB = $v\,\delta t$
(b) $r\,\delta\theta = v\,\delta t$ and so $\delta\theta = v\,\delta t/r$
(c) See Lesson 3 where relative velocity is considered. The vector diagram is used to find the velocity at B relative to the velocity at A and hence the change in velocity.
(d) Change in velocity in time $\delta t = v\,\delta\theta = v^2\,\delta t/r$ and so $a$ = change in velocity/$\delta t = v^2/r$.
(e) The direction of the acceleration is towards the centre of the circle. The line XZ along which the acceleration is directed bisects the angle between the velocities at A and at B. This bisection means that XZ must be directed towards the centre of the circle.
(f) $F = ma = mv^2/r$

**7.3** (a) Tension in the string.
(b) On a flat road, friction between the road and the tyres. See SAQ 7.7 for a banked road where the force can be provided by the component of the weight of the car in the horizontal direction.
(c) Gravitational forces.

## Answers II

**7.4** $F = mv^2/r = 0.20 \times 4.0^2/0.80 = 4.0$ N

**7.5** The centripetal force required is $mv^2/r$. At the top of the path the weight $mg$ of the object acts towards the centre. Thus $T + mg = mv^2/r$ at the top of the path. At the bottom of the path the weight acts away from the centre and thus $T - mg = mv^2/r$. Hence the maximum tension occurs at the bottom of the path and the minimum at the top. At the bottom of the path $T = (0.50 \times 5.0^2/1.0) + 0.50 \times 9.8 = 17.4$ N. At the top of the path $T = (0.50 \times 5.0^2/1.0) - 0.5 \times 9.8 = 7.6$ N.

**7.6** This arrangement is sometimes known as the conical pendulum. Figure A7.1 shows the forces acting on the object. $T\sin(90 - \theta) = mg$, hence $T\cos\theta = mg$, for the vertical forces. The horizontal force is $T\cos(90 - \theta)$ or $T\sin\theta$. This provides the centripetal force and so $T\sin\theta = mv^2/r$.
(a) Using the vertical force equation $T = mg/\cos\theta = 0.20 \times 9.8/\cos 30° = 2.3$ N.
(b) Using the centripetal equation $v^2 = (rT\sin\theta)/m = (0.60 \times 2.3 \times \sin 30°)/0.20$ and so $v$ = 1.9 m $s^{-1}$.

**7.7** See Figure A7.2. (a) $R\cos\theta = mg$ (b) $R\sin\theta = mv^2/r$ (c) Dividing equation (b) by equation (a) gives $\tan\theta = v^2/rg$.

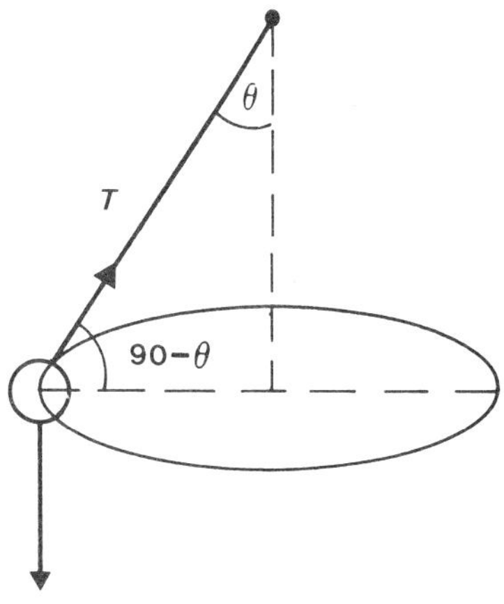


Figure A7.1

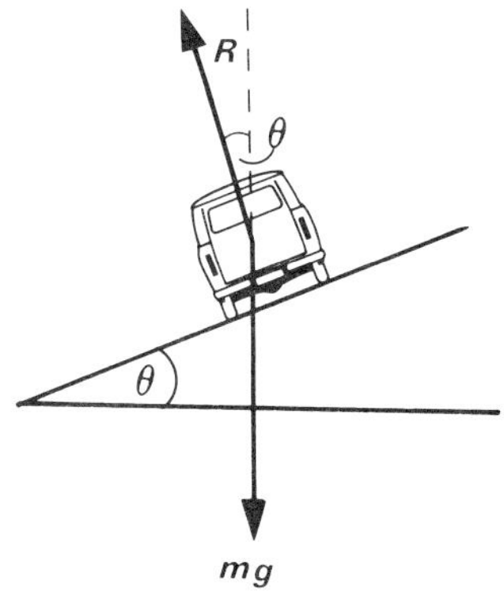


Figure A7.2

**7.8** $\tan\theta = v^2/rg = (80 \times 10^3/3600)^2/(80 \times 9.8)$ and hence $\theta = 32.2°$.

**7.9** $a = v^2/r = 4 \times 9.8 = v^2/300$, hence $v = 108\,\text{m}\,\text{s}^{-1}$.

# Lesson 8 Angular motion

## Aims

After completing this lesson you should be able to

- define and use the terms *angular displacement*, *angular velocity*, *angular acceleration*, and *frequency*
- derive and use the equations relating linear and angular motions
- derive and use the equations for uniformly accelerated angular motion.

## Prerequisites

Lesson 2 on linear motion.

## STUDY

**I Read** about the definitions of angular displacement, angular velocity, angular acceleration and frequency. References: *B* PP ch. 2, *B* ST1 ch. 4, *D* ch. 7, *N* ch. 2, *W et al.* (ch. 7), *Wh&H* ch. 6.

**Self-assessment questions**

**8.1** With linear motion (see Lesson 2) it is possible to distinguish between *distance* and *displacement*, with the distance being the total distance covered, regardless of the path followed, and displacement being the straight-line distance between two points. Distinguish between *angle* and *angular displacement* in a similar manner.

**8.2** The relationship between arc length $s$, the radius $r$ and the angle subtended by the arc $\theta$ is (see Figure 8.1)

$$s = r\theta$$

Use this equation to define the *radian*.

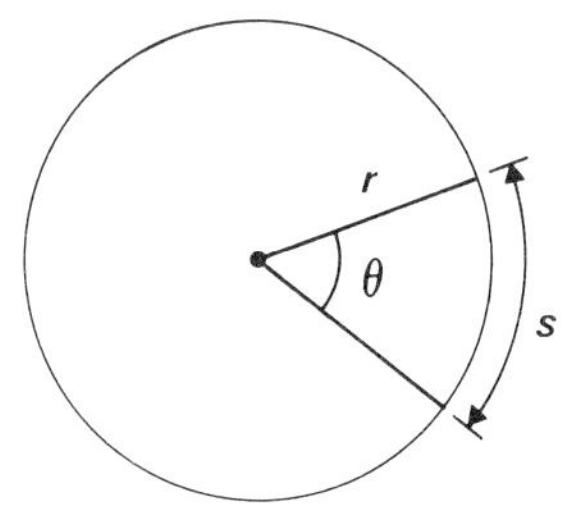


Figure 8.1

**8.3** How many radians is the angular displacement when there is a rotation through (a) a quarter of a revolution, (b) a full revolution?

**8.4** Define *angular velocity* and state its unit.

**8.5** If the frequency of revolution is $n$, i.e. the number of complete revolutions per second, what is the angular velocity?

**8.6** Define *angular acceleration* and state its unit.

**8.7** A car tyre has a diameter of 560 mm. What is (a) the distance moved by a point on the tyre tread when it rotates through 1 rad, (b) the angular velocity of the wheel if it completes 10 revolutions per second?

**8.8** The angular velocity of a wheel changes from 5.0 rad $s^{-1}$ to 12 rad $s^{-1}$ in 20 s. What is the average angular acceleration over that time?

**II** **Read** about the derivation of the relationships between angular and linear velocities, angular and linear accelerations.
References: *B* PP ch. 2, *B* ST1 ch. 2, *D* ch. 7, *N* ch. 2, *W et al.* (ch. 7), *Wh&H* ch. 6. Only *B* ST1 and *Wh&H* give both relationships.

**Self-assessment questions**

**8.9** Consider a point on the rim of a wheel of radius $r$ moving with a constant angular velocity $\omega$.
(a) In a time $t$ what is the angle $\theta$ swept out?
(b) The linear speed $v$ is the distance covered round the arc divided by the time taken. What is the distance covered when the angle swept out is $\theta$?
(c) Hence arrive at a relationship between $v$ and $\omega$.

**8.10** A wheel has a radius of 120 mm and rotates at 20 revolutions per second. What is the linear velocity at the circumference of the wheel?

***8.11** A car has a constant linear velocity of 40 m $s^{-1}$. If the wheels of the car have a radius of 350 mm what is their angular velocity?

**8.12** Consider the motion of a point on the rim of a wheel when the angular velocity is changing from $\omega_0$ to $\omega_1$ in a time $t$.
(a) What is the average angular acceleration $\alpha$ over that time?
(b) The point on the rim will have both a centripetal acceleration (see Lesson 7) and a tangential acceleration. If in this time $t$ the linear velocity of the point changes from $v_0$ to $v_1$, what is the tangential linear acceleration?
(c) Using the relationship $v = r\omega$, where $r$ is the radius of the wheel, derive a relationship between the tangential linear acceleration and the angular acceleration.

***8.13** The angular velocity of a car wheel changes from 4.0 rad $s^{-1}$ to 20 rad $s^{-1}$ in 25 s. If the wheel has a radius of 350 mm, what is (a) the average angular acceleration and (b) the average tangential linear acceleration of a point on the rim of the wheel?

**III** **Refer** to the derivations of the equations for uniformly accelerated linear motion in Lesson 2. The next group of questions concerns the equations for uniformly accelerated *angular* motion, which are derived in a comparable way.

**Self-assessment questions**

**8.14** Derive the equation $\omega_1 = \omega_0 + \alpha t$ from the definition of angular acceleration, given that there is constant angular acceleration $\alpha$ and that the object's angular velocity changes from $\omega_0$ to $\omega_1$ in time $t$.

**8.15** The average angular velocity for the object described in SAQ 8.14 is $\frac{1}{2}(\omega_0 + \omega_1)$. But the average angular velocity is $\theta/t$, where $\theta$ is the angular displacement in time $t$. Hence derive the relationship $\theta = \omega_0 t + \frac{1}{2}\alpha t^2$.

**8.16** Using the relationships derived in SAQs 8.14 and 8.15, derive the relationship $\omega_1^2 = \omega_0^2 + 2\alpha\theta$.

***8.17** A wheel has an angular velocity of 20 rad $s^{-1}$ before brakes are applied and the wheel is brought to rest with a constant retardation in 12 s. What is (a) the average angular retardation and (b) the number of revolutions the wheel makes in coming to rest?

***8.18** The armature of an electric motor is running at 50 revolutions per second. When the current to the motor is switched off it comes to rest, with uniform retardation, in 10 s. What is (a) the average angular retardation and (b) the number of revolutions the armature makes in coming to rest?

***8.19** A flywheel rotates through 40 complete revolutions in attaining an angular velocity of 30 rad $s^{-1}$ from rest. What is (a) the angular acceleration and (b) the time taken to attain this angular velocity, if the motion is one of constant angular acceleration?

# SELF-ASSESSMENT

## Key points

**Angular displacement** is the angular difference between two positions, the unit used being the **radian**. One radian is the angle subtended at the centre of a circle by an arc equal in length to the radius.

$$\text{Arc length } s = r\theta$$

where $r$ is the radius of the arc and $\theta$ the angle subtended by the arc.
**Angular velocity** is the rate of change of angular displacement with time, symbol $\omega$ and unit rad $s^{-1}$.

$$\text{Average angular velocity } \omega = \frac{\theta}{t}$$

where $\theta$ is the angular displacement in time $t$.
**Angular acceleration** is the rate of change of angular velocity with time, symbol $\alpha$ and unit rad $s^{-2}$.

$$\text{Average angular acceleration } \alpha = \frac{(\omega_1 - \omega_0)}{t}$$

where $\omega_0$ is the initial angular velocity and $\omega_1$ the angular velocity after a time $t$.
The angular velocity is related to the frequency of rotation $n$ by

$$\omega = 2\pi n$$

The relationships between linear and angular motion are

$$v = r\omega \quad \text{and} \quad a = r\alpha$$

The **equations of motion** for a constant angular acceleration are:

$$\omega_1 = \omega_0 + \alpha t$$

$$\theta = \omega_0 t + \tfrac{1}{2}\alpha t^2$$

$$\omega_1^2 = \omega_0^2 + 2\alpha\theta$$

## Answers I

**8.1** *Angle* is the total angle covered, regardless of path. Thus an object could oscillate back and forth and the angle would be the *total* angle covered. *Angular displacement* is the angle between the initial and final positions. Unless the rotation is in just one sense and plane it will not be the same as the angle covered.

**8.2** See the key points.

**8.3** (a) $\frac{1}{2}\pi$ (b) $2\pi$ (It is customary not to fully work out these angles but to quote them in terms of $\pi$.)

**8.4** See the key points.

**8.5** One complete revolution is $2\pi$ radians, hence $n$ revolutions is an angular displacement of $2\pi n$ rad. This is the angular displacement in 1 s, hence the angular velocity $\omega = 2\pi n$.

**8.6** See the key points.

**8.7** (a) $s = r\theta = 280 \times 1 = 280$ mm (b) $\omega = 2\pi n = 2\pi \times 10 = 20\pi$ rad $s^{-1}$

**8.8** $\alpha = (\omega_1 - \omega_0)/t = (12 - 5)/20 = 0.35$ rad $s^{-2}$

## Answers II

**8.9** (a) $\theta = \omega t$ (b) $s = r\theta$ (c) $v = s/t = r\theta/t = r\omega$

**8.10** $v = r\omega = r \times 2\pi n = 0.120 \times 2\pi \times 20 = 15.1$ m $s^{-1}$

**8.11** $v = r\omega$, hence $\omega = v/r = 40/0.350 = 114$ rad $s^{-1}$

**8.12** (a) $\alpha = (\omega_1 - \omega_0)/t$ (b) $a = (v_1 - v_0)/t$ (c) $a = (r\omega_1 - r\omega_2)/t = r\alpha$

**8.13** (a) $\alpha = (\omega_1 - \omega_0)/t = (20 - 4)/25 = 0.64$ rad $s^{-2}$ (b) $a = r\alpha = 0.350 \times 0.64 = 0.22$ m $s^{-2}$

## Answers III

**8.14** $\alpha = (\omega_1 - \omega_0)/t$, hence $\alpha t = \omega_1 - \omega_0$ and so $\omega_1 = \omega_0 + \alpha t$.

**8.15** $\theta/t = \frac{1}{2}(\omega_1 + \omega_0)$, hence $\theta = \frac{1}{2}\omega_1 t + \frac{1}{2}\omega_0 t$. But $\omega_1 = \omega_0 + \alpha t$, hence $\theta = \frac{1}{2}(\omega_1 + \alpha t)\, t + \frac{1}{2}\omega_0 t = \omega_0 t + \frac{1}{2}\alpha t^2$.

**8.16** $\omega_1^2 = (\omega_0 + \alpha t)^2 = \omega_0^2 + 2\omega_0\alpha t + \alpha^2 t^2 = \omega_0^2 + 2\alpha(\omega_0 t + \frac{1}{2}\alpha t^2) = \omega_0^2 + 2\alpha\theta$

**8.17** (a) $\omega_1 = \omega_0 + \alpha t$, hence $0 = 20 + 12\alpha$ and $\alpha = -1.67$ rad $s^{-2}$.
(b) $\theta = \omega_0 t + \frac{1}{2}\alpha t^2 = 20 \times 12 - \frac{1}{2} \times 1.67 \times 12^2 = 119.8$ rad. Number of revolutions $= 119.8/2\pi = 19.1$.

**8.18** (a) $\omega_1 = \omega_0 + \alpha t$, hence as $\omega_0 = 2\pi n$, we have $0 = 2\pi \times 50 + 10\alpha$ and $\alpha = -31.4$ rad $s^{-2}$.
(b) $\theta = \omega_0 t + \frac{1}{2}\alpha t^2 = 2\pi \times 50 \times 10 - \frac{1}{2} \times 31.4 \times 10^2 = 1572$ rad. Number of revolutions $= 1572/2\pi = 250.1$.

**8.19** (a) $\omega_1^2 = \omega_0^2 + 2\alpha\theta$, hence $\alpha = (\omega_1^2 - \omega_0^2)/2\theta = (30^2 - 0)/(2 \times 40 \times 2\pi) = 1.8$ rad $s^{-2}$. (b) $\omega_1 = \omega_0 + \alpha t$, hence $t = (\omega_1 - \omega_0)/\alpha = (30 - 0)/1.8 = 17$ s.

# Lesson 9 Gravitation

## Aims

After completing this lesson you should be able to
- state Kepler's laws
- state and use Newton's law of gravitation
- derive the relationship between the acceleration due to gravity and the universal gravitational constant, and the variation of the acceleration with both height and depth from the earth's surface.

## Prerequisites

Lesson 4 on Newton's laws and Lesson 7 on motion in a circle.

# STUDY

**I Read** about Kepler's laws and Newton's law of gravitation.
References: *B* PP ch. 1, ch. 2, *B* ST5 ch. 1, *D* ch. 7, *N* ch. 2, *W et al.* ch. 36(28), *Wh&H* ch. 20. *B* PP gives some of the historical background.

**Self-assessment questions**

**9.1** State Kepler's three laws.

**9.2** State Newton's law of gravitation.

**9.3** State the basic principles of a method that can be used to measure the universal gravitational constant.

**9.4** Consider a planet moving in a circular orbit.
(a) If the planet has a mass $m$ and speed $v$ and the orbit has a radius $r$, what is the centripetal force necessary for the circular motion?
(b) If the planet takes a time $T$ to complete one orbit at a constant speed $v$, what is the centripetal force in terms of $T$?
(c) The centripetal force is provided by the gravitational force of attraction between the sun, mass $M$, and the planet. What is this force?
(d) Hence derive a relationship between the time taken to complete an orbit, $T$, and the radius $r$ of the orbit.

***9.5** The earth orbits the sun with an almost circular orbit. If the radius of this orbit is $1.5 \times 10^{11}$ m and the time taken for a complete orbit is 365 days, determine the mass of the sun. ($G = 6.7 \times 10^{-11}$ N m$^2$ kg$^{-2}$)

***9.6** The moon orbits the earth with an almost circular orbit. If the radius of this orbit is $3.84 \times 10^8$ m and the time taken for a complete orbit is 27.3 days, determine the mass of the earth. ($G = 6.7 \times 10^{-11}$ N m$^2$ kg$^{-2}$)

***9.7** An artificial satellite is placed in a stationary orbit above the earth's surface, i.e. it always remains above the same position on the earth's surface. Calculate the height of such a satellite above the earth's surface if the earth has a mass of $5.98 \times 10^{24}$ kg and a radius of $6.37 \times 10^6$ m. ($G = 6.7 \times 10^{-11}$ N m$^2$ kg$^{-2}$)

**II Read** about the acceleration due to gravity, its relationship with the universal gravitational constant and the derivation of the variation of the acceleration with both height above and depth below the surface of the earth.
References: *B* ST5 ch. 1, *D* ch. 7, *N* ch. 2, *W et al.* (ch. 28), *Wh&H* ch. 20.

**Self-assessment questions**

**9.8** Consider an object of mass $m$ at the surface of the earth where the acceleration due to gravity is $g$.
(a) What is the force acting on the object when it accelerates with acceleration $g$?
(b) If we assume that the earth, mass $M$, acts as if its entire mass were concentrated at its centre, and the earth has a radius $r$, what is the force acting on the object according to the law of gravitation?
(c) The answers to (a) and (b) must be equal. Hence arrive at a relationship between the acceleration due to gravity and the universal gravitational constant.

***9.9** Calculate the acceleration due to gravity at the surface of the moon if the moon has a radius of 1839 km and a mass of $7.36 \times 10^{22}$ kg. ($G = 6.7 \times 10^{-11}$ N m$^2$ kg$^{-2}$)

***9.10** If the earth is assumed to have a constant density, obtain its value knowing that the acceleration due to gravity at the earth's surface is 9.8 m s$^{-2}$ and the radius of the earth is $6.37 \times 10^6$ m.
($G = 6.7 \times 10^{-11}$ N m$^2$ kg$^{-2}$)

***9.11** The earth is elliptical, with the polar radius being $6.357 \times 10^6$ m and the equatorial radius $6.378 \times 10^6$ m. What is the resulting difference in the acceleration due to gravity at the pole and that at the equator if the earth has a mass of $6.0 \times 10^{24}$ kg? ($G = 6.7 \times 10^{-11}$ N m$^2$ kg$^{-2}$)

***9.12** (a) How does the acceleration due to gravity vary with distance from the centre of the earth when the distance is greater than the radius of the earth?
(b) Consider a point a distance $h$ above the earth's surface. How does the acceleration due to gravity at that height, $g'$, compare with the acceleration $g$ at the surface of the earth? Express your answer in terms of $h$ and the earth's radius $r$.
(c) If $h$ is very small compared with $r$ so that powers of $(h/r)$ greater than the first can be neglected, show that $g' = g(1 - 2h/r)$.

***9.13** The acceleration due to gravity at a point below the earth's surface is due entirely to that part of the earth's mass that is below the point concerned. Assume the earth to have a uniform density and derive the relationship $g' = (a/r)g$, where $g'$ is the acceleration due to gravity at a radius $a$ and $g$ the acceleration due to gravity at the surface where the radius is $r$.

***9.14** Sketch a graph showing how, if the earth is assumed to have a constant density, the acceleration due to gravity varies with distance from the centre of the earth to distances greater than the earth's radius.

# SELF-ASSESSMENT

## Key points

**Kepler's laws** can be stated as:
1 Each planet moves in an ellipse which has the sun at one focus.
2 The line joining the sun to the moving planet sweeps out equal areas in equal times.
3 The squares of the periods of revolution of the planets about the sun are proportional to the cubes of their mean distances from it, i.e. $T^2 \propto r^3$ or $r^3/T^2$ = a constant.
**Newton's law of gravitation:** Every particle of matter attracts every other particle with a force which is directly proportional to the product of their masses and inversely proportional to the square of their distances apart.

$$F = \frac{Gm_1m_2}{r^2}$$

$G$ is the universal gravitational constant.

## Answers I

**9.1** See the key points.

**9.2** See the key points.

**9.3** The gravitational force between objects in a laboratory is very small and hence difficult to measure. The method used by Cavendish consisted of a beam of negligible mass supported by a fine suspension wire (Figure A9.1). From each end of the beam a small lead sphere was suspended. Two larger lead spheres were placed on either side of the beam, each close to one of the suspended spheres. The resulting forces of attraction caused the beam to rotate slightly, twisting the suspension wire. The angle of deflection of the beam was then measured. *B* ST5 ch. 1 shows a photograph of the original apparatus, *N* ch. 2 describes a slightly different method and *B* PP ch. 1 mentions another method.

**9.4** (a) $F = mv^2/r$ (b) $v = 2\pi r/T$, hence $F = m(2\pi r/T)^2/r = 4\pi^2 mr/T^2$ (c) $F = GMm/r^2$ (d) $T^2 = (4\pi^2/GM)r^3$, i.e. $T^2 \propto r^3$. This is Kepler's third law.

**9.5** $F = mv^2/r = GMm/r^2$ and so $M = v^2r/G$. But $v = 2\pi r/T$ and so $M = 4\pi^2r^3/(GT^2) = 4\pi^2 \times (1.5 \times 10^{11})^3/(6.7 \times 10^{-11} \times 365^2 \times 24^2 \times 3600^2) = 2.0 \times 10^{30}$ kg.

**9.6** As in 9.5 above we have $M = 4\pi^2r^3/(GT^2) = 4\pi^2 \times (3.84 \times 10^8)^3/(6.7 \times 10^{-11} \times 27.3^2 \times 24^2 \times 3600^2) = 6.0 \times 10^{24}$ kg.

**9.7** $F = mv^2/r = GMm/r^2$ and $v = 2\pi r/T$, hence $r^3 = GMT^2/4\pi^2$. The stationary condition means that the orbit time is the same as the time taken for one revolution of the earth, i.e. 1 day. Hence $r^3 = 6.7 \times 10^{-11} \times 5.98 \times 10^{24} \times (24 \times 3600)^2/4\pi^2$ and so $r = 42.31 \times 10^6$ m. Hence distance above surface $= 42.31 \times 10^6 - 6.37 \times 10^6 = 35.94 \times 10^6$ m.

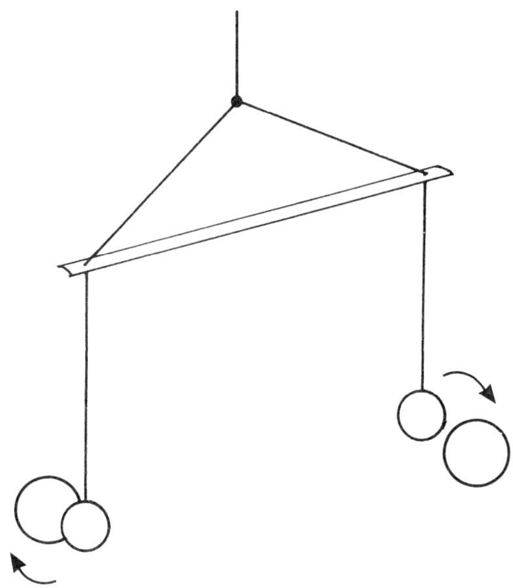

**Figure A9.1**

## Answers II

**9.8** (a) $F = mg$ (b) $F = GMm/r^2$ (c) $g = GM/r^2$

**9.9** $g = GM/r^2 = 6.7 \times 10^{-11} \times 7.36 \times 10^{22}/(1.839 \times 10^6)^2 = 1.46\ \text{m s}^{-2}$

**9.10** $g = GM/r^2 = G(\frac{4}{3}\pi r^3\rho)/r^2$ and so $\rho = 3g/(4\pi rG) = 3 \times 9.8/(4 \times \pi \times 6.37 \times 10^6 \times 6.7 \times 10^{-11}) = 5480\ \text{kg m}^{-3}$

**9.11** $g_p - g_e = (GM/r_p^2) - (GM/r_e^2) = 0.065\ \text{m s}^{-2}$.
This is not the only cause of a difference in the accelerations due to gravity at the equator and at the pole. Because of the earth's rotation, there is a centripetal force on bodies at the equator which does not occur at the poles.

**9.12** (a) $g$ is inversely proportional to $r^2$.
(b) $g' = GM/(r + h)^2$, also $g = GM/r^2$. Hence $g' = gr^2/(r + h)^2$.

(c) $g' = g(1 + h/r)^{-2}$. Using the binomial expansion and neglecting all but the first power term gives $g' = g(1 - 2h/r)$.

**9.13** $g = GM/r^2 = G(\frac{4}{3}\pi r^3 \rho)/r^2$. Similarly $g' = G(\frac{4}{3}\pi a^3 \rho)/a^2$. Hence $g/r = g'/a$ and so $g' = (a/r)g$.

**9.14** For radii up to the radius of the earth $g$ is proportional to the radius, for radii greater than the radius of the earth $g$ is proportional to $1/r^2$. Figure A9.2 shows the graph.

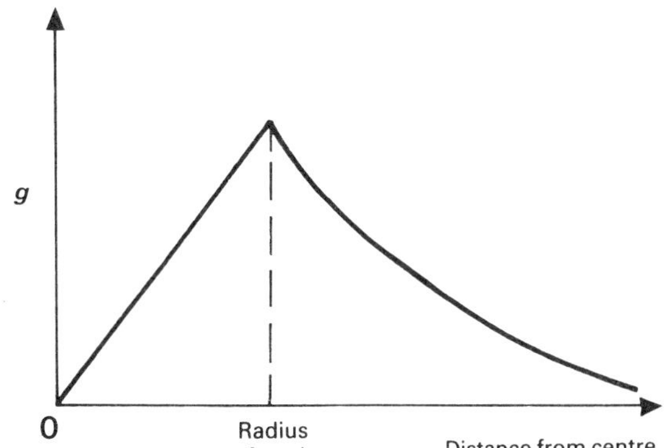


**Figure A9.2**

# Lesson 10 Gravitational fields

## Aims

After completing this lesson you should be able to:

- define gravitational *field strength*
- derive and use the relationship for the field strength due to a spherical mass
- define gravitational *potential* and use it in problems involving energy transformations
- derive and use the relationship for the potential near a sphere
- calculate escape velocities.

## Prerequisites

Lessons 5 and 9, and for one question a knowledge of calculus is assumed.

## STUDY

**I Read** about the definition of gravitational field strength and the derivation of the relationship for the field strength due to a spherical mass.
References: *B* PP ch. 11, *B* ST5 ch. 1, *D* ch. 7, *W et al.* ch. 36(28), *Wh&H* ch. 20.

### Self-assessment questions

**10.1** Define *gravitational field strength* and state its unit.

**10.2** How is the gravitational field strength related to the acceleration due to gravity?

**10.3** What is the gravitational field strength at the surface of the moon where the acceleration due to gravity is $1.6\,\mathrm{m\,s^{-2}}$?

**10.4** Consider a sphere of mass $M$ and assume that it behaves as though all the mass were concentrated at a point at its centre.
(a) What is the force on an object of mass $m$ a distance $r$ from the centre of the sphere, where $r$ is greater than the radius of the sphere?
(b) What is the gravitational field strength at this distance $r$?

***10.5** The gravitational field strength at the surface of the earth, radius $R$, is $9.8\,\mathrm{N\,kg^{-1}}$. What is the gravitational field strength at distances from the centre of the earth of (a) $2R$ and (b) $3R$?

***10.6** What is the gravitational field strength at a distance of $40 \times 10^6\,\mathrm{m}$ from the centre of the earth? Take the product of $G$ and the mass of the earth to be $4.0 \times 10^{14}\,\mathrm{N\,m^2\,kg^{-1}}$. (Radius of the earth = $6.4 \times 10^6\,\mathrm{m}$)

**II Read** about the definition of gravitational potential, the derivation of the relationship for the potential near a sphere, and the solution of problems involving potential, in particular the calculation of escape velocities.
References: *B* PP ch. 11, *B* ST5 ch. 1, *D* ch. 7, *N* ch. 2, *W et al.* ch. 38(29), *Wh&H* ch. 20.

### Self-assessment questions

**10.7** Define *gravitational potential* and state its unit.

**10.8** What is the work done in moving an object of mass 2 kg through a distance of 5 m in the opposite direction to a gravitational field of 10 N $kg^{-1}$, if the field is constant over the distance?

**10.9** Consider a sphere of mass $M$ and assume that it behaves as though all the mass were concentrated at a point at its centre. Now consider a small mass $m$ placed a distance $x$ from the centre of the sphere, with $x$ being greater than the radius of the sphere.
(a) What is the gravitational force acting on this mass $m$?
(b) What is the direction of this force?
(c) Would work have to be done to move the object towards the sphere?
(d) What is the work done when the object moves a small distance $\delta x$ towards the sphere? If an external agent has to do work, i.e. supply energy, to move the object closer to the sphere, then the work done is given a positive sign. If the external agent has to exert a restraining force to prevent the object moving closer to the sphere, then moving it closer means the work done is negative. Give the appropriate sign to the work done in this case.
(e) Use integral calculus to determine the work done in bringing the mass $m$ from infinity to a distance $r$ from the centre of the sphere.
(f) Hence obtain the equation for the potential at a distance $r$.

***10.10** A spacecraft returning to the earth from a trip to the moon is coasting through space without using its rocket motors. If it had a velocity of 5000 m $s^{-1}$ when 26 000 km from the centre of the earth, what would be its velocity when 20 000 km away? Take the value of $G$ multiplied by the mass of the earth to be $4.0 \times 10^{14}$ N $m^2$ $kg^{-1}$.

***10.11** Calculate the energy needed to raise a rocket from rest on the earth's surface to a distance of 600 km above the earth's surface and for it to have an outward velocity of 2000 m $s^{-1}$ at that point. The rocket has a mass of 1000 kg and the value of $G$ multiplied by the mass of the earth can be taken as $4.0 \times 10^{14}$ N $m^2$ $kg^{-1}$. Radius of the earth = $6.4 \times 10^6$ m.

***10.12** What kinetic energy must an object have at the surface of the earth if it is to escape to infinity from the earth? Hence derive a relationship for the escape velocity for objects to leave the earth.

***10.13** What is the escape velocity from the surface of the earth where the field strength is 9.8 N $kg^{-1}$? Radius of the earth = $6.4 \times 10^6$ m.

# SELF-ASSESSMENT

## Key points

A **gravitational field** is said to exist at a point if a force is exerted on an object placed at that point by virtue of its mass. The **gravitational field strength** $g$ at a point is defined as the force per unit mass at the point, i.e. $g = F/m$. The unit of field strength is N $kg^{-1}$. The field strength due to a spherical mass $M$ is, at a distance $r$ from the centre of the mass, where $r$ is greater than the radius of the sphere,

$$g = \frac{GM}{r^2}$$

The **gravitational potential** at a point, $V$, can be defined as the potential energy possessed by a unit mass at that point. By convention the zero of potential energy is taken to be at infinity. Hence the potential at a point can be defined as the work done in bringing unit mass from infinity to the point concerned. The potential $V$ at a distance $r$ from a spherical mass $M$ is, where $r$ is greater than the radius of the sphere,

$$V = -\frac{GM}{r}$$

The unit of potential is J $kg^{-1}$.

## Answers I

**10.1** See the key points.

**10.2** Field strength = $F/m$. But a force $F$ can cause an acceleration, in this case the acceleration due to gravity. Using $F = ma$, the acceleration is $F/m$. But this is the same as the field strength. Hence numerically the field strength and the acceleration due to gravity are the same. The units in which they are expressed differ.

**10.3** Gravitational field strength = 1.6 N $kg^{-1}$

**10.4** (a) $F = GMm/r^2$ (b) $g = F/m = GM/r^2$

**10.5** (a) $9.8/2^2 = 2.45$ N $kg^{-1}$ (b) $9.8/3^2 = 1.1$ N $kg^{-1}$

**10.6** $g = GM/r^2 = 4.0 \times 10^{14}/(40 \times 10^6)^2 = 0.25$ N $kg^{-1}$

## Answers II

**10.7** See the key points.

**10.8** $W = Fs = (gm)s = 10 \times 2 \times 5 = 100$ J

**10.9** (a) $F = GMm/x^2$
(b) Towards the centre of the sphere.

(c) Work is needed to move the object *away* from the sphere, work is done *by* the mass in moving towards the sphere and does not have to be supplied by any external agent.
(d) $-F\delta x = -(GMm/x^2)\delta x$

(e) $W = \int_r^\infty -\frac{GMm}{x^2}\,dx = -\frac{GMm}{r}$

(f) $V = W/m = -GM/r$

**10.10** k.e. gain = p.e. loss, hence $\frac{1}{2}mv_2^2 - \frac{1}{2}mv_1^2 = -(mV_2 - mV_1) = GMm(1/r_2 - 1/r_1)$. Hence $v_2^2 - v_1^2 = 2 \times 4.0 \times 10^{14}\,[(1/20 \times 10^6) - (1/26 \times 10^6)] = 9.23 \times 10^6$. Hence $v_2 = 5851\text{ m s}^{-1}$.

**10.11** Difference in potential energy $= mV_2 - mV_1$. Hence total energy required $= \frac{1}{2}mv^2 + mV_2 - mV_1 = \frac{1}{2}mv^2 - GMm(1/r_2 - 1/r_1) = \frac{1}{2} \times 1000 \times 2000^2 + 4.0 \times 10^{14} \times 1000 \times [-1/(7.0 \times 10^6) + 1/(6.4 \times 10^6)] = 7.4 \times 10^9$ J.

**10.12** $\frac{1}{2}mv^2 = GMm/R$, where $R$ is the radius of the earth, hence $v^2 = 2GM/R$. But $g = GM/R^2$ and so $v^2 = 2gR$.

**10.13** $v^2 = 2gR = 2 \times 9.8 \times 6.4 \times 10^6$ and so $v = 11.2 \times 10^3\text{ m s}^{-1}$

# Lesson 11 Simple harmonic motion

## Aims

After completing this lesson you should be able to

- define simple harmonic motion (s.h.m.)
- derive and use expressions for the (a) displacement–time, (b) velocity–displacement and (c) acceleration–displacement of an object performing s.h.m.
- explain the energy transformations occurring during oscillations and relate the energy of an oscillator to its amplitude.

## Prerequisites

A knowledge of linear, circular and angular motion, Newton's laws and energy is required, i.e. Lessons 2, 4, 5, 7 and 8. For one question a knowledge of calculus is required, there is however an alternative version of the question which does not require it.

## STUDY

**I Read** about the definition of simple harmonic motion and the equations of motion for it.
References: *B* PP ch. 4, *B* ST6 ch. 5, *D* ch. 8, *N* ch. 2, *W et al.* ch. 17, ch. 18 (19), *Wh&H* ch. 10.

**Self-assessment questions**

**11.1** Define simple harmonic motion.

**11.2** The following questions are concerned with a derivation of equations for s.h.m. by considering its relation with circular motion.

(a) Consider a point P moving round a circle of radius $r$ with a constant angular velocity $\omega$ (Figure 11.1). Describe in general terms how the horizontal projection of the radius OP, i.e. ON or $x$, varies with time.
(b) The point N moves back and forth along the diameter, the acceleration of N being the component of the acceleration of P in that direction. What is the centripetal acceleration of P, in terms of the angular velocity of P? Hence, what is the acceleration component along the diameter?
(c) How is the displacement of N from O, i.e. $x$, related to $\theta$?
(d) Use your answers to (b) and (c) to arrive at an equation relating the acceleration of N along the diameter to its displacement $x$ from O.

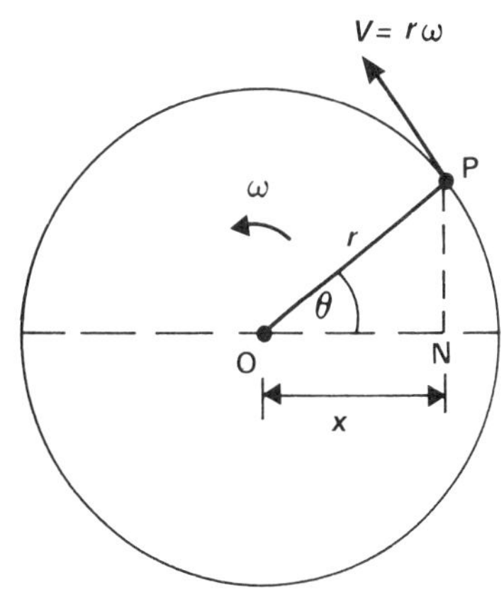


**Figure 11.1**

**11.3** This continues the previous question to give the velocity variation with displacement.
(a) The velocity of N along the diameter is given by the component of the velocity of P which is in that direction. If P has the velocity $v$ tangential to the circle, what is the component of $v$ along the diameter?
(b) Since $v = r\omega$ the component can be expressed in terms of the angular velocity. Use the equation $\sin^2\theta + \cos^2\theta = 1$ and the fact that the displacement $x = r\cos\theta$ to obtain a relationship between the velocity component, $r$, $\omega$ and $x$. This describes the velocity variation with displacement.

**11.4** This continues the previous two questions.
(a) If the point P makes $n$ complete oscillations per second, what is the relationship between $n$ and $\omega$?
(b) If $T$ is the time taken to complete one oscillation, this being known as the *periodic time*, what is the relationship between $T$ and $n$, and hence $\omega$?

**11.5** This is an alternative to SAQ 11.2 and SAQ 11.3 and involves the use of calculus. This version starts from the displacement equation; a version starting from the acceleration equation is given in *N* ch. 2.
(a) The variation of the displacement $x$ with time can be represented by the equation

$$x = r\cos\omega t$$

But velocity $v = \mathrm{d}x/\mathrm{d}t$. Hence derive an equation relating velocity and time.
(b) $\sin^2\theta + \cos^2\theta = 1$
Hence rearrange your answer to (a) to obtain the relationship

$$v = \pm\,\omega\sqrt{(r^2 - x^2)}$$

(c) Acceleration $a = \mathrm{d}v/\mathrm{d}t$. Hence differentiate your answer to (a) and obtain an equation relating the acceleration and the displacement.

**11.6** A particle, P, oscillates with simple harmonic motion along a line between two extremes A and B (see Figure 11.2) with O being the centre of the motion.
(a) What, at any instant, can be said about the direction of the acceleration?
(b) When is the acceleration (i) a maximum, (ii) zero?
(c) When is the velocity (i) a maximum, (ii) zero?

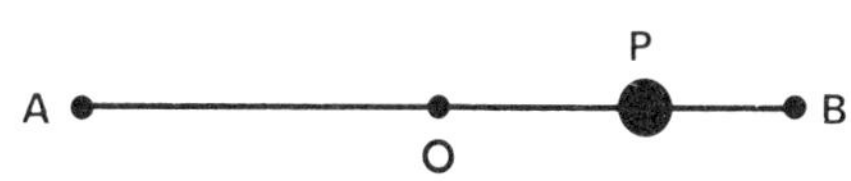


**Figure 11.2**

***11.7** A particle oscillating with simple harmonic motion has a velocity of 20 mm s$^{-1}$ when 60 mm from its equilibrium position and 10 mm s$^{-1}$ when at 120 mm. What is (a) the amplitude, (b) the frequency and (c) the maximum velocity of the oscillating particle?

***11.8** The displacement of an object oscillating with simple harmonic motion is given by $x = 10\cos 3\pi t$ where $x$ is the displacement in millimetres and $t$ the time in seconds. What is (a) the amplitude, (b) the periodic time and (c) the maximum velocity of the oscillating object?

**II** **Read** about the energy transformations involved in s.h.m. and the energy of an oscillator.
References: *B* PP ch. 4, *B* ST6 ch. 5, *D* ch. 8, *N* ch. 2, *W et al.* ch. 18(19), *Wh&H* ch. 10.

**Self-assessment questions**

**11.9** (a) What is the maximum value of the kinetic energy of an object of mass $m$ oscillating with simple harmonic motion?
(b) When is the kinetic energy a maximum?
(c) When is the kinetic energy zero? What happens to the energy at this point in the motion?
(d) If the oscillating object neither gains nor loses energy during its oscillation, after being set in motion, what must be the total energy due to the oscillation?
(e) Express the total energy in terms of $m$, $\omega$ and the maximum displacement $r$.

***11.10** A particle of mass 10 g oscillates with simple harmonic motion, having a periodic time of 0.5 s. What is the energy of the oscillating particle if the amplitude of the oscillation is 100 mm?

# SELF-ASSESSMENT

## Key points

**Simple harmonic motion** (s.h.m.) is said to occur if the acceleration of a body is directly proportional to its displacement $x$ from a fixed point and always directed towards that point.

$$\text{acceleration } a = -(\text{constant})x$$

The following equations describe the motion:

$$a = -\omega^2 x$$
$$x = r\cos\omega t$$
$$v = -\omega r\sin\omega t$$
$$v = \pm\,\omega\sqrt{(r^2 - x^2)}$$
$$T = \frac{1}{n} = \frac{2\pi}{\omega}$$

$\omega$ is the angular velocity, $r$ the amplitude, $T$ the periodic time, $n$ the frequency, $v$ the velocity and $t$ the time. The energy of a body oscillating with s.h.m. is given by

$$\text{energy} = \tfrac{1}{2}m\omega^2r^2$$

## Answers I

**11.1** See the key points.

**11.2** (a) The point N moves back and forth along the diameter. Thus the value of $x$, starting from N at the centre, increases to a maximum, then decreases back to zero, then increases in the opposite direction to a maximum before again decreasing back to zero.
(b) $a = v^2/r$ and as $v = r\omega$ we have $a = r\omega^2$. Component $= r\omega^2 \cos\theta$.
(c) $x = r\cos\theta$.
(d) $a = \omega^2x$. Because the acceleration is always directed towards O, in the opposite direction to which $x$ increases, the equation is written as $a = -\omega^2x$.

**11.3** (a) Component $= v\sin\theta$.
(b) Component $= r\omega\sin\theta = \pm r\omega\sqrt{(1 - \cos^2\theta)} = \pm r\omega\sqrt{(1 - (x/r)^2)} = \pm\omega\sqrt{(r^2 - x^2)}$

**11.4** (a) $\omega = 2\pi n$ (b) $T = 1/n = 2\pi/\omega$

**11.5** (a) $v = -\omega r\sin\omega t$ (b) $v = \pm\omega\sqrt{(r^2 - x^2)}$, derived as in 11.3(b) above. (c) $a = -\omega^2 r\cos\omega t = -\omega^2x$

**11.6** (a) The acceleration is always directed towards O.
(b) (i) At maximum displacement A or B. (ii) At O when the displacement is zero.
(c) (i) At O when the displacement is zero. (ii) At A or B when the displacement is a maximum.

**11.7** (a) $v = \pm\omega\sqrt{(r^2 - x^2)}$. If we consider just the displacement on one side of the equilibrium position then we can consider, say, just the positive sign for $v$. Thus $20 = \omega\sqrt{(r^2 - 60^2)}$ and $10 = \omega\sqrt{(r^2 - 120^2)}$. Dividing the first equation by the second gives $2 = \sqrt{(r^2 - 60^2)}/\sqrt{(r^2 - 120^2)}$. Squaring and rearranging the terms gives $4r^2 - 57\,600 = r^2 - 3600$ and so $3r^2 = 54\,000$ and $r = 134.2$ mm.
(b) Substituting for $r$ in the first velocity expression above gives $20 = \omega\sqrt{(134.2^2 - 60^2)}$ and so $\omega = 0.167$ rad s$^{-1}$. Hence as $\omega = 2\pi n$, $n = 0.0266$ Hz. Note that 1 hertz = 1 vibration per second.
(c) The maximum velocity occurs at the equilibrium position when $x = 0$. Thus $v = \pm\omega\sqrt{(r^2)} = \pm\omega r = \pm 0.167 \times 134.2 = \pm 22.4$ mm s$^{-1}$.

**11.8** (a) 10 mm (b) $\omega = 3\pi = 2\pi/T$ and so $T = 0.67$ s
(c) $v = -\omega r\sin\omega t$ and so the maximum velocity is when the sine term is $\pm 1$. Hence maximum velocity is $\omega r = 3\pi \times 10 = 94.2$ mm s$^{-1}$.

## Answers II

**11.9** (a) k.e. $= \frac{1}{2}mv^2$. As the maximum velocity is $\omega r$ (see previous question), the maximum kinetic energy is $\frac{1}{2}m(r\omega)^2$.
(b) The kinetic energy is a maximum when the velocity is a maximum, i.e. when the object is passing through the equilibrium position.
(c) The kinetic energy is zero when the velocity is zero, i.e. when the object is at the extreme ends of its oscillation. When this happens the kinetic energy has become transformed into potential energy.
(d) The entire oscillation is a continual transformation back and forth from kinetic energy to potential energy. Hence the total energy equals the maximum kinetic energy, or the maximum potential energy.
(e) Energy $= \frac{1}{2}mr^2\omega^2$. This means that the energy is proportional to the square of the amplitude $r$.

**11.10** Energy $= \frac{1}{2}mr^2\omega^2$. As $\omega = 2\pi/T$, energy $= \frac{1}{2} \times 0.010 \times 0.100^2 \times (2\pi/0.5)^2 = 7.9 \times 10^{-3}$ J.

# Lesson 12 Mechanical oscillations

## Aims

After completing this lesson you should be able to

- derive and use expressions for the frequency and periodic time of oscillation of objects performing simple harmonic motion
- describe the effects of damping, forced oscillations and resonance.

## Prerequisites

Lesson 11 on simple harmonic motion.

# STUDY

**I** **Read** about the derivation of the equations for the frequencies and periodic times for oscillating objects such as the simple pendulum and a mass on the end of a spring.
References: *B* PP ch. 4, *B* ST6 ch. 5, *D* ch. 8, *N* ch. 2, *W et al.* ch. 18(19), *Wh&H* ch. 10.

**Self-assessment questions**

**12.1** Figure 12.1 shows a mass tethered by a spring to a rigid support. When the mass is pulled so that it has a displacement $x$ from its equilibrium position, then the spring is extended by $x$ and there is a restoring force.

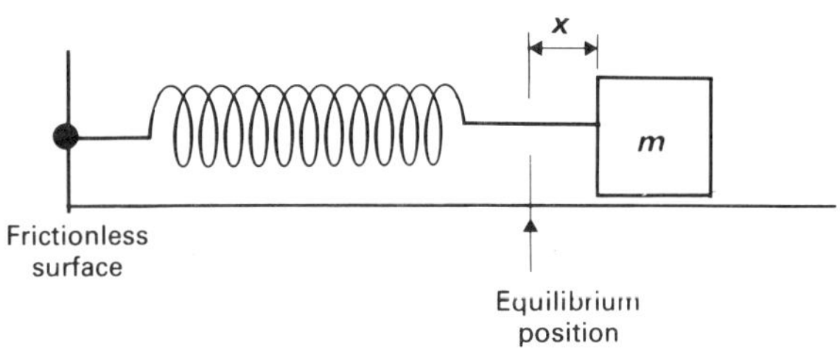


**Figure 12.1**

(a) If the spring obeys Hooke's law then the force $F$ is directly proportional to the extension $x$, i.e. $F = kx$ where $k$ is a constant called the spring or force constant. What is the acceleration of the mass if it is released when at extension $x$?
(b) What is the direction of the acceleration?
(c) Is the oscillation s.h.m.?
(d) For s.h.m. $a = -\omega^2 x$, hence for the mass on the spring what is the value of $\omega$?
(e) As $\omega = 2\pi/T$, derive an equation for the periodic time of the oscillation.

**12.2** When a mass is suspended from a vertical spring, stretching of the spring occurs and an extension $e$ is produced.
(a) What is the force causing the extension?
(b) For a spring obeying Hooke's law the extension is directly proportional to the restoring force and the relationship can be expressed as $F = ke$. Hence obtain, for the static situation, $mg = ke$.
(c) The mass is then pulled down a further distance $x$. What is the restoring force?
(d) What is the acceleration of the mass when it is released?
(e) Hence derive an equation for the periodic time of the oscillation if the motion is assumed to be simple harmonic.

***12.3** (a) Consider the situation shown in Figure 12.2(a), where two identical springs are in series. The mass applied to the end of the springs will stretch each spring by the same amount as if it were applied to each separately. Derive an equation for the periodic time of oscillation of the mass.
(b) Figure 12.2(b) shows two identical springs in parallel. What is the force acting on each spring? Derive an equation for the periodic time of oscillation of the mass.

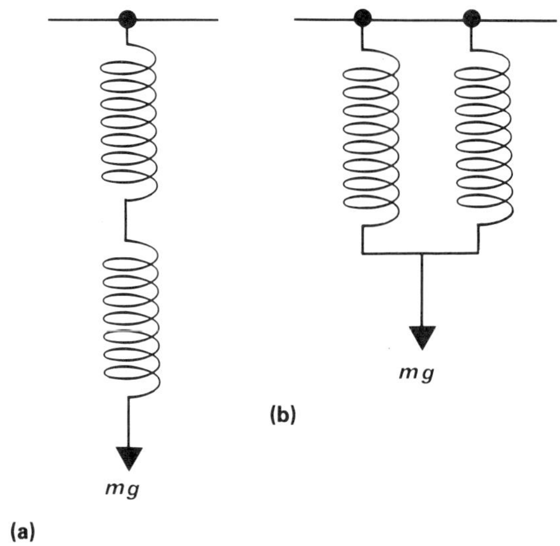


**Figure 12.2**

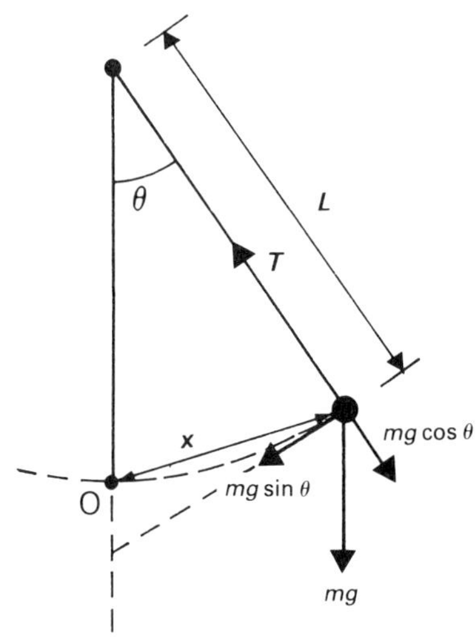


**Figure 12.3**

**12.4** Consider the simple pendulum shown in Figure 12.3.
(a) What is the restoring force when the pendulum bob is at a displacement $x$ from the equilibrium position?
(b) For small values of $\theta$, we have $\sin \theta$ approximating to $x/L$. Hence derive an expression for the periodic time of oscillation of a simple pendulum for small angles.

***12.5** A simple pendulum has a bob with a mass of 20 g. If it oscillates with a periodic time of 2.0 s and an amplitude of 100 mm, what is (a) the length of the pendulum, (b) the maximum velocity of the bob, (c) the maximum kinetic energy? ($g = 9.8\ \text{m s}^{-2}$)

***12.6** An object of mass 100 g is attached to the lower end of a vertically suspended spring. This causes the spring to extend by 20 mm. The mass is then pulled down a further 10 mm and released. What is (a) the periodic time of the oscillation and (b) the maximum kinetic energy of the mass? ($g = 9.8\ \text{m s}^{-2}$)

***12.7** An object of mass 120 g hangs from a vertically suspended spiral spring. When the mass is pulled down to 100 mm below its equilibrium position and then released it oscillates with a simple harmonic motion having a periodic time of 1.5 s. What is (a) the velocity of the mass when it passes through its equilibrium position, (b) the time taken to move to 100 mm above the equilibrium position, (c) the total energy of the oscillating system?

**II** **Read** about damping, forced oscillations and resonance.
References: *B* PP ch. 4, *B* ST6 ch. 5, *D* ch. 8, *W et al.* ch. 19(20), *Wh&H* ch. 11.

**Self-assessment questions**

**12.8** Sketch the form of the displacement–time graphs for oscillations which are (a) lightly damped, (b) critically damped and (c) heavily damped.

***12.9** Consider a simple pendulum with a frequency of oscillation $n$, and suppose such a pendulum to be attached to a support which can be set into continuous oscillation. The pendulum is said to be subject to *forced vibration*.
(a) Sketch a graph showing how, for different degrees of damping, the amplitude of oscillation of the pendulum will depend on the forcing frequency. Consider forcing frequencies both smaller and larger than the frequency $n$.
(b) Explain the term *resonance*.

****12.10** For slight damping of an oscillating system the amplitude of the oscillation decays exponentially. If the initial amplitude is $a$ then the amplitude after a time $t$ is $ae^{-kt}$, where $k$ is the decay constant (not to be confused with the force constant).
(a) Sketch a graph showing how the displacement varies with time for such a system.
(b) If for such a system the amplitude diminishes to half its initial value in 20.5 s, what is the value of $k$?

# SELF-ASSESSMENT

## Key points

To determine whether an object is performing s.h.m.:
1 Consider the body to be given some displacement $x$.
2 Determine the restoring force acting to bring the body back to its equilibrium position.
3 Use $F = ma$ to determine the acceleration.
4 If the motion is simple harmonic the acceleration will follow the relationship

$$\text{acceleration} = -(\text{constant})x$$

In order to determine the periodic time of such an oscillation, equate the constant in the above expression to $\omega^2 = (2\pi/T)^2$.
In practice oscillations are damped, i.e. energy is taken from the oscillating system. With **slight damping** there is an oscillation but the amplitude decreases with time. With **critical damping** there is no real oscillation, and the time taken for the displacement to become effectively zero is a minimum. With **heavy damping** there is only a very slow return of the system to the initial equilibrium position.
**Forced vibration** is when a continuous oscillation is applied by some external agent to an oscillator. When the forcing vibration frequency equals the natural frequency of oscillation of the system then **resonance** is said to occur. When this happens the energy of the system becomes a maximum.

## Answers I

**12.1** (a) $F = kx = ma$ and hence $a = (k/m)x$. Because the acceleration is in the opposite direction to that in which $x$ increases, a minus sign is included to give $a = -(k/m)x$.
(b) The acceleration is always directed towards the equilibrium position.
(c) Yes, because the acceleration = $-(\text{constant})x$.
(d) $\omega = \sqrt{(k/m)}$ (e) $\omega = 2\pi/T$, so $T = 2\pi\sqrt{(m/k)}$.

**12.2** (a) The weight $mg$.
(b) For equilibrium $ke = mg$.
(c) $F = k(e + x) - mg$. This is the resultant upwards force.
(d) $F = ma = ke + kx - mg = kx$ *upwards*, hence $a = -(k/m)x$.
(e) Thus $\omega^2 = k/m$ and as $\omega = 2\pi/T$ then $T = 2\pi\sqrt{(m/k)}$.

**12.3** (a) There will be twice the extension that would have occurred with just the single spring. For one of the springs we have $F = ke = mg$. But the total extension is $2e$, or $2mg/k$, and so $mg = (k/2) \times$ total extension. The arrangement is thus behaving as a single spring with a spring constant of $\frac{1}{2}k$. Thus using the answer obtained in the previous question $T = 2\pi\sqrt{(m/\frac{1}{2}k)}$.
(b) Force on each spring $= \frac{1}{2}mg$ so the extension is only half as much as would have occurred with a single spring. Hence total extension $= \frac{1}{2}mg/k$ and so $mg = (2k) \times$ total extension. The arrangement is behaving as a single spring with a spring constant of $2k$. Hence $T = 2\pi\sqrt{(m/2k)}$.

**12.4** (a) $mg \sin\theta$ (b) $F = mg\,(x/L) = ma$ and so $a = (g/L)x$. We need to include a minus sign because the direction of the acceleration is opposite to the direction in which $x$ increases. Hence $a = -(g/L)x$ and so $T = 2\pi\sqrt{(L/g)}$.

**12.5** (a) $T = 2\pi\sqrt{(L/g)}$ and so $L = gT^2/4\pi^2 = 9.8 \times 2.0^2/4\pi^2 = 0.99$ m.
(b) $\omega = 2\pi/T = 2\pi/2.0 = \pi$ rad s$^{-1}$. Maximum velocity $= \omega r$ (see answer to SAQ 11.7) $= \pi \times 0.100 = 0.31$ m s$^{-1}$.
(c) Maximum kinetic energy is when velocity is maximum and so is $\frac{1}{2} \times 0.020 \times 0.31^2 = 9.6 \times 10^{-4}$ J.

**12.6** (a) $F = ke = mg$ and so $k = 0.100 \times 9.8/0.020 = 49$ N m$^{-1}$. $T = 2\pi\sqrt{(m/k)} = 2\pi\sqrt{(0.100/49)} = 0.28$ s.
(b) Maximum k.e. $=$ total energy $= \frac{1}{2}m\omega^2r^2$ (see previous lesson) and as $\omega = 2\pi/T$ the maximum k.e. $= \frac{1}{2} \times 0.100 \times (2\pi/0.28)^2 \times 0.010^2 = 2.5 \times 10^{-3}$ J.

**12.7** (a) This is the maximum velocity and is therefore $\omega r$ (see answer to SAQ 11.7), thus, as $\omega = 2\pi/T$, velocity $= (2\pi/1.5) \times 0.100 = 0.42$ m s$^{-1}$.
(b) This is the time to move to the maximum displacement and so is half a complete oscillation, i.e. 0.75 s.
(c) Total energy $= \frac{1}{2}m\omega^2r^2 = \frac{1}{2} \times 0.120 \times (2\pi/1.5)^2 \times 0.100^2 = 0.0105$ J

## Answers II

**12.8** See Figure A12.1.

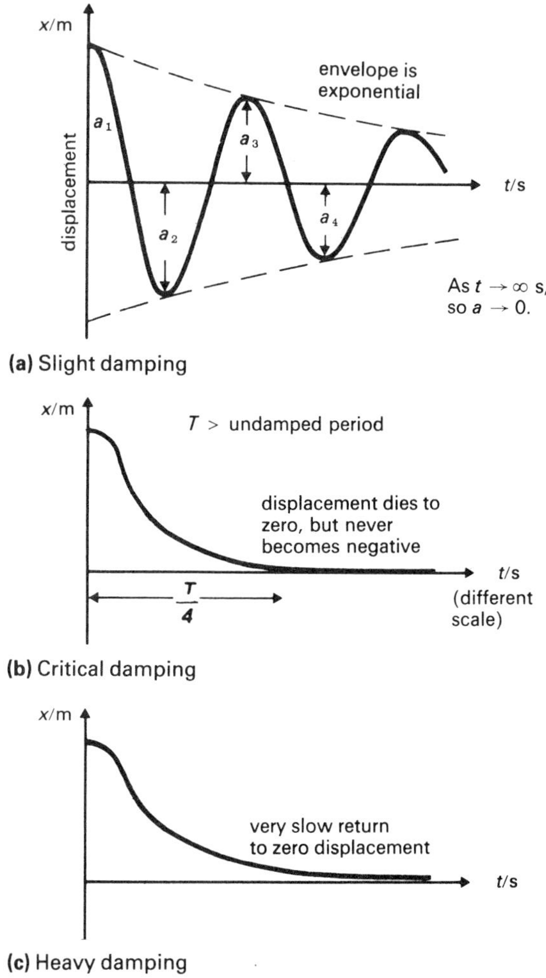


**Figure A12.1**

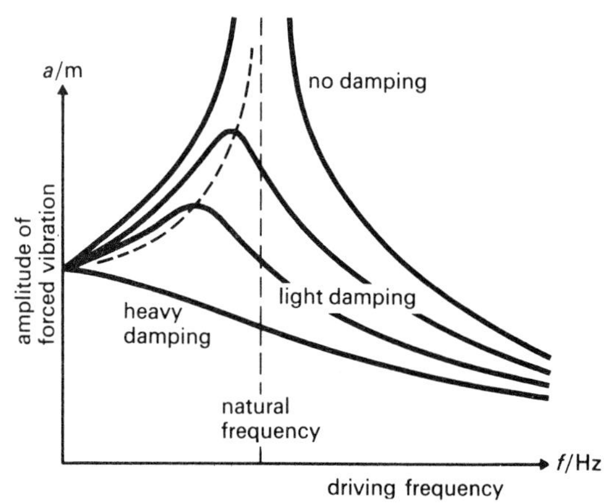


**Figure A12.2 Response of damped systems to varying driving frequencies**

**12.9** (a) See Figure A12.2.
(b) See the key points. When resonance occurs there is maximum energy transfer from the driving force to the oscillating system. For a damped system the amplitude of the oscillation reaches a maximum when the forcing frequency is just less than the natural oscillation frequency of the system. See *B* PP ch. 4 for some interesting examples of forced vibration of systems.

**12.10** (a) See Figure A12.1(a). For exponential decay

$$\frac{a_1}{a_2} = \frac{a_2}{a_3} = \frac{a_3}{a_4} = \text{etc.}$$

where these represent the amplitudes of successive maximum displacements of the oscillation.
(b) $a_1 = ae^{-kt}$, hence $(a_1/a) = e^{-kt}$ and so $\ln(a_1/a) = -kt$. Thus $\ln(0.5) = -20.5k$ and $k = 0.0338\,\text{s}^{-1}$.

## CHAPTER 2

# Material properties

## Lesson 1 Elasticity

### Aims

After completing this lesson you should be able to

- state and use *Hooke's law*
- define and know how to calculate *stress*, *strain*, *limit of proportionality*, *Young's modulus* and *elastic limit*
- describe stress–strain graphs using the above quantities
- distinguish between *brittle* and *ductile* materials
- solve problems on the stresses produced when thermal expansion or contraction is restricted
- calculate the energy stored in a stretched or compressed material.

### Prerequisites

A basic understanding of force and energy, e.g. Lesson 5 in Chapter 1, and for one group of questions a knowledge of linear expansion of solids due to temperature.

## STUDY

**I Read** about Hooke's law, stress, strain and the terms used with stress–strain graphs, also the difference between ductile and brittle materials.
References: *B* PP ch. 9, *B* ST2 ch. 1, *D* ch. 2, *N* ch. 5, *W et al.* ch. 14 (13), *Wh&H* ch. 17.

**Self-assessment questions**

**1.1** *Hooke's law* can be stated as: the extension of a piece of material is proportional to the force or tension in the material. For a material obeying this law, if the extension is $x$ when the force is $F$, what is the extension when the force is (a) $2F$ and (b) $3F$? What form is the graph of force against extension for such a material?

**1.2** Hooke's law can be expressed in the form of an equation $F = kx$, where $k$ is a constant called the *force constant*. For a strip of steel a force of 1500 N was found to give an extension of 0.10 mm. If Hooke's law is obeyed, what is the value of the force constant for the steel?

**1.3** Define *stress* and *strain*, stating their units.

**1.4** What is the stress in a strip of steel with a cross-sectional area of 50 $mm^2$ when acted on by a force of 2000 N?

**1.5** What is the strain in a strip of polythene of length 120 mm which extends by 40 mm?

**1.6** Explain what is meant by the *limit of proportionality* in relation to (a) a force–extension graph and (b) a stress–strain graph.

**1.7** Define *Young's modulus of elasticity* and state its unit.

***1.8** A load of 0.60 kg is hung from the end of a vertical, suspended wire of length 2.0 m and diameter 0.43 mm. If the wire extends by 0.40 mm what is the value of Young's modulus if the limit of proportionality has not been exceeded? ($g = 9.8\ m\ s^{-2}$)

***1.9** The following are force–extension data for a 2.0 m length of copper wire of diameter 0.90 mm.
(a) Plot the stress–strain graph.
(b) Determine the stress at the limit of proportionality.
(c) Determine Young's modulus.

| | | | | | | | | |
|---|---|---|---|---|---|---|---|---|
| Force/N | 0 | 10 | 20 | 30 | 40 | 50 | 60 | 70 |
| Extension/mm | 0 | 0.25 | 0.50 | 0.75 | 1.00 | 1.25 | 1.58 | 2.75 |

***1.10** What is the maximum load which can be hung from a steel wire of diameter 0.50 mm if the strain must not exceed 0.001 and the material has a Young's modulus of $2.0 \times 10^{11}$ Pa?

***1.11** The rubber cord of a catapult has a cross-sectional area of 4.0 $mm^2$ and an unstretched length of 100 mm. What is (a) the tension in the rubber and (b) the force needed to stretch it when the rubber cord is pulled back until its length has been doubled? The two arms of the catapult are 70 mm apart and Young's modulus for the rubber is $5.0 \times 10^8$ Pa.

*1.12 A copper wire and a steel wire, each of length 1.0 m and diameter 1.0 mm, are joined at one end to form a composite wire 2.0 m long. What are the strains in the two wires when a force causes the composite wire to stretch by 0.002 mm? (Young's modulus for copper = $1.2 \times 10^{11}$ Pa, for steel $2.0 \times 10^{11}$ Pa)

**1.13** A material is said to show *elastic* behaviour when the removal of stress results in the material returning to its original shape; it is said to show *plastic* behaviour when it does not return to its original shape but shows permanent deformation. Sketch stress–strain graphs for a material, showing both the loading and unloading results, when it is (a) stretched within its elastic region and (b) stretched to within its plastic region.

**1.14** Define *elastic limit*.

**1.15** Materials which show a considerable amount of plastic deformation before they break are called *ductile* materials; those materials which show little if any plastic deformation are called *brittle* materials. Sketch typical stress–strain graphs for such materials. Give an example of a brittle and a ductile material.

**II Read** about the stresses produced when thermal expansion or contraction is restricted.
References: *B* ST2 ch. 1, *D* ch. 4, *N* ch. 5.

**Self-assessment questions**

**1.16** The linear expansivity (or coefficient of expansion) $\alpha$ is given by

$$\alpha = \frac{L_\theta - L_0}{L_0\theta}$$

where $L_0$ is the original length and $L_\theta$ the length after a temperature change of $\theta$.
(a) Consider a rod of length $L_0$ which is restrained from expanding when the temperature rises by $\theta$. What would have been the length of the rod if it had been allowed to expand?
(b) We can consider the stress in the rod to be that which would have occurred if the expanded-length rod had been compressed to the length $L_0$. What is the strain?
(c) Simplify your answer to (b) by considering $L_0\alpha\,\theta$ to be small when compared with $L_0$. Hence, assuming the limit of proportionality is not exceeded, determine the stress.
(d) What is the force exerted by the material on its restraining supports if the cross-sectional area of the rod is $A$?

*1.17 Calculate the forces exerted on restraining supports which prevent a steel girder, cross-sectional area 80 $cm^2$, from expanding when the temperature rises by 15 °C. (Young's modulus = $2.0 \times 10^{11}$ Pa, linear expansivity = $1.2 \times 10^{-5}\,K^{-1}$)

**III Read** about the derivation and use of the equation for the energy stored in a stretched or compressed material.
References: *B* PP ch. 9, *B* ST3 ch. 1, *D* ch. 2, *W et al.* ch. 16 (8), *Wh&H* ch. 17.

**Self-assessment questions**

**1.18** (a) If a force $F$ stretches a wire by $\delta x$, what is the work done?
(b) If Hooke's law is obeyed, the extension is proportional to the force causing the stretching. What therefore is the work done in stretching a wire by $x$? The force will not be constant during this extension; Figure 1.1 shows the force–extension graph.

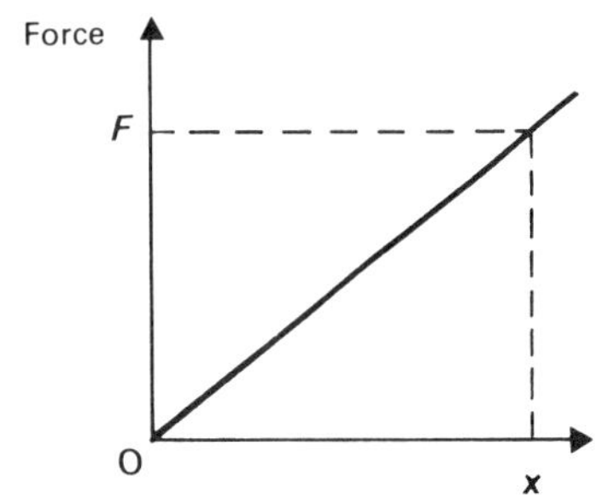


**Figure 1.1**

(c) The work done in stretching the material is stored in it as strain energy. What is the strain energy stored per unit volume of the material?

*1.19 A vertical steel wire 2.0 m long and of diameter 1.0 mm has a load of 5.0 kg applied to its lower end. What is the energy stored in the wire? (Young's modulus = $2.0 \times 10^{11}$ Pa, $g = 9.8\,m\,s^{-2}$)

*1.20 How can the energy needed to stretch a material to some extension beyond its elastic limit be determined from its stress–strain graph?

# SELF-ASSESSMENT

## Key points

**Hooke's law:** the extension is proportional to the force; alternatively, strain is proportional to stress.
**Stress** = force/area, **strain** = extension/original length.

Up to the **limit of proportionality** the strain is proportional to the stress, the stress divided by the strain being **Young's modulus of elasticity**.
Up to the **elastic limit** a material returns to its original shape when the stretching force is removed, the material being said to show **elastic** behaviour. Beyond the elastic limit the material shows a permanent change in shape when the force is removed, the material being said to show **plastic** behaviour.
The **energy stored** in a stretched or compressed material is $\frac{1}{2}$ (stress × strain) per unit volume of material.

## Answers I

**1.1** (a) $2x$ (b) $3x$ See Figure 1.1 on page 35.

**1.2** $k = F/x = 1500/(0.10 \times 10^{-3}) = 1.5 \times 10^7\ \mathrm{N\,m^{-1}}$

**1.3** See the key points; stress has units of $\mathrm{N\,m^{-2}}$ or the pascal (Pa) where $1\ \mathrm{N\,m^{-2}} = 1$ Pa. Strain is a ratio and so has no units.

**1.4** Stress $= F/A = 2000/(50 \times 10^{-6}) = 4.0 \times 10^7$ Pa

**1.5** Strain $= x/L = 40/120 = 0.3$

**1.6** (a) The limit at which the force is proportional to the extension and so the limit of the straight-line part of the graph.
(b) The limit at which the stress is proportional to the strain and so the limit of the straight-line part of the graph.

**1.7** See the key points; the unit is Pa.

**1.8** Young's modulus $E$ = stress/strain = $(F/A) \div (x/L) = FL/Ax = 0.60 \times 9.8 \times 2.0/(\frac{1}{4}\pi \times 0.43^2 \times 10^{-6} \times 0.40 \times 10^{-3}) = 2.02 \times 10^{11}$ Pa

**1.9** (a) See Figure A1.1. $A = 6.36 \times 10^{-7}\ \mathrm{m^2}$, therefore stress for the 10 N force is $1.57 \times 10^7$ Pa. The strain for the 0.25 mm extension is $0.25/2000 = 1.25 \times 10^{-4}$.
(b) About $8 \times 10^7$ Pa.
(c) The strain at which the limit occurs is $6.4 \times 10^{-4}$, hence $E$ = stress/strain $= 8 \times 10^7/6.4 \times 10^{-4} = 1.3 \times 10^{11}$ Pa.

**1.10** $E$ = stress/strain and so $F = EA \times$ strain $= 2.0 \times 10^{11} \times \frac{1}{4}\pi \times (0.50 \times 10^{-3})^2 \times 0.001 = 39.3$ N, and in terms of mass hung on the wire, 4.01 kg.

**1.11** (a) $E$ = stress/strain = $FL/Ax$, hence $F = EAx/L = 5.0 \times 10^8 \times 4.0 \times 10^{-6} \times (100/100) = 2000$ N.
(b) See Figure A1.2. Force required $= 2T\cos\theta$. As $\sin\theta = 35/100$ then $\theta = 20.5°$ and so force $= 2 \times 2000 \times \cos 20.5° = 3750$ N.

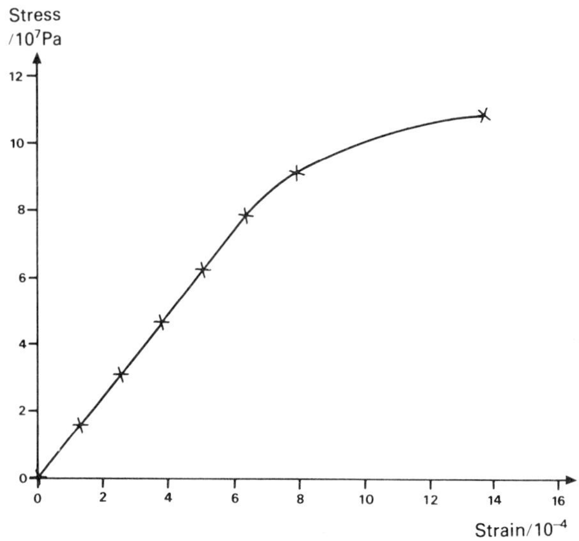


**Figure A1.1**

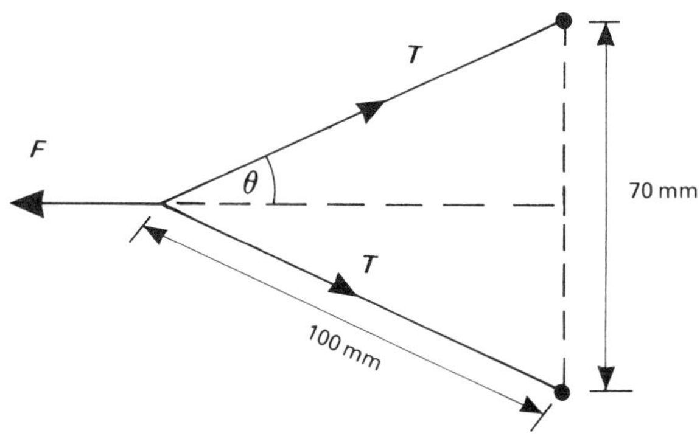


**Figure A1.2**

**1.12** $E$ = stress/strain = $FL/Ax$, hence $x = FL/AE$. Thus total extension $x_t = x_c + x_s = (FL/AE_c) + (FL/AE_s)$ and so $F = (Ax_t/L)/(1/E_c + 1/E_s)$. Therefore the strain in the copper $= x_c/L = F/AE_c = (x_t/LE_c)/(1/E_c + 1/E_s) = 1.25 \times 10^{-3}$. This is an extension of $1.25 \times 10^{-3}$ m in the 1.0 m length. Thus the extension for the steel is $0.75 \times 10^{-3}$ m and so the strain in the steel is $0.75 \times 10^{-3}$.

**1.13** See Figure A1.3.

**1.14** See the key points.

**1.15** See Figure A1.4. Glass and concrete are examples of brittle materials; mild steel is ductile.

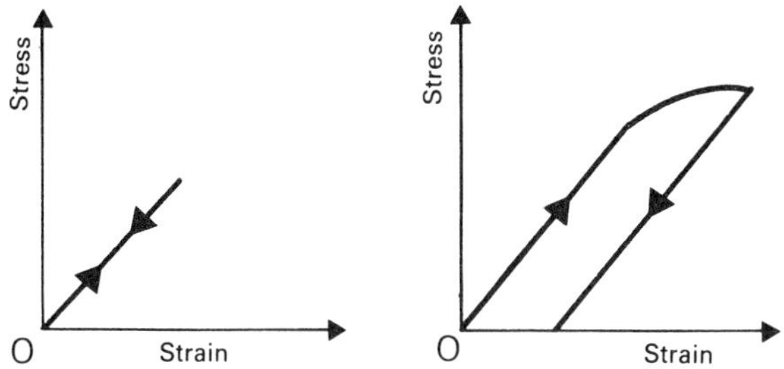


**(a)** Elastic behaviour **(b)** Plastic deformation

**Figure A1.3**

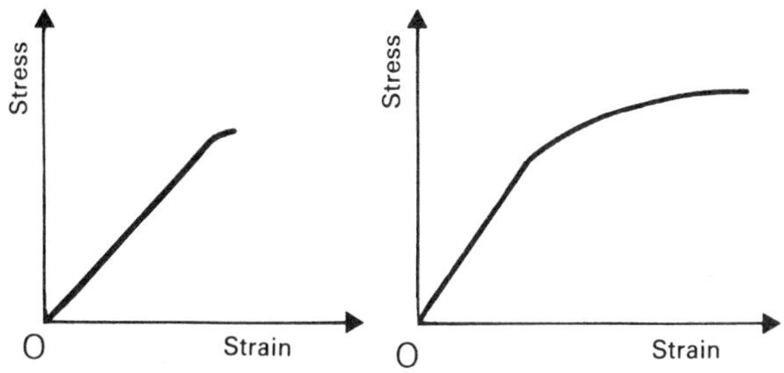


**(a)** Brittle **(b)** Ductile

**Figure A1.4**

## Answers II

**1.16** (a) $L_0 + L_0\theta\alpha$ (b) $L_0\theta\alpha/(L_0 + L_0\theta\alpha)$ (c) strain = $\theta\alpha$, hence stress = $E\theta\alpha$ (d) $F = AE\theta\alpha$

**1.17** $F = AE\theta\alpha = 80 \times 10^{-4} \times 2.0 \times 10^{11} \times 15 \times 1.2 \times 10^{-5} = 2.9 \times 10^5$ N

## Answers III

**1.18** (a) $F\delta x$ (b) It is the sum of all the $F\delta x$ terms. This sum is the area under the force–extension graph up to the ordinate for $x$. Area = $\frac{1}{2}Fx$ (where $F$ is the force at extension $x$) = work done. Another way of considering it is as the average force multiplied by the extension. (c) $W = \frac{1}{2}Fx/AL = \frac{1}{2}(F/A)(x/L) = \frac{1}{2}$(stress × strain)

**1.19** Energy = volume × $\frac{1}{2}$(stress × strain) = volume × $\frac{1}{2}$ × stress × (stress/$E$) = $AL \times \frac{1}{2}(F/A)^2/E = \frac{1}{2}LF^2/AE$ = $\frac{1}{2} \times 2.0 \times (5.0 \times 9.8)^2/(\frac{1}{4}\pi \times 1.0^2 \times 10^{-6} \times 2.0 \times 10^{11})$ = 0.015 J

**1.20** The area under the graph up to the ordinate for the extension concerned gives the energy per unit volume.

# Lesson 2 Surface tension

## Aims

After completing this lesson you should be able to

- define surface tension and surface energy and show the relationship between them
- define the angle of contact and account for the shapes of liquid surfaces
- derive and use the relationship for the excess pressure inside a spherical surface
- determine the relationship for capillary rise and use it in problems.

## Prerequisites

A basic knowledge of pressure in fluids, the relationship $p = h\rho g$, and Lesson 5 in Chapter 1 on energy.

## STUDY

**I Read** about surface tension and surface energy, their definitions and the relationship between them.

References: *B* PP ch. 9, *D* ch. 9, *N* ch. 6, *W et al.* ch. 16 (15), *Wh&H* ch. 18.

### Self-assessment questions

**2.1** Define *surface tension* and state its unit.

**2.2** Define *free surface energy* and state its unit.

**2.3** Consider a liquid film in a wire frame, as in Figure 2.1.

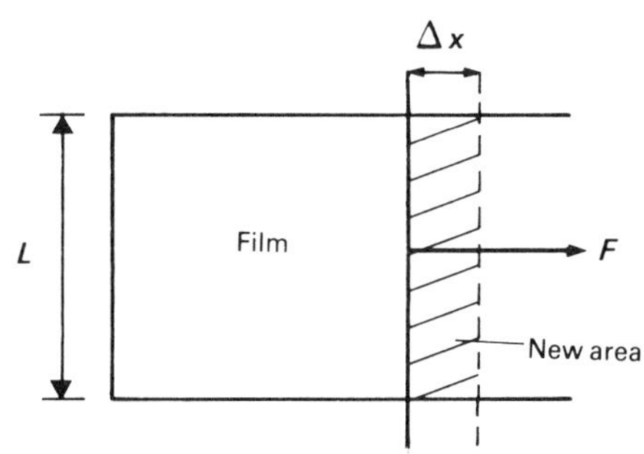


**Figure 2.1**

(a) What is the work done when the force $F$ moves through a distance $\Delta x$?
(b) What is this force in terms of the surface tension $\gamma$? Remember that the film of liquid has two surfaces.
(c) We can arrive at the work done by considering the energy required to increase the surface area of the liquid film. What is the increase in surface area?
(d) Hence, in terms of free surface energy $\sigma$, what is the increase in surface energy?
(e) By considering your answers to (b) and (d) when constant temperature conditions exist, arrive at the relationship between $\gamma$ and $\sigma$.

***2.4** A circular ring of mean radius 4.0 cm is horizontal in the surface of a liquid. What is the force required to pull the ring clear of the liquid? Assume that when the ring breaks free the surface tension forces act vertically downwards.

***2.5** What is the work done against surface tension forces in blowing a soap bubble of diameter 20 mm if the surface energy of the liquid used is $3.0 \times 10^{-2}\,\mathrm{J\,m^{-2}}$? (Surface area of a sphere $= 4\pi r^2$)

***2.6** What is the energy needed to split a spherical drop of water of diameter 6.0 mm into six equal-size droplets, given that the surface energy of water is $7.0 \times 10^{-2}\,\mathrm{J\,m^{-2}}$? (Surface area of a sphere $= 4\pi r^2$, volume $= \frac{4}{3}\pi r^3$)

**II Read** about the shape of liquid surfaces and the angle of contact.
References: *D* ch. 9, *N* ch. 6, *W et al.* (ch. 15), *Wh&H* ch. 18.

**Self-assessment questions**

**2.7** Define the *angle of contact.*

**2.8** (a) Water has zero angle of contact with a perfectly clean glass surface. What is the shape adopted by water when dropped onto such a surface?
(b) Water on a greasy glass surface has an angle of contact of about 10°. What is the shape adopted by water when dropped onto such a surface?
(c) Mercury has an angle of contact of about 140° on a clean glass surface. What is the shape adopted by mercury when dropped onto such a surface?

**2.9** The shapes of liquid surfaces in contact with a vertical glass surface can be explained in terms of two forces acting on liquid molecules near to the glass, these being the attractive force between neighbouring liquid molecules (called the *cohesive* force) and the attractive force between the liquid and the glass molecules (called the *adhesive* force). Explain what happens when (a) the adhesive force is larger than the cohesive force and (b) the cohesive force is larger than the adhesive force.

**III Read** about the derivation and use of the relationship for the excess pressure inside a spherical surface.
References: *D* ch. 9, *N* ch. 6, *W et al.* (ch. 15), *Wh&H* ch. 18.

**Self-assessment questions**

**2.10** Consider the equilibrium of one half of a bubble in a liquid, as in Figure 2.2.

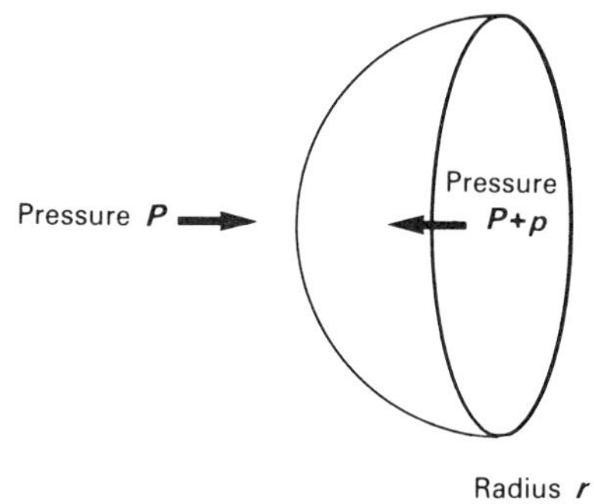


**Figure 2.2**

(a) What are the forces acting on the hemisphere?
(b) The force due to the external pressure $P$ acts over the entire spherical face of the drop, however the component of the force at right angles to the central plane is $P \times$ the area of that plane. The surface tension force at right angles to this plane is given by the product of $\gamma$ and the circumference of the hemisphere. Hence derive a relationship between the pressure difference between the inside and the outside of the bubble, $\gamma$ and the bubble radius.

**2.11** What is the excess pressure in a soap bubble of radius 10 mm if the surface tension of the soap solution is $2.5 \times 10^{-2}\,\mathrm{N\,m^{-1}}$?

***2.12** Two soap bubbles of different diameters are connected by a tube. Why does the smaller bubble gradually collapse and the larger bubble grow bigger?

***2.13** A soap bubble of radius 2.0 cm and another soap bubble of radius 3.0 cm are brought together so that the combined bubble has a common interface of radius $r$, as in Figure 2.3. What is the value of $r$? Hint: consider the excess pressure across the interface between the two bubbles.

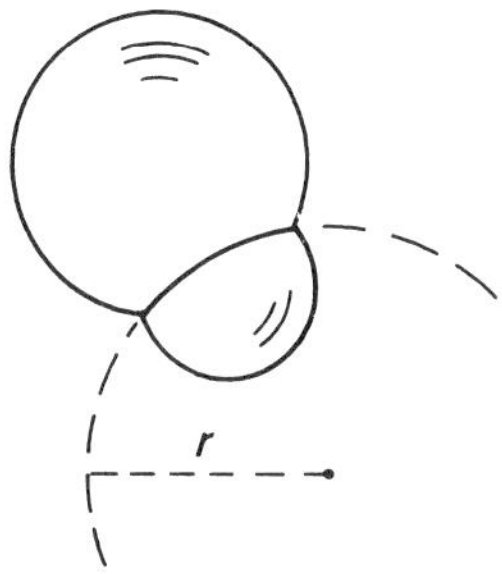


Figure 2.3

*2.14 A drop of water is squeezed between two glass plates to form a thin film, as in Figure 2.4. If the water film has a thickness of 0.002 mm and an area of 400 $mm^2$, what is the force needed to pull the plates apart? Take the surface tension of the water to be $7 \times 10^{-2}$ N $m^{-1}$.

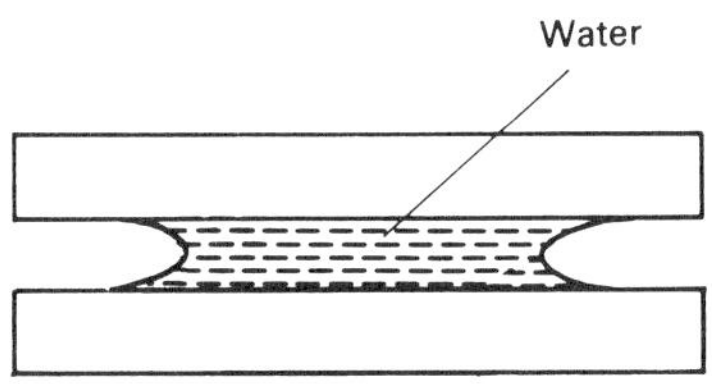


Figure 2.4

**IV Read** about the rise, or depression, of liquids in capillary tubes.
References: *B* PP ch. 9, *D* ch. 9, *N* ch. 6, *W et al.* (ch. 15), *Wh&H* ch. 18.

**Self-assessment questions**

**2.15** Explain, in terms of pressure differences, why water rises up a capillary tube and why the level of a mercury surface is depressed in a capillary tube.

**2.16** Consider the situation shown in Figure 2.5 where a liquid, such as water, has risen up a capillary tube to a height $h$, the tube having a radius $r$ and the liquid a density $\rho$ and surface tension $\gamma$.

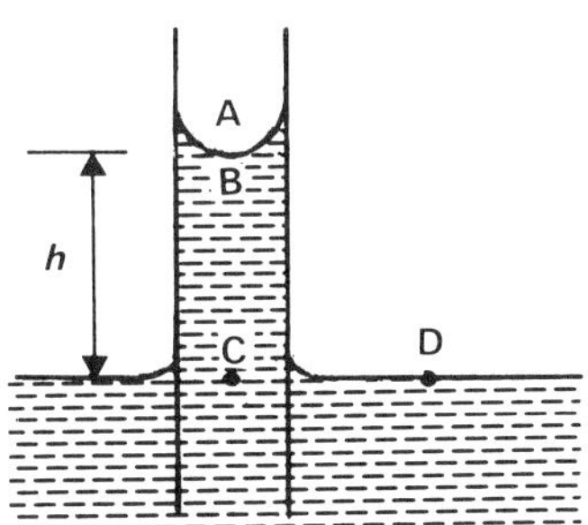


Figure 2.5

(a) What is the pressure difference across the liquid surface in the tube, i.e. between A and B, if the radius of the liquid surface is assumed to be the same as that of the tube?
(b) How is the pressure at C related to that at D?
(c) What is the pressure difference between B and C due to the column of liquid of height $h$?
(d) If the pressures at A and at D are assumed to be the same, i.e. atmospheric pressure, derive the relationship $h = 2\gamma/(r\rho g)$.

***2.17** How high does water rise up a capillary tube of radius 0.50 mm when held vertically with its lower end dipping into water? Surface tension of water = $7.0 \times 10^{-2}$ N $m^{-1}$ and density = 1000 kg $m^{-3}$. ($g$ = 9.8 m $s^{-2}$)

***2.18** A capillary tube of radius 0.30 mm is held vertically with its lower end dipping into water. If the length of the tube above the water is 20 mm, what happens? Surface tension of water = $7.0 \times 10^{-2}$ N $m^{-1}$ and density = 1000 kg $m^{-3}$. ($g$ = 9.8 m $s^{-2}$)

***2.19** A capillary tube of radius 0.60 mm is held vertically with its lower end dipping into water. What pressure must be applied to the upper end of the tube for the water level inside the tube to be the same as that outside the tube? Surface tension of water = $7.0 \times 10^{-2}$ N $m^{-1}$.

***2.20** Describe an experiment to measure the surface tension of water.

# SELF-ASSESSMENT

## Key points

**Surface tension** $\gamma$ can be defined as the force per unit length acting in the surface of a liquid perpendicular to one side of a line in the surface, i.e. $\gamma = F/L$, and has the unit of N $m^{-1}$.
The **free surface energy** $\sigma$ of a liquid is the energy required to create unit area of new surface, i.e. $\sigma = W/A$, and has the unit of J $m^{-2}$.
$\sigma$ and $\gamma$ are numerically equal. Both decrease with an increase in temperature.
The **angle of contact** between a liquid and a solid surface is the angle between the tangent planes to the two surfaces when measured through the liquid.
The excess pressure inside a spherical liquid surface of radius $r$ is $2\gamma/r$.
The height $h$ to which a liquid rises up a capillary tube of radius $r$ is $h = 2\gamma \cos\theta/(r\rho g)$, where $\theta$ is the angle of contact. For water in a clean glass tube $\theta$ is taken as 0°.

## Answers I

**2.1** See the key points.

**2.2** See the key points.

**2.3** (a) $F\,\Delta x$ (b) $2\gamma L$ (c) $2L\,\Delta x$ (d) $2\sigma L\,\Delta x$ (e) $2\gamma L\,\Delta x = 2\sigma L\,\Delta x$, hence $\gamma = \sigma$.

**2.4** There are two sides to the ring and hence $2\gamma \times 2\pi r = F$.

**2.5** Surface area created, bearing in mind there are two surfaces, is $2 \times 4\pi r^2$, hence energy $= \sigma A = 3.0 \times 10^{-2} \times 2 \times 4\pi \times (10 \times 10^{-3})^2 = 7.5 \times 10^{-5}$ J.

**2.6** The volume remains constant, thus $6.0^3 = 6r^3$ where $r$ is the radius in mm of one of the six droplets. Thus $r = 3.3$ mm. Change in surface area $= 6 \times 4\pi \times (3.3 \times 10^{-3})^2 - 4\pi \times (6.0 \times 10^{-3})^2\,\mathrm{m}^2$. Hence energy required $= 4\pi \times 10^{-6} \times 7.0 \times 10^{-2} \times (3.3^2 \times 6 - 6.0^2) = 2.6 \times 10^{-5}$ J.

## Answers II

**2.7** See the key points.

**2.8** (a) It spreads out completely and wets the glass, no drop being formed. (b) See Figure A2.1(a). (c) See Figure A2.1(b).

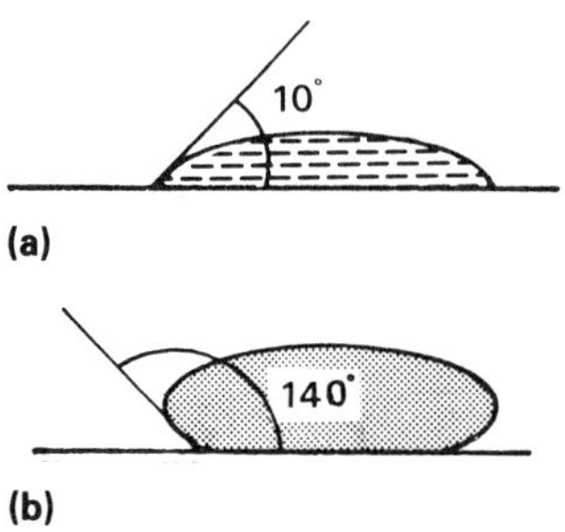


**Figure A2.1**

**2.9** (a) See Figure A2.2(a). (b) See Figure A2.2(b). In both cases the resulting liquid surface is at right angles to the resultant force.

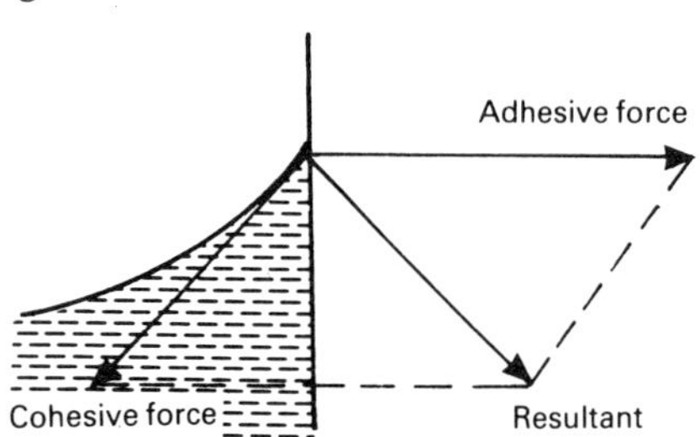


**Figure A2.2(a)**

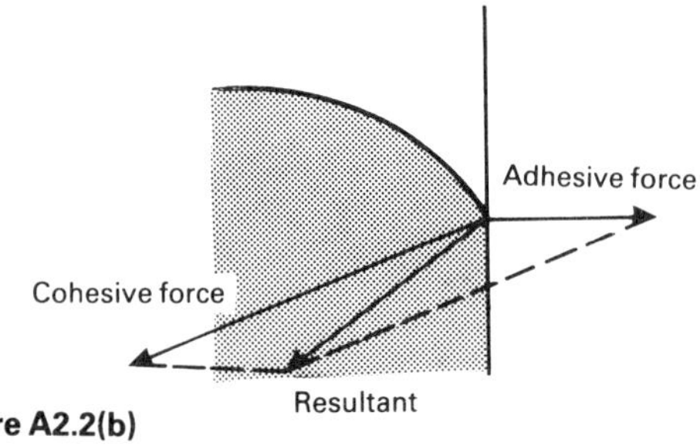


**Figure A2.2(b)**

## Answers III

**2.10** (a) External and internal pressures and surface tension. (b) $\pi r^2(P + p) = \pi r^2 P + 2\pi r\gamma$, hence $p = 2\gamma/r$.

**2.11** Excess pressure $= 4\gamma/r = 4 \times 2.5 \times 10^{-2}/(10 \times 10^{-3}) = 10$ Pa. The bubble has two surfaces.

**2.12** Because the pressure inside the smaller bubble is greater than that inside the larger bubble and the pressures tend to equalize.

**2.13** Excess pressure in the 2.0 cm bubble is $4\gamma/r = 4\gamma/(2.0 \times 10^{-2}\,\mathrm{m})$. Excess pressure in the 3.0 cm bubble is $4\gamma/(3.0 \times 10^{-2}\,\mathrm{m})$. Thus the pressure difference across the interface is $4\gamma/(2.0 \times 10^{-2}\,\mathrm{m}) - 4\gamma/(3.0 \times 10^{-2}\,\mathrm{m})$. The resulting radius is $4\gamma$/pressure $= 5.9$ cm. Because the pressure is always greater on the side of the small bubble, the interface is always concave to the smaller bubble.

**2.14** Force = excess pressure × area $= (2\gamma/r)A = 2 \times 7 \times 10^{-2} \times 400 \times 10^{-6}/(0.001 \times 10^{-3}) = 56$ N.

## Answers IV

**2.15** The pressure under the curved water surface in the tube is less than the pressure above it, which is approximately the same as the pressure at the surface of the water outside the tube (i.e. atmospheric). To equalize the pressures in the water inside and outside the tube at the same level, the water has to rise up the inside of the tube. With mercury the pressure under the curved surface in the tube is greater than the pressure above it, and so the level has to be depressed.

**2.16** (a) Pressure at A minus pressure at B $= 2\gamma/r$.
(b) They must be the same because they are at the same level.
(c) Pressure at C minus pressure at B $= h\rho g$.
(d) Subtracting the equation in (c) from that in (a) gives pressure at C minus pressure at A $= h\rho g - 2\gamma/r$. But pressure at C equals pressure at D equals pressure at A, hence $0 = h\rho g - 2\gamma/r$ and so $h = 2\gamma/(r\rho g)$.

**2.17** $h = 2\gamma/(r\rho g) = 2 \times 7.0 \times 10^{-2}/(0.50 \times 10^{-3} \times 1000 \times 9.8) = 0.0286$ m or 28.6 mm

**2.18** The water would, if possible, rise up the tube to a height of 47.6 mm, but the length of the tube does not permit this. The result is that a spherical surface is formed at the top of the tube with a radius $R$ different from that of the tube. The pressure difference across this surface is $2\gamma/R$. This must be equal to $H\rho g$, where $H$ is the 20 mm height, i.e. $H\rho g = 2\gamma/R$ and so $R = 2\gamma/(H\rho g) = 2 \times 7.0 \times 10^{-2}/(20 \times 10^{-3} \times 1000 \times 9.8) = 7.14 \times 10^{-4}$ m = 0.714 mm. The contact angle $\theta$ between the water and the glass is given by $\cos\theta = r/R$ (see Figure A2.3) and so $\theta = 45.6°$. If the tube had been long enough the contact angle would have been zero.

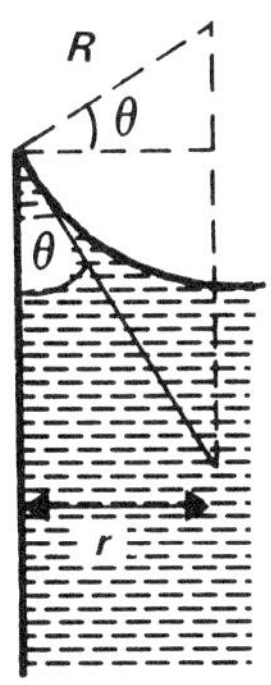


**Figure A2.3**

**2.19** Pressure = $h\rho g$, where $h$ is the height to which the liquid would otherwise have risen. Therefore pressure = $2\gamma/r = 2 \times 7.0 \times 10^{-2}/(0.60 \times 10^{-3}) = 233$ Pa.

**2.20** A number of methods could be described:
(a) The 'force to break a film' method, or *chemical balance method*. This involves measuring the force needed to pull a wire frame free of the surface of the liquid, as for example in SAQ 2.4.
(b) The *capillary rise method*. This involves measuring the height to which the liquid rises in a capillary tube of known radius, then using the equation developed in the answer to SAQ 2.16.
(c) The measurement of the pressure required to blow a bubble in the liquid, or *Jaeger's method*. This is the only method that is very convenient for the measurement of surface tension over a range of temperature. As the pressure is increased inside a tube dipping into the liquid, a bubble grows at the end of the tube, and as it does so the pressure rises to a maximum and then falls away as the bubble breaks away from the tube. The maximum pressure occurs when the radius of the bubble is a minimum, this occurring when the radius is about the same as that of the tube. Making this assumption the surface tension can be determined. The pressure is measured with a simple U-tube manometer.
The references quoted for this lesson give more details of all these methods.

# Lesson 3 Viscosity

## Aims

After completing this lesson you should be able to
- describe *laminar* flow of a fluid past a surface
- define the coefficient of viscosity and use it in problems
- state *Stoke's law* and use it in problems
- state *Poiseuille's equation* and use it in problems.

## Prerequisites

Lesson 4 of Chapter 1 on Newton's laws.

## STUDY

**I Read** about the laminar flow of a fluid past a surface, the velocity gradient and the coefficient of viscosity.

References: *B* PP ch. 9, *B* ST2 ch. 4, *D* ch. 10, *N* ch. 7, *W et al.* (ch. 11), *Wh&H* ch. 19.

**Self-assessment questions**

**3.1** What is *laminar* (or *streamline*) flow?

**3.2** When the flow of water through a pipe is laminar, what can be said about the layer of water in contact with the walls of the pipe?

**3.3** With a fluid flowing past a solid surface there is said to be a 'velocity gradient'. Define this.

**3.4** When adjacent layers of a material are displaced laterally over each other, as in Figure 3.1, the material is said to be subject to *shear stress*. Define shear (sometimes known as *tangential*) stress.

**3.5** Define the coefficient of viscosity and state its unit.

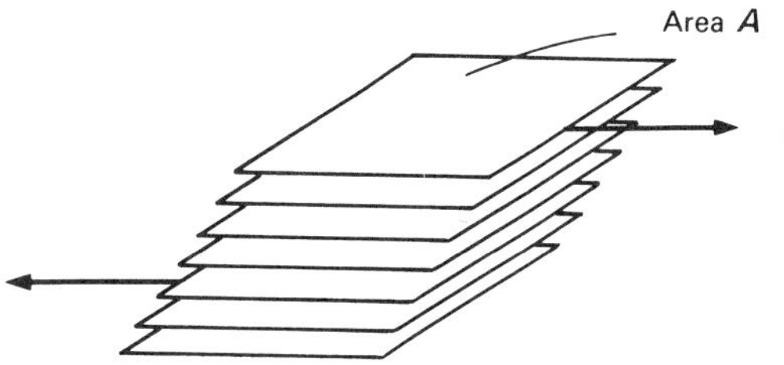


**Figure 3.1 Shearing a material**

**3.6** What is meant by a *Newtonian* fluid?

**3.7** Water flows past a horizontal plate of area 1.2 $m^2$. If the water adjacent to the plate has a velocity gradient of 10 $s^{-1}$ and the coefficient of viscosity is $1.3 \times 10^{-3}$ N s $m^{-2}$, what is the force acting on the plate?

**II Read** about Stokes' law and Poiseuille's equation, and their use in problems.
References: *B* PP ch. 9, *B* ST2 ch. 4, *D* ch. 10, *N* ch. 7, *W et al.* (ch. 11), *Wh&H* ch. 19.

**Self-assessment questions**

**3.8** When a sphere falls in a fluid, what forces act on it if the motion is laminar?

**3.9** The viscous drag force on a sphere falling in a fluid is given by *Stokes' law* as $6\pi\eta rv$, where $r$ is the radius of the sphere and $v$ its velocity. The other forces acting on the sphere are that of gravity, $mg$, and the upthrust due to the fluid being displaced by the presence of the sphere, this force being the weight of fluid displaced and so $V\rho_f g$, where $\rho_f$ is the density of the fluid and $V$ the volume of the sphere.
(a) When the sphere starts from rest, what are the sizes and directions of the forces acting on it?
(b) Why does the sphere have an initial acceleration?
(c) How do the forces acting on the sphere change as the velocity of the sphere increases?
(d) State the condition for there to be no resultant force acting on the sphere.
(e) What is the motion of the sphere when there is no resultant force acting on it?

**3.10** Define *terminal* velocity.

***3.11** Calculate the terminal velocity of raindrops falling in air if the raindrops can be considered to be spheres of diameter 1 mm and the flow is laminar. Coefficient of viscosity of air = $1.8 \times 10^{-5}$ N s $m^{-2}$, density of air = 1.3 kg $m^{-3}$, density of water = 1000 kg $m^{-3}$, $g$ = 9.8 m $s^{-2}$.

***3.12** An oil has a coefficient of viscosity of 2.4 N s $m^{-2}$ and a density of 940 kg $m^{-3}$. What will be the terminal velocity of a steel sphere of radius 2.0 mm falling freely in the oil? Density of the steel = 7800 kg $m^{-3}$, $g$ = 9.8 m $s^{-2}$. Assume laminar flow.

***3.13** Estimate the minimum time that would be taken for particles of sand sprinkled onto the surface of water in a beaker, filled to a depth of 120 mm, to reach the bottom. Assume that the sand particles are spherical and have a diameter of 0.10 mm, the viscosity of water is $1.1 \times 10^{-3}$ N s $m^{-2}$, the density of water is 1000 kg $m^{-3}$, the density of the sand is 2000 kg $m^{-3}$ and $g$ is 9.8 m $s^{-2}$.

**3.14** The rate of flow of fluid in a tube is given by *Poiseuille's equation*, for laminar flow, as

$$\frac{V}{t} = \frac{\pi r^4 p}{8\eta L}$$

where $V$ is the volume flowing in time $t$, $r$ is the radius of the tube, $L$ its length and $p$ the pressure difference between the two ends of this length of tube.
(a) If the pressure difference between the ends of a pipe is doubled, how does the rate of fluid flow through the pipe change?
(b) If the radius of the pipe is doubled, for the same pressure gradient how does the rate of fluid flow through the pipe change?

***3.15** In an experiment to measure the coefficient of viscosity of water, 22.2 $cm^3$ of water was found to flow in one minute through a capillary tube of internal diameter 1.07 mm and length 252 mm when the pressure difference between the ends was that due to a column of water of height 175 mm. If the density of water is 1000 kg $m^{-3}$ and $g$ is 9.81 m $s^{-2}$, what is the coefficient of viscosity of the water?

***3.16** A horizontal tube consists of two parts joined together end to end. One part is 100 mm long and has a radius of 1.00 mm and the other is 140 mm long with a radius of 0.50 mm. When water flows steadily along the tube there is a pressure difference between the tube ends equivalent to that due to 120 mm of water. What are the pressure differences across each of the component tubes, in terms of millimetres of water?

# SELF-ASSESSMENT

## Key points

With fluid flow past a surface there is a velocity gradient produced in the layers of fluid near to the surface. The

**shear**, or **tangential**, **stress** is the force per unit area, the direction of the force being in the plane of the area. The **coefficient of viscosity** $\eta$ is given by

$$\eta = \frac{\text{tangential stress}}{\text{velocity gradient}} = \frac{F/A}{dv/dy}$$

and has the unit of N s m$^{-2}$ or Pa s.

**Stokes' law** for streamline flow past a sphere states that the viscous drag force is $6\pi\eta rv$.

For streamline flow through a pipe, **Poiseuille's equation** applies:

$$\frac{V}{t} = \frac{\pi r^4 p}{8\eta L}$$

where $V$ is the volume flowing in time $t$, $r$ is the radius of the tube, and $p$ is the pressure difference between the ends of the tube of length $L$.

## Answers I

**3.1** *Laminar flow* is when the flow is orderly and can be considered as sheets of fluid moving parallel to a surface. A *streamline* is defined as a line, the tangent to which at any point gives the direction of the fluid motion at that point. *Streamline flow* is when the streamlines in a fluid do not change with time.

**3.2** It is stationary.

**3.3** The velocity gradient is the rate of change in velocity with distance in a direction perpendicular to the velocity.

**3.4** $F/A$; see the key points.

**3.5** See the key points.

**3.6** A *Newtonian fluid* is one where the tangential stress is directly proportional to the velocity gradient. This means that $\eta$ is independent of the tangential stress.

**3.7** $F = \eta A(dv/dy) = 1.3 \times 10^{-3} \times 1.2 \times 10 = 1.56 \times 10^{-2}$ N. This is the force on the fluid and, by Newton's third law of motion, also the force on the surface.

## Answers II

**3.8** Downwards the gravitational force; upwards the upthrust, equal to the weight of fluid displaced, and the viscous drag force.

**3.9** (a) Downwards $mg$, independent of velocity, and upwards, also independent of velocity, the upthrust $V\rho_f g$. Also upwards is the viscous drag $6\pi\eta rv$. Since this is proportional to the velocity it is zero when the sphere just starts to fall.
(b) Because the sphere is denser than the fluid, so there is a resultant force in the downwards direction.
(c) The viscous drag force increases while the others remain constant. At some particular velocity, called the *terminal* velocity, these forces all balance and there is no resultant force.
(d) $mg = V\rho_f g + 6\pi\eta rv$
(e) constant velocity

**3.10** The terminal velocity is the constant velocity that occurs in the case of an object falling in a fluid when there is no resultant force acting on it.

**3.11** Using the equation given in the answer to SAQ 3.9(d) and substituting $\frac{4}{3}\pi r^3\rho_s$ for $m$ and $\frac{4}{3}\pi r^3$ for $V$, where $\rho_s$ is the density of the sphere and $r$ its radius, then $\frac{4}{3}\pi r^3(\rho_s - \rho_f)g = 6\pi\eta rv$ and so we have $v = 2r^2(\rho_s - \rho_f)g/9\eta = 2 \times (0.5 \times 10^{-3})^2(1000 - 1.3) \times 9.8/(9 \times 1.8 \times 10^{-5}) = 30.2$ m s$^{-1}$.

**3.12** As in 3.11 above, $v = 2r^2(\rho_s - \rho_f)g/9\eta = 2 \times (2.0 \times 10^{-3})^2(7800 - 940) \times 9.8/(9 \times 2.4) = 0.025$ m s$^{-1}$.

**3.13** Assume that the sand reaches its terminal velocity so quickly that it can be assumed to fall the entire 120 mm with that velocity. Hence, using the equation in 3.11 above, $v = 2r^2(\rho_s - \rho_f)g/9\eta = 2 \times (0.05 \times 10^{-3})^2 \times (2000 - 1000) \times 9.8/(9 \times 1.1 \times 10^{-3}) = 0.049$ m s$^{-1}$. Therefore the time is $s/v = (120 \times 10^{-3})/0.049 = 2.4$ s.

**3.14** (a) Doubled. (b) Increased by a factor of 16.

**3.15** Using the equation given in SAQ 3.14 and rearranging, $\eta = \pi r^4 pt/(8LV) = \pi(1.07 \times 10^{-3})^4 \times 0.175 \times 1000 \times 9.8 \times 60/(8 \times 252 \times 10^{-3} \times 22.2 \times 10^{-6}) = 9.47 \times 10^{-3}$ N s m$^{-2}$.

**3.16** Using Poiseuille's equation for each component tube with $(V/t)$ and $\eta$ the same for both tubes, we have $pr^4/L$ = a constant. Thus $p_1 \times (1.00^4/100) = p_2 \times (0.50^4/140)$ and so $p_1 = 0.0446p_2$. But $p_1 + p_2 = 120$ and so $p_1 = 5.12$ mm and $p_2 = 114.88$ mm.

# Lesson 4 Structure of matter

## Aims

After completing this lesson you should be able to

- define the terms *atoms* and *molecules* and state their orders of magnitude, also the *Avogadro constant* and the *mole*
- describe force–separation and potential energy–separation graphs for intermolecular forces
- use such graphs to explain properties of solids
- explain surface tension and the latent heat of vaporization in terms of the breaking of intermolecular bonds.

## Prerequisites

Lessons 1 and 2 in this Chapter, Lesson 5 in Chapter 1, and a basic knowledge of latent heat, e.g. Lesson 3 in Chapter 3.

## STUDY

**I Read** about atoms and molecules, their size, the Avogadro constant, the mole, and calculations involving these quantities.
References: *B* PP ch. 9, *B* ST2 ch. 5, *D* ch. 1, *N* ch. 5, *Wh&H* ch. 16.

### Self-assessment questions

**4.1** Define the terms *atom* and *molecule*.

**4.2** Roughly how big is an atom?

**4.3** In atomic terms how does a compound differ from an element?

**4.4** Atomic and molecular masses give the masses of atoms and molecules when compared with the mass of the carbon-12 atom which is given the value of exactly 12. What is the atomic mass of (a) hydrogen, (b) oxygen, (c) copper?

**4.5** The atomic mass of carbon is 12 while that of copper is 63.5. Thus if 12 g of carbon contains $6.02 \times 10^{23}$ atoms, how many atoms will there be in 63.5 g of copper?

**4.6** The *Avogadro constant* is defined as being numerically equal to the number of atoms in exactly 12 g of carbon-12. The *mole* is the amount of a substance which contains as many particles as there are carbon atoms in 12 g of carbon-12.
(a) What is the mass of 1 mole of hydrogen ($H_2$)?
(b) The water molecule consists of two hydrogen atoms and one oxygen atom ($H_2O$). What is the molecular mass of the water molecule?
(c) What is the mass of $6.02 \times 10^{23}$ molecules of water?
(d) What is the mass of one mole of water?

**4.7** How many molecules are there in 5 g of water?

**II Read** about the force–separation and potential energy–separation graphs for intermolecular forces and the uses made of such graphs.
References: *B* PP ch. 3, *D* ch. 1, *N* ch. 5, *W et al.* ch. 14(13), *Wh&H* ch. 16.

### Self-assessment questions

**4.8** Figure 4.1 shows a force–separation graph for intermolecular forces between two molecules.

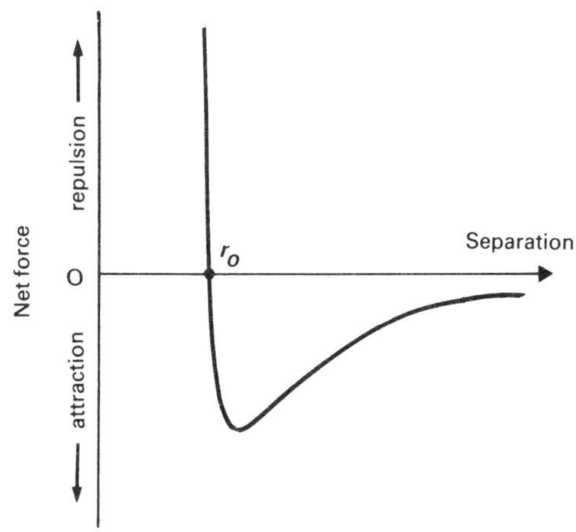


**Figure 4.1**

(a) What is the net force when the separation of the two molecules is less than $r_0$?
(b) What is the net force when the separation is greater than $r_0$?
(c) Why is $r_0$ the equilibrium separation of the two molecules?
(d) Explain how the force between two molecules changes as they are gradually brought closer together from a separation much greater than $r_0$ to one less than $r_0$.

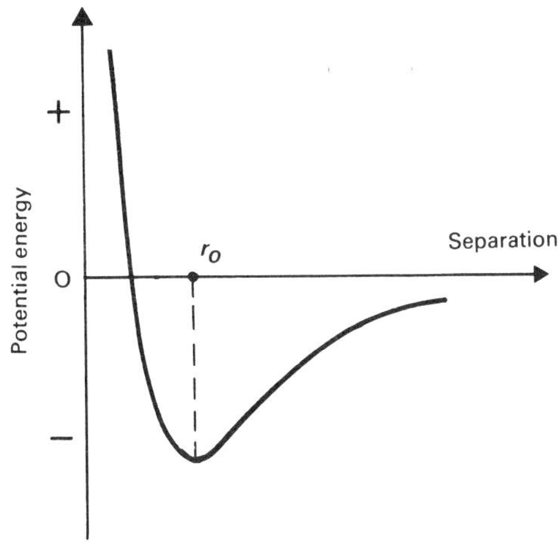


**Figure 4.2**

**4.9** Figure 4.2 shows the potential energy–separation graph for intermolecular forces between two molecules.
(a) What is the separation of the molecules for minimum potential energy?
(b) What, from Figure 4.1, is the net force on the molecules at the minimum potential energy position?
(c) What is the significance of the potential energy having positive values at small separations but negative values at larger separations?

***4.10** This question concerns the use of the force–separation graph in Figure 4.1 to explain elasticity.
(a) When a solid is stretched, why does it spring back when the force is removed?
(b) When a solid is compressed, why does it spring back when the force is removed?
(c) Explain how Hooke's law relates to the shape of the graph near the separation $r_0$.
(d) Identify on the graph the point at which the solid will break under tensile forces.

***4.11** This question concerns the use of the potential energy–separation graph in Figure 4.2 to explain thermal expansion.
(a) At absolute zero molecules are stationary and have no thermal energy. What point on the graph corresponds to the separation at this temperature?
(b) At higher temperatures the molecules have thermal energy and oscillate between two positions. Explain how these positions can be identified on the graph with a specified amount of thermal energy.
(c) Explain why the amplitude of the oscillations increases as the thermal energy supplied increases.
(d) Explain why solids expand when the thermal energy is increased.

**II Read** about surface tension and latent heat in relation to intermolecular forces.
References: *B* PP ch. 9, *D* ch. 9, *N* ch. 6, *W et al.* (ch. 15), *Wh&H*, ch. 18.

**Self-assessment questions**

**4.12** A molecule in the surface of a liquid has more energy than a molecule deeper in the liquid. Justify this statement qualitatively.

***4.13** Why does the temperature of a liquid decrease when the surface area is increased? Hint: this follows on from the previous question.

***4.14** If the surface tension of water is $7.0 \times 10^{-2}$ N m$^{-1}$ and water molecules have an effective diameter of $3.0 \times 10^{-10}$ m, what is the extra free energy possessed by a water molecule in the surface of water?

***4.15** This follows from the previous question. If within the body of the liquid a molecule has 10 near neighbours with which it has bonds, use the data given in SAQ 4.14 to arrive at a value for the energy needed to break a bond.

***4.16** When a liquid evaporates a molecule breaks all its bonds with other molecules in the liquid and escapes from the liquid. Explain why the values of the latent heat of vaporization and the free surface energy of a liquid are related.

# SELF-ASSESSMENT

## Key points

An **atom** is the smallest neutral particle that represents an element.
A **molecule** of a substance is the smallest particle of it that can exist under normal conditions.
**Atomic** and **molecular masses** give the masses of atoms and molecules relative to the mass of the carbon-12 atom.
The **Avogadro constant** is numerically equal to the number of atoms in exactly 12 g of carbon-12 and has the value $6.02 \times 10^{23}$.
At large distances of separation the forces between molecules are negligible; however if the molecules are brought closer there is a net force of attraction between them. At still closer distances the net force becomes repulsive. At the equilibrium position the net force is zero, also the potential energy of the system is a minimum.

Molecules in the surface of a liquid have about twice the potential energy that molecules in the body of the liquid have, due to their having only half the intermolecular bonds. It is this energy which accounts for the **free surface energy. Vaporization** is concerned with the breaking of molecular bonds and hence the values of the free surface energy and latent heat are related.

## Answers I

**4.1** See the key points.

**4.2** About 1 to 3 × $10^{-10}$ m.

**4.3** An element consists of only one type of atom, a compound consists of at least two different types of atom.

**4.4** (a) 1 (b) 16 (c) 63.5

**4.5** The same, i.e. 6.02 × $10^{23}$.

**4.6** (a) The molecular mass of the hydrogen is 2 since it is composed of two atoms of hydrogen. Hence the mass of 1 mole is 2 g. (b) 18 (c) 18 g (d) 18 g

**4.7** One mole is 18 g and contains 6.02 × $10^{23}$ molecules, therefore 5 g contains 1.67 × $10^{23}$ molecules.

## Answers II

**4.8** (a) repulsive (b) attractive (c) Because the net force is zero. (d) At large distances the force is negligible. As the molecules are brought closer there is a net attractive force which increases to a maximum value before diminishing to zero at $r_0$. Further bringing together of the molecules results in a net repulsive force.

**4.9** (a) $r_0$ (b) zero (c) Negative values mean that work has to be done to separate the molecules. Positive values mean that work has to be done to move the molecules closer together.

**4.10** (a) The attractive forces between the molecules increase.
(b) The repulsive forces between the molecules increase.
(c) According to Hooke's law the extension is directly proportional to the force. This means that the part of the force–separation graph on either side of $r_0$ should be a straight line. It is, very nearly.
(d) When the material is stretched the separation of the molecules increases as a result of the force increasing. However, the material can be considered to have broken when, although the separation increases, the force decreases. This corresponds to the point of maximum attractive force on the graph.

**4.11** (a) Separation $r_0$.
(b) See Figure A4.1. The height above the minimum represents the thermal energy.
(c) The amplitude increases because the potential energy–separation curve is opening out.
(d) Expansion occurs because the mean separation of the molecules increases, due to the asymmetry of the curve.

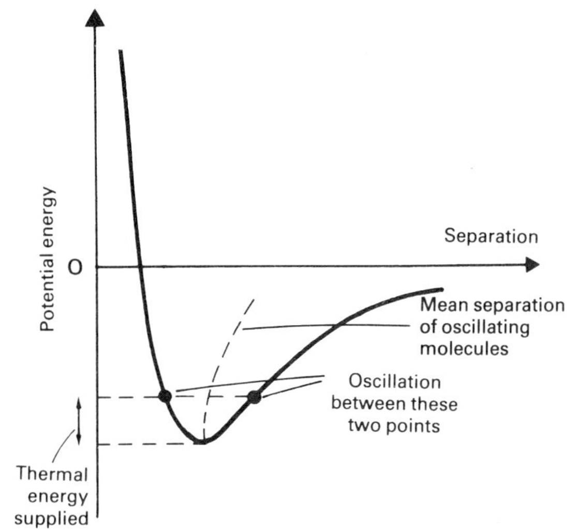


**Figure A4.1**

## Answers III

**4.12** Consider a molecule in the body of the liquid. It is in a constant state of motion, continually making and breaking bonds with neighbouring molecules all around it. A molecule in the surface has neighbours only below and to the sides of it, so it makes only about half the number of bonds that the molecule deeper in the liquid makes. Since a bond is a state of negative potential energy which requires energy to be used to break it, the surface molecule has more 'free' energy.

**4.13** When the surface area is increased the number of molecules in the surface increases and so the total free energy has to increase, this increase coming from a lowering of the kinetic energy of the molecules in the liquid.

**4.14** The free surface energy is numerically equal to the surface tension, i.e. the energy to produce 1 $m^2$ of surface is 7.0 × $10^{-2}$ J. In an area of 1 $m^2$ the number of molecules is $1/(\frac{1}{4}\pi\, 3.0^2 \times 10^{-20}) = 1.4 \times 10^{19}$. Thus the extra energy possessed by a molecule is $7.0 \times 10^{-2}/(1.4 \times 10^{19}) = 5.0 \times 10^{-21}$ J.

**4.15** The extra free energy comes from the number of bonds made by a molecule decreasing by half, i.e. from 10 to 5. Thus the energy needed to break a bond is $\frac{1}{2} \times 5.0 \times 10^{-21}/5 = 5.0 \times 10^{-22}$ J. The extra $\frac{1}{2}$ comes from the fact that breaking one bond releases 2 molecules.

**4.16** The free surface energy is a measure of the energy needed to break about half the bonds for the molecules in a surface area of 1 $m^2$. The latent heat is a measure of the energy needed for a specified mass of molecules to break their bonds.

## CHAPTER 3

# Heat

## Lesson 1 Temperature scales

### Aims

After completing this lesson you should be able to

- state the *zeroth law of thermodynamics*
- explain how a temperature scale is determined
- describe properties that can be used for temperature measurement
- describe the ideal-gas scale and the thermodynamic scale.

## STUDY

**I Read** about the zeroth law and how a temperature scale is determined.
References: *B* ST3 ch. 2, *D* ch. 4, ch. 19, *N* ch. 8, *W. et al.* ch. 9(9), *Wh&H.* ch. 21. Also *B* PP ch. 3, ch. 8, as useful background and extension reading.

**Self-assessment questions**

**1.1** State the zeroth law of thermodynamics.

**1.2** Explain how a temperature scale can be defined, using the terms *thermometric property, fixed points* and *fundamental interval.*

**1.3** A temperature scale is defined using the e.m.f. developed by a thermocouple as the thermometric property. If at the zero of the scale the e.m.f. is 2.0 μV and at the 100 point it is 3.2 μV, what is the temperature on such a scale when the e.m.f. is 2.7 μV?

**II Read** about the thermometric properties used, the absolute scale and the kelvin scale.
References: *B* ST3 ch. 2, *D* ch. 4, *N* ch. 8, *W et al.* ch. 9(9), *Wh&H* ch. 21.

**Self-assessment questions**

***1.4** What are the characteristics which make a property useful as a thermometric property?

**1.5** State five properties that are used as thermometric properties.

***1.6** Why might scales derived on the basis of different thermometric properties differ in the value they give to a particular temperature, even when using the same fixed points?

**1.7** The ideal-gas scale uses an ideal gas as the working substance and the property measured is the pressure when the volume is kept constant. What are the two fixed points used and the temperature values given to those points?

**1.8** The fundamental temperature scale is the thermodynamic scale, the unit of temperature on this scale being the kelvin (K). What are the values of the normal ice-melting and water-boiling points on this scale?

***1.9** The resistance $R_\theta$ of a coil of wire at a temperature $\theta$, on the ideal-gas scale, is given by

$$R_\theta = R_0(1 + \alpha\theta + \beta\theta^2)$$

where $R_0$ is the resistance at 0 °C, $\alpha = 4.0 \times 10^{-3}\ \mathrm{K}^{-1}$ and $\beta = -6.0 \times 10^{-7}\ \mathrm{K}^{-2}$. If the resistance of this coil of wire were used to define a resistance temperature scale, what would be the difference between the temperature value assigned to an object by this scale and the temperature of 40 °C indicated by the ideal-gas scale, assuming both scales coincided at 0 °C and 100 °C?

## SELF-ASSESSMENT

### Key points

The **zeroth law of thermodynamics** states that if two bodies A and B are separately in thermal equilibrium with a third body C, then A and B are in thermal equilibrium with each other.
When two bodies are in thermal equilibrium with each other then they are at the same **temperature**.
To define a temperature scale a thermometric property of a body is chosen, two fixed points are selected and the thermometric property measured at these two points.

Then the temperature $\theta$ when the thermometric property has some other value, $X_\theta$, is

$$\theta = \frac{(X_\theta - X_1)}{(X_2 - X_1)} N + \theta_1$$

where $X_2$ is the value of the property at the upper fixed point, $X_1$ the value at the lower fixed point, $N$ the number of scale divisions chosen to divide the interval between the fixed points, and $\theta_1$ the value given to the lower fixed point.

The **ideal-gas scale** uses absolute zero as the lower fixed point, giving it the value 0, and for the upper fixed point uses the triple point of pure air-free water, the value 273.16 being given to it. With such a scale we thus have

$$T = \frac{p_T}{p_{tr}} \times 273.16$$

where $p_T$ is the pressure measured at temperature $T$ and $p_{tr}$ the pressure at the triple point. In practice real gases only approach the ideal-gas behaviour when extrapolated to zero pressure. By definition the values at the fixed points have been chosen to agree with those on the thermodynamic scale, and so are quoted in kelvin.

### Answers I

**1.1** See the key points. This law is important as without it we could not measure temperature; it tells us when two bodies are at the same temperature.

**1.2** See the key points.

**1.3** $\theta = (2.7 - 2.0) \times 100/(3.2 - 2.0) = 58.3$

### Answers II

**1.4** Repeatability; a unique value at a particular temperature; sensitivity to temperature changes over the range concerned.

**1.5** Length of a column of mercury in glass; electrical resistance; pressure of a gas when the volume is kept constant; e.m.f. of a thermocouple; saturation vapour pressure of a liquid; colour of the light emitted by a hot body; intensity of the radiation emitted by a hot body; etc.

**1.6** While, by definition, scales can be made to agree at the fixed points, between the fixed points a linear relationship between the property and the temperature is assumed for each property, so the different scales might not agree. For instance, a scale based on the length of a mercury column in a glass capillary tube would indicate that the e.m.f. of a thermocouple did not vary linearly with temperature. However, if the thermocouple e.m.f. had been chosen as the thermometric property, it would have been linear and the length of the mercury column would not have varied linearly with temperature.

**1.7** The absolute zero, value 0, and the triple point of water, value 273.16.

**1.8** 273 K, 373 K, to three-figure accuracy.

**1.9** For the resistance scale with the same fixed points and fundamental interval, then

$$\theta = \frac{R_\theta - R_0}{(R_{100} - R_0)} \times 100$$

$$= \frac{(\alpha\theta + \beta\theta^2)}{(100\alpha + 100^2\beta)} \times 100$$

$$= 40.37\,°\mathrm{C}$$

Hence the difference in value is 0.37 K.

# Lesson 2 Heat capacity

### Aims

After completing this lesson you should be able to

- define heat capacity, *specific* heat capacity and *molar* heat capacity
- solve problems involving the above quantities
- explain methods used for the measurement of specific heat capacities, explaining how heat losses are reduced and determined.

### Prerequisites

Rate of conversion of electrical energy to heat as $IV$, the mole concept and a basic knowledge of energy is required, thus Lesson 5 of Chapter 1 and Lesson 4 of Chapter 2 would be useful.

# STUDY

**I Read** about the definitions of heat capacity, specific heat capacity and molar heat capacity, and the solution of problems involving these quantities.
References: *B* PP ch. 3, *B* ST3 ch. 2, *D* ch. 4, *N* ch. 9, *W et al.* ch. 11(9), *Wh&H* ch. 23.

**Self-assessment questions**

**2.1** Define *heat capacity*, *specific heat capacity* and *molar heat capacity*, stating their units in each case.

**2.2** If the specific heat capacity of copper is $4.0 \times 10^2$ J kg$^{-1}$ K$^{-1}$, what heat has to be supplied to a mass of 100 g to raise the temperature by 20 K?

**2.3** What is the molar heat capacity of copper if the atomic mass is 63.5 and the specific heat capacity is 400 J kg$^{-1}$ K$^{-1}$?

**2.4** An aluminium container has a mass of 150 g and contains 0.8 kg of water. What heat is required to raise the temperature of the container and water by 60 K? (Specific heat capacity of aluminium = $8.8 \times 10^2$ J kg$^{-1}$ K$^{-1}$, specific heat capacity of water = $4.2 \times 10^3$ J kg$^{-1}$ K$^{-1}$.)

**2.5** A container has a heat capacity of 300 J K$^{-1}$. What heat has to be supplied to the empty container to raise its temperature by 40 K? What heat would have had to be supplied if the container had contained 0.6 kg of water? (Specific heat capacity of water = $4.2 \times 10^3$ J kg$^{-1}$ K$^{-1}$.)

**2.6** A 2 kW electric immersion heater is used to heat 20 kg of water in a copper tank. If the tank has a heat capacity of 600 J K$^{-1}$ and all the heat from the heater is used to heat the water and the tank, how long will it take to heat the water from 20 °C to 80 °C? (Specific heat capacity of water = $4.2 \times 10^3$ J kg$^{-1}$ K$^{-1}$.)

**2.7** If 0.5 kg of cold water at 15 °C is mixed with 1.0 kg of hot water at 60 °C, what will be the resulting temperature if you can assume that no heat enters or leaves the mixture?

**2.8** A piece of aluminium, of mass 30 g, and at a temperature of 100 °C, is dropped into 0.20 kg of water in a copper container, both being at 20 °C. What is the resulting temperature of the water? Assume that no heat enters or leaves the system. The copper container has a heat capacity of 40 J K$^{-1}$, the water a specific heat capacity of $4.2 \times 10^3$ J kg$^{-1}$ K$^{-1}$ and the aluminium a specific heat capacity of 880 J kg$^{-1}$ K$^{-1}$.

***2.9** A bullet travelling at 400 m s$^{-1}$ hits a target and is brought to rest. What was the kinetic energy of the bullet on impact with the target, in terms of its mass *m*? If 50% of this energy goes into heating up the bullet, what will be the resulting temperature rise? The bullet has a specific heat capacity of 130 J kg$^{-1}$ K$^{-1}$.

***2.10** A beam of electrons, travelling at $1.2 \times 10^6$ m s$^{-1}$, hits a metal target. If every second $1.0 \times 10^{14}$ electrons hit the target and all their energy is converted into heat, how long will it take for the target temperature to increase from 18 °C to 80 °C? The target has a heat capacity of 0.15 J K$^{-1}$ and the mass of an electron is $9.0 \times 10^{-31}$ kg.

**II Read** about the measurement of specific heat capacity and how heat losses are reduced, determined or eliminated from the calculation.
References: *B* ST3 ch. 2, *D* ch. 4, *N* ch. 9, *W et al.* ch. 11, *Wh&H* ch. 23. *B* PP ch. 3 gives details from a relatively recent paper on the measurement of a specific heat capacity.

**Self-assessment questions**

**2.11** State two methods by which heat losses are minimized in a calorimetric experiment.

**2.12** *Newton's law of cooling* states that the rate of loss of heat from a body is proportional to its excess temperature over the surroundings. Bearing in mind that the rate of loss of heat for an object of constant heat capacity is proportional to the rate of change of temperature, explain how you could carry out an experiment to test this law.

**2.13** This question is concerned with the cooling correction that can be applied in calorimetric experiments in order to compensate for heat losses during the experiment.
(a) During an experiment a record was kept of the temperature variation with time, Figure 2.1 showing the resulting graph. What was the rate of cooling at about 26–27 °C?
(b) What is the average rate of cooling during the temperature change from 20 °C to 27 °C?
(c) Hence what is the cooling correction and the temperature that would have been attained in the absence of heat losses?

**2.14** A common method used for the measurement of the specific heat of a solid is the method of mixtures. The solid, of measured mass, is heated to a known temperature $\theta_3$, generally that of boiling water, and then dried and dropped into a liquid, often water, of known

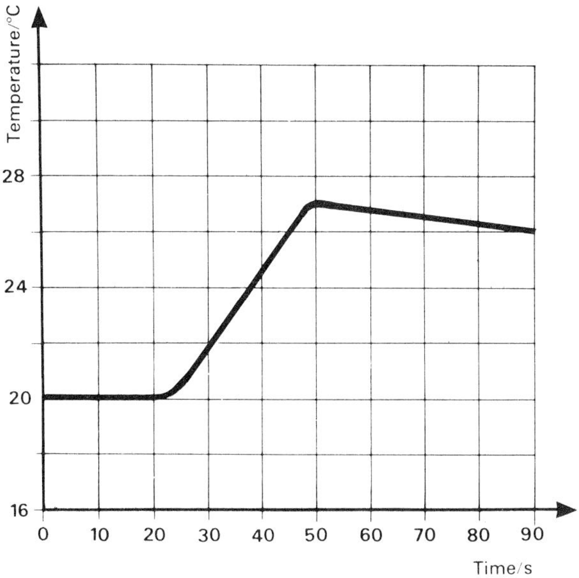


**Figure 2.1**

mass, in a calorimeter of known mass and specific heat capacity. The temperature of the liquid is taken prior to the solid being immersed in it, $\theta_1$, and then monitored during the mixing of the solid with the liquid so that the maximum temperature attained can be determined, $\theta_2$.
(a) What is the heat lost by the cooling solid?
(b) What is the heat gained by the liquid?
(c) What is the heat gained by the calorimeter?
(d) Hence deduce an expression for the specific heat of the solid.
(e) How could the above be changed so that the specific heat of the liquid in the calorimeter is determined?

**2.15** A block of metal of mass 150 g at 100 °C is dropped into a calorimeter containing water at 18 °C. If the mass of the water is 100 g and the calorimeter has a heat capacity of 40 J $kg^{-1}$, and the temperature of the resulting mixture is 34 °C, what is the specific heat capacity of the solid? (Specific heat capacity of water = $4.2 \times 10^3$ J $kg^{-1}$ $K^{-1}$.)

**2.16** The specific heat capacity of a metal can be found by an electrical method in which an electric heater is inserted into a hole in the metal and a thermometer is placed in another hole.
(a) If the current $I$ through the heater and the potential difference $V$ across it are measured for a time $t$, what is the electrical energy transformed into heat in that time?
(b) The rise in temperature $\theta$ of the block of metal during the time interval is measured. What is the heat received by the metal block in that time?
(c) If the block is sufficiently insulated to assume that no heat is lost or gained by the system, derive an equation for the specific heat capacity of the metal.
(d) One method that is used to compensate for heat losses is to start the heating of the block at a temperature as far below the room temperature as the temperature at which the experiment finishes is above room temperature. Explain this method.

***2.17** A metal block of mass 1.0 kg is heated by an electric heater which supplies 50 W for one minute. The initial rate of rise in temperature is 2.0 K $min^{-1}$. When the heater is switched off the rate of cooling is 0.15 K $min^{-1}$. What is the specific heat capacity of the metal?

***2.18** Describe the continuous-flow method of determining the specific heat capacity of a liquid.
(a) Why is the heat capacity of the apparatus not required?
(b) Explain how the heat loss that occurs can be eliminated from the final calculation of the specific heat capacity.

***2.19** In the continuous-flow method for the determination of the specific heat capacity of a liquid, an electric element is used which dissipates 50.1 W. When the rate of flow of the liquid over the heater is 60 g $min^{-1}$, the inflow and outflow temperatures are 15.1 °C and 18.6 °C. When the experiment is repeated with the same inflow and outflow temperatures the electric element has to dissipate 58.1 W when the rate of flow is 102 g $min^{-1}$. Calculate the specific heat capacity of the liquid.

# SELF-ASSESSMENT

## Key points

The **heat capacity** of a body is the quantity of heat required to produce unit rise in temperature, i.e. $C = \Delta Q/\Delta\theta$. The **specific heat capacity** is the heat rquired to produce unit rise in temperature in unit mass of the material, i.e. $c = \Delta Q/(m\,\Delta\theta)$. The **molar heat capacity** is the heat required to produce unit rise in temperature in one mole of the substance concerned, i.e. $C_m = \Delta Q/(\mu\,\Delta\theta)$, where $\mu$ is the number of moles.

## Answers I

**2.1** See the key points. Heat capacity has the unit J $K^{-1}$, specific heat capacity J $kg^{-1}$ $K^{-1}$, and molar heat capacity J $mol^{-1}$ $K^{-1}$.

**2.2** $\Delta Q = mc\,\Delta\theta = 0.100 \times 4.0 \times 10^2 \times 20 = 800\,\text{J}$

**2.3** One mole has a mass of 63.5 g, see Lesson 4, Chapter 2. Thus $\Delta Q = mc\,\Delta\theta = 0.0635 \times 400 \times 1 = 25.4\,\text{J}$. Hence $C_m = 25.4\,\text{J mol}^{-1}\,\text{K}^{-1}$. For many solids the molar heat capacity is about $25\,\text{J mol}^{-1}\,\text{K}^{-1}$, this fact being known as *Dulong and Petit's law*.

**2.4** For the container $\Delta Q = mc\,\Delta\theta = 0.150 \times 8.8 \times 10^2 \times 60 = 7920\,\text{J}$. For the water $\Delta Q = 0.8 \times 4.2 \times 10^3 \times 60 = 201\,600\,\text{J}$. Hence the total heat required is $2.10 \times 10^5\,\text{J}$.

**2.5** $\Delta Q = C\,\Delta\theta = 300 \times 40 = 1.2 \times 10^4\,\text{J}$. For the water $\Delta Q = mc\,\Delta\theta = 0.6 \times 4.2 \times 10^3 \times 40 = 1.0 \times 10^5\,\text{J}$ and so the total is $1.1 \times 10^5\,\text{J}$.

**2.6** $\Delta Q = 600 \times 60 + 20 \times 4.2 \times 10^3 \times 60 = 5.07 \times 10^6\,\text{J}$. Time $= 5.07 \times 10^6/2000 = 2.5 \times 10^3\,\text{s}$, or about 42 minutes.

**2.7** $0.5 \times 15 + 1.0 \times 60 = 1.5\,\theta$ and so $\theta = 45\,°\text{C}$.

**2.8** $0.030 \times 880 \times (100 - \theta)$ = heat lost by the aluminium. Heat gained by the container plus the water $= 40 \times (\theta - 20) + 0.20 \times 4200 \times (\theta - 20)$. Heat gained equals heat lost, and so $\theta = 23.7\,°\text{C}$.

**2.9** k.e. $= \frac{1}{2}mv^2 = \frac{1}{2}m \times 400^2$. $\Delta Q = mc\,\Delta\theta = m \times 130 \times \Delta\theta$ and so $\frac{1}{2}m \times 400^2 \times 50/100 = m \times 130 \times \Delta\theta$ and so $\Delta\theta = 308\,°\text{C}$.

**2.10** Total kinetic energy $= n \times \frac{1}{2}mv^2 = 1.0 \times 10^{14} \times \frac{1}{2} \times 9.0 \times 10^{-31} \times (1.2 \times 10^6)^2 = 6.48 \times 10^{-5}\,\text{J}$ per second. $\Delta Q = C\,\Delta\theta = 0.15 \times (80 - 18) = 9.3\,\text{J}$. Hence time $= 9.3/(6.48 \times 10^{-5}) = 1.44 \times 10^5\,\text{s}$.

## Answers II

**2.11** The use of insulation, polishing to reduce radiation loss, surrounding the calorimeter with an evacuated jacket, starting the experiment at a temperature as far below the temperature of the surroundings as the final temperature will be above.

**2.12** The body could be a beaker of hot water, the temperature being determined as a function of time. A graph of temperature against time can then be plotted. The rate of cooling at a particular temperature is the slope of the graph at that temperature. By measuring a number of slopes at different temperatures the relationship between rate of cooling and temperature above the surroundings can be determined.

**2.13** (a) $1/40\,\text{K s}^{-1}$
(b) If we assume that the rate of cooling is proportional to the temperature excess above the surroundings, at 20 °C it is zero and at 27 °C it is $1/40\,\text{K s}^{-1}$. The average is thus half this, i.e. $0.0125\,\text{K s}^{-1}$.
(c) It took 30 s and so the correction is $30 \times 0.0125 = 0.375\,\text{K}$ and the temperature attained should be 27.375 °C.

**2.14** (a) $\Delta Q = mc(\theta_3 - \theta_2)$ (b) $\Delta Q = m_l c_l(\theta_2 - \theta_1)$ (c) $\Delta Q = m_c c_c(\theta_2 - \theta_1)$ (d) Heat lost = heat gained, hence $c = (m_l c_l + m_c c_c)(\theta_2 - \theta_1)/m(\theta_3 - \theta_2)$
(e) The solid used could be of known specific heat capacity.

**2.15** Using the expression derived in 2.14(d) above, $c = 743\,\text{J kg}^{-1}\,\text{K}^{-1}$.

**2.16** (a) $IVt$ (b) $\Delta Q = mc\theta$ (c) $c = IVt/m\theta$
(d) The heat gained when below the temperature of the surroundings equals the heat lost above that temperature.

**2.17** In 1 minute the temperature rise is 2.0 K. At the end of the heating the rate of cooling is $0.15\,\text{K min}^{-1}$, therefore the average rate of cooling is $0.075\,\text{K min}^{-1}$, and the temperature gain should have been 2.075 K. Hence, using the expression derived in 2.16(c) above, $c = 50 \times 60/(1.0 \times 2.075) = 1.4 \times 10^3\,\text{J kg}^{-1}\,\text{K}^{-1}$.

**2.18** See any of the references quoted for a description. The salient points are as follows. There is a constant rate of flow of the liquid over a heating coil in a tube. The inflow and the outflow temperatures are measured for a particular flow rate and electrical power. The tube is generally surrounded by a vacuum jacket and a constant-temperature jacket (a water-filled tube).
(a) Because the temperature of the apparatus does not change during the experiment. For this reason it is vital that readings are not taken until steady-state conditions exist.
(b) By repeating the experiment with a different flow rate and electrical power but the same inflow and outflow temperatures. The heat loss in each case will be the same and can then be eliminated between two simultaneous equations of heat flow.

**2.19** For first flow rate: $I_1V_1 = m_1c(\theta_2 - \theta_1) + h$, where $m_1$ is the mass flowing per second and $h$ the heat loss per second. Thus $50.1 = 0.001\,c \times (18.6 - 15.1) + h$. For second flow rate: $58.1 = (0.102/60)c \times (18.6 - 15.1) + h$. Subtracting the first equation from this eliminates $h$ and gives $c = 3265\,\text{J kg}^{-1}\,\text{K}^{-1}$.

# Lesson 3 Latent heat

## Aims

After completing this lesson you should be able to

- explain what latent heat is and what is meant by a *change of phase*
- define *specific latent heat* and solve problems involving it
- describe experiments by which latent heats can be measured
- relate latent heat to intermolecular forces.

## Prerequisites

Lesson 2 on heat capacity. Lesson 4 of Chapter 2 on the structure of matter is a useful cross-reference for study section III below.

# STUDY

**I Read** about latent heat, change of phase and the definition of specific latent heat.
References: *B* ST3 ch. 2, *D* ch. 4, *N* ch. 9, *W et al.* (ch. 9), *Wh&H* ch. 27.

**Self-assessment questions**

**3.1** What is meant by a 'change of phase'?

**3.2** What is *latent heat* and how is it related to a change of phase?

**3.3** Define *specific latent heat* and state its units.

**3.4** Sketch a graph of temperature against time for a material that is heated when solid, changes phase and becomes a liquid, and is then heated as a liquid. Explain the varying gradients of the graph.

**3.5** The specific latent heat of fusion for water is $3.3 \times 10^5\,\mathrm{J\,kg^{-1}}$. How much heat is required to change 200 g of ice at its melting point to water at the same temperature?

**3.6** How much heat is needed to heat 200 g of water at 20 °C and turn it into steam at 100 °C if the specific heat capacity of water is $4.2 \times 10^3\,\mathrm{J\,kg^{-1}\,K^{-1}}$ and the specific latent heat of vaporization is $2.3 \times 10^6\,\mathrm{J\,kg^{-1}}$?

***3.7** An electric heater of power 2 kW is used to heat 0.8 kg of water in a container of heat capacity $300\,\mathrm{J\,K^{-1}}$. If the water is initially at 18 °C, how long will it take to heat it to the boiling point of 100 °C? What mass of water will have boiled away in 4 minutes? (Specific heat capacity = $4.2 \times 10^3\,\mathrm{J\,kg^{-1}\,K^{-1}}$, specific latent heat of vaporization = $2.3 \times 10^6\,\mathrm{J\,kg^{-1}}$.)

***3.8** Steam at 100 °C is passed into 200 g of water initially at 20 °C in a container of heat capacity $50\,\mathrm{J\,K^{-1}}$. Neglecting heat losses, what mass of steam will have been passed into, and condensed in, the water when the water temperature has risen to 35 °C? (Specific heat capacity = $4.2 \times 10^3\,\mathrm{J\,kg^{-1}\,K^{-1}}$, specific latent heat of vaporization = $2.3 \times 10^6\,\mathrm{J\,kg^{-1}}$.)

***3.9** A lead pellet fired from a gun hits a sheet of steel and melts. If the initial temperature of the pellet was 20 °C and lead melts at 327 °C, what is the minimum speed the pellet must have if 60% of the kinetic energy of the pellet is used to heat it on impact? (Specific heat capacity of lead = $1.3 \times 10^2\,\mathrm{J\,kg^{-1}\,K^{-1}}$, specific latent heat of fusion = $2.1 \times 10^4\,\mathrm{J\,kg^{-1}}$.)

**II Read** about experiments to measure latent heats.
References: *D* ch. 4, *N* ch. 9, *Wh&H* ch. 27.

***3.10** Explain how the method of mixtures can be used for the determination of the specific latent heat of fusion of water. Explain how a cooling correction can be avoided.

***3.11** Describe an electrical method for the determination of the specific latent heat of vaporization of water. Explain how heat losses can be both minimized and determined.

***3.12** In an experiment to determine the specific latent heat of vaporization by an electrical method, a current of 3.5 A flows through the heating coil when the potential difference across it is 8.6 V. In 3 minutes 6.3 g of the liquid is vaporized. When the experiment is repeated with a current of 4.2 A and a potential difference of 11.9 V, the mass vaporized in 3 minutes is 10.8 g. What is (a) the specific latent heat of vaporization and (b) the heat loss per second?

**III Read** about the relationship between latent heat and intermolecular forces and bond-breaking.
References: *B* PP ch. 9, *D* ch. 4. Lesson 4 in Chapter 2 could also be referred to.

**Self-assessment question**

****3.13** The *bonding energy* for two molecules is the amount of energy that has to be supplied to pull them apart to what amounts to an infinite separation of the molecules.
(a) If a molecule has $n$ neighbours with which it has bonds and each bond has a bonding energy of $E_0$, what is the total energy needed to break this molecule away from all its bonds?
(b) What would be the energy needed to break all the bonds in 1 mole of these molecules? Remember that when a bond is broken for one molecule another molecule also has a bond broken.
(c) An energy of $2.3 \times 10^6$ J is needed to vaporize 1 kg of water. What is the energy needed for one mole of water? (1 mole of water has a mass of 18 g.)
(d) Hence arrive at a value for the bonding energy of water molecules taking $n$ as 10. (Avogadro's number $N_0$ is $6.02 \times 10^{23}$.)

# SELF-ASSESSMENT

## Key points

The **specific latent heat** of transformation from one phase to another is the energy needed per unit mass of the substance for the phase change to occur, without any change of temperature occurring, i.e. $l_p = L/m$ where $l_p$ is the specific latent heat for the phase change, $L$ the energy required and $m$ the mass of the substance.

## Answers I

**3.1** A change from one form of molecular configuration to another, e.g. solid to liquid or liquid to vapour.

**3.2** Latent heat is the heat needed for the change of phase to occur. The term *latent* is present because the energy causes no change in temperature but is used to break bonds between molecules and possibly to increase the volume.

**3.3** See the key points; the unit is J $kg^{-1}$.

**3.4** See Figure A3.1. The slope of AB is related to the specific heat capacity of the material when in the solid phase. BC shows no temperature change although energy continues to be supplied. This is because a phase change is occurring. The slope CD is related to the specific heat capacity of the liquid.

**3.5** $L = ml_m$ where $l_m$ is the specific latent heat of fusion. Thus $L = 200 \times 10^{-3} \times 3.3 \times 10^5 = 6.6 \times 10^4$ J.

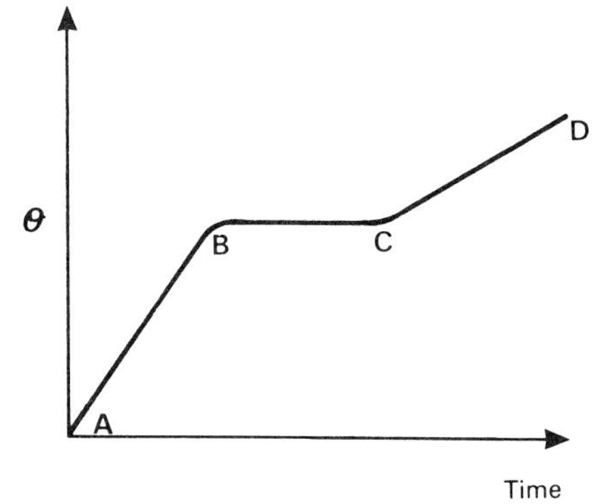


**Figure A3.1**

**3.6** $\Delta Q = mc\,\Delta\theta + ml_v = 0.200 \times 4.2 \times 10^3 \times 80 + 0.200 \times 2.3 \times 10^6 = 5.27 \times 10^5$ J

**3.7** $\Delta Q = mc\,\Delta\theta = 0.8 \times 4.2 \times 10^3 \times 82 + 300 \times 82 = 3.00 \times 10^5$ J. Hence time $= 3.00 \times 10^5/2000 = 150$ s. Therefore time for boiling $= 4 \times 60 - 150 = 90$ s. Energy supplied $= 90 \times 2000 = 1.8 \times 10^5$ J. Mass evaporated $= 1.8 \times 10^5/2.3 \times 10^6 = 0.078$ kg.

**3.8** Heat gained by water and container $= 0.200 \times 4.2 \times 10^3 \times 15 + 50 \times 15 = 1.34 \times 10^4$ J. Heat lost by steam in condensing and then cooling to 35 °C $= 2.3 \times 10^6 \times m + m \times 4.2 \times 10^3 \times 65 = 2.57 \times 10^6 m$. Hence, as heat gained = heat lost, $m = 5.21 \times 10^{-3}$ kg.

**3.9** Heat required $= m \times 1.3 \times 10^2 \times 307 + m \times 2.1 \times 10^4 = 6.09 \times 10^4 m$. This has to be supplied by $(60/100)\frac{1}{2}mv^2$, hence $v = 451$ m $s^{-1}$.

## Answers II

**3.10** The significant points are that pieces of ice are in water to bring them to 0 °C, taken out and dried, then dropped into water in a calorimeter and the temperature monitored. The water should initially be, say, 4 degrees above the room temperature and then sufficient ice added to bring the temperature to the same amount below room temperature. The mass of the water has been determined initially and again after the ice is added. Then heat lost by the water, and calorimeter, equals the heat gained by the ice melting and then the resulting water being warmed to the mixture temperature.

**3.11** The significant points are that a steady state must be reached and then the mass vaporizing in a measured time can be determined. There must be a steady state as there is then no need to know the heat capacity of the container. Then $IVt = ml_v + Q$ where $Q$ is the heat loss in time $t$. Repeating the experiment with a different power gives $I_1V_1t = m_1l_v + Q$ and the heat loss $Q$ can be determined. It is the same in both experiments because the temperature is the same and the same time is considered.

**3.12** Using the equations given in 3.11 above: $3.5 \times 8.6 \times 3 \times 60 = 6.3 \times 10^{-3}l_v + Q$ and $4.2 \times 11.9 \times 3 \times 60 = 10.8 \times 10^{-3}l_v + Q$, where $Q$ is the heat loss in 3 minutes.
(a) Eliminating $Q$ by subtracting the first equation from the second gives $l_v = 7.95 \times 10^5\ \text{J kg}^{-1}$.
(b) Substituting the $l_v$ value in the first equation gives $Q = 409.5$ J and hence heat loss per second = 2.28 W.

### Answers III

**3.13** (a) $nE_0$ (b) $\frac{1}{2}N_0nE_0$ (c) $4.14 \times 10^4$ J (d) $1.38 \times 10^{-20}$ J

# Lesson 4 The gas laws

### Aims

After completing this lesson you should be able to

- state and use the ideal-gas laws
- define the *specific gas constant* and the *universal gas constant* and solve problems involving the gas laws with these constants.

### Prerequisites

Pressure = force/area, the thermodynamic temperature scale (Lesson 1) and the mole (Lesson 4 of Chapter 2).

## STUDY

**I Read** about the three gas laws and their combination into the form $pV/T$ = a constant.
References: *B* PP ch. 9, *B* ST2 ch. 3, *D* ch. 19, *N* ch. 10, *W et al.* ch. 1(1), *Wh&H* ch. 24.

#### Self-assessment questions

**4.1** State *Boyle's law*.

**4.2** A fixed mass of gas has a volume of $0.04\ \text{m}^3$ under a pressure of $1.0 \times 10^5$ Pa. What will be its volume under a pressure of $5.0 \times 10^5$ Pa if the temperature does not change?

**4.3** What is the relationship for an ideal gas of constant mass between (a) the volume and the temperature when the pressure is constant, (b) the pressure and the temperature when the volume is constant, if the temperatures are measured on the thermodynamic scale?

**4.4** Define an ideal gas.

**4.5** A fixed mass of gas under a constant pressure has its temperature raised from 27 °C to 54 °C. By what factor does the volume change?

**4.6** A gas cylinder of constant volume contains gas under a pressure of $5.0 \times 10^5$ Pa at 15 °C. What will be the pressure at 35 °C?

***4.7** A barometer tube has a tube length of 1000 mm above the external mercury level, the internal height of the mercury in the tube being 750 mm. When the barometer tube is pressed down into the mercury so that the tube length above the external mercury level is 950 mm, the internal height of the mercury in the tube becomes 747 mm. What is the atmospheric pressure?

***4.8** A closed vessel contains a gas which exerts a pressure of $2.0 \times 10^4$ Pa at a temperature of 150 K. At what temperature will it exert a pressure of $10 \times 10^4$ Pa?

**II Read** about the specific gas constant and the universal gas constant and the solution of problems involving the mass of a gas.
References: *B* ST2 ch. 3, *D* ch. 19, *N* ch. 10, *W et al.* ch. 13(10), *Wh&H* ch. 24.

#### Self-assessment questions

**4.9** The gas equation can be written as $pV/T$ = a constant. On what factors does this constant depend?

**4.10** Considering 1 mole of different ideal gases, what can be said about the value of the constant in the equation $pV/T$ = a constant?

**4.11** Define the *universal gas constant* and write the ideal-gas equation in terms of this constant when $\mu$ moles of gas are present.

*4.12 Calculate the volume of 1 mole of gas at s.t.p. (standard temperature and pressure, i.e. 273 K and $1.01 \times 10^5$ Pa). The universal gas constant $R$ = 8.31 J mol$^{-1}$ K$^{-1}$.

4.13 Another way of writing the ideal-gas equation is in the form $pV = mrT$, where $m$ is the mass of the gas concerned and $r$ the value of the gas constant for unit mass of the gas concerned. This constant is known as the *specific gas constant*. Calculate the value of this constant for hydrogen, if at s.t.p. hydrogen has a density of 0.09 kg m$^{-3}$.

*4.14 A cylinder of volume 0.05 m$^3$ contains oxygen at a temperature of 18 °C and a pressure of $3.0 \times 10^5$ Pa. What is the mass of gas in the cylinder? The specific gas constant for oxygen is 260 J kg$^{-1}$ K$^{-1}$.

*4.15 Two gas containers of equal volume are joined by a tube of negligible volume and are filled with air at s.t.p. What mass of air will be in the two containers if each has a volume of $2.0 \times 10^{-3}$ m$^3$? If one of the containers is then heated to 80 °C while the other remains at 0 °C, what mass of air will pass through the connecting tube? For air the specific gas constant is 287 J kg$^{-1}$ K$^{-1}$.

# SELF-ASSESSMENT

## Key points

**Boyle's law:** If mass and $T$ are constant, $pV$ = a constant.
**Charles' law:** If mass and $p$ are constant, $V/T$ = a constant.
**Pressure law:** If mass and $V$ are constant, $p/T$ = a constant.
The temperature $T$ is on the thermodynamic scale and can be found by adding 273 to Celsius scale values.
The three laws can be combined as $pV/T$ = a constant. This constant has the same value if one mole of any ideal gas is considered: $pV/T = R$ where $R$ is the **universal gas constant**. If $\mu$ moles are present then $pV/T = \mu R$. The gas equation can also be written as $pV/T = mr$, where $m$ is the mass and $r$ the **specific gas constant**. The value of this constant depends on the gas concerned.

## Answers I

**4.1** See the key points.

**4.2** $p_1V_1 = p_2V_2$ and so $V_2 = 1.0 \times 10^5 \times 0.04/5.0 \times 10^5 = 0.008$ m$^3$

**4.3** (a) $V \propto T$ (b) $p \propto T$ See the key points.

**4.4** One that obeys the gas laws.

**4.5** $V_1/T_1 = V_2/T_2$ and so $V_2/V_1 = 327/300 = 1.09$

**4.6** $p_1/T_1 = p_2/T_2$ and so $p_2 = 308 \times 5.0 \times 10^5/288 = 5.4 \times 10^5$ Pa

**4.7** The fact that the height changes means there must be air, or some other gas, above the mercury in the tube. The air pressure is $(A - 750)$ in mm of mercury, where $A$ is the atmospheric pressure. Thus $p_1V_1 = p_2V_2$ gives $(A - 750)250 = (A - 747)203$ and so $A$ = 763 mm of mercury. In some situations it is quite common to quote pressures in terms of the pressure that would be exerted by a specified height of a column of mercury.

**4.8** $p_1/T_1 = p_2/T_2$ and so $T_2 = 10.0 \times 10^4 \times 150/(2.0 \times 10^4) = 750$ K

## Answers II

**4.9** The mass of the gas and the particular gas concerned.

**4.10** It is the same.

**4.11** $pV = \mu RT$. The universal gas constant is the value of $pV/T$ for one mole of the gas and is the same for all ideal gases.

**4.12** $pV = RT$ and so $V = 8.31 \times 273/(1.01 \times 10^5) = 0.0225$ m$^3$. One mole of any ideal gas has this volume at s.t.p.

**4.13** $\rho = m/V$ and so $p = \rho rT$ and so $r = 1.01 \times 10^5/(0.09 \times 273) = 4.1 \times 10^3$ J kg$^{-1}$ K$^{-1}$

**4.14** $pV = mrT$ and so $m = 3.0 \times 10^5 \times 0.05/(260 \times 291) = 0.20$ kg

**4.15** $pV = mrT$ and thus for one container $m = 1.01 \times 10^5 \times 2.0 \times 10^{-3}/(287 \times 273) = 2.58 \times 10^{-3}$ kg. Thus the total mass is $5.16 \times 10^{-3}$ kg. For the container at 80 °C we have $m_1 = p \times 2.0 \times 10^{-3}/(287 \times 353) = 1.97 \times 10^{-8}\,p$. For the container at 0 °C we have $m_2 = p \times 2.0 \times 10^{-3}/(287 \times 273) = 2.55 \times 10^{-8}\,p$. The pressure $p$ must be the same in both containers because they are connected together. The total mass remains unchanged and so $1.97 \times 10^{-8}p + 2.55 \times 10^{-8}p = 5.16 \times 10^{-3}$ and so $p = 1.14 \times 10^5$ Pa. Hence $m_2 - m_1 = (2.55 - 1.97)10^{-8}p = 6.60 \times 10^{-4}$ kg.

# Lesson 5 Thermal behaviour of ideal gases

## Aims

After completing this lesson you should be able to
- state the *first law of thermodynamics*
- define the principal molar heat capacities and derive the relationship between them
- define *isothermal* and *adiabatic* processes and state and use the pressure–volume relationships for them.

## Prerequisites

Lesson 2 on heat capacity, Lesson 4 on ideal gases and Lesson 5 in Chapter 1 on work, energy and power. Calculus is needed for two of the questions.

# STUDY

**I Read** about the first law of thermodynamics, the concept of internal energy, the principal molar heat capacities and the relationship between them.
References: *B* PP ch. 9, *D* ch. 19, *N* ch. 10, *W et al.* ch. 13(9), *Wh&H* ch. 25.

### Self-assessment questions

**5.1** State the first law of thermodynamics.

**5.2** Explain what is meant by *internal energy* in the context of the first law.

**5.3** Consider a gas enclosed in a cylinder by a piston. If the piston has an area $A$ (see Figure 5.1) and is acted on by a pressure $p$, what is the work done when the piston moves a small distance $\Delta x$? What is this work in terms of the volume change? Assume that the change in volume is so small that the pressure remains constant.

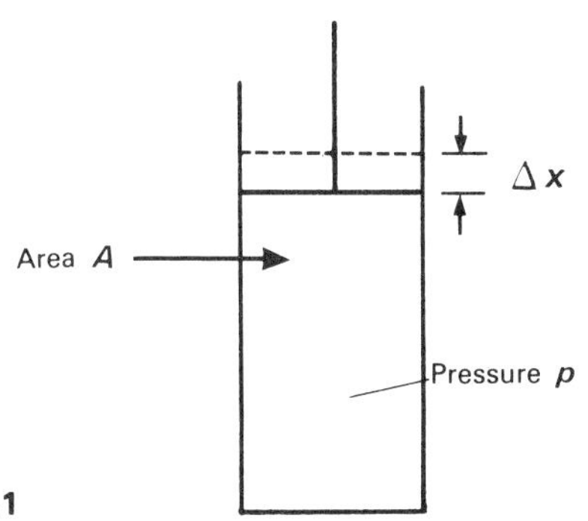


**Figure 5.1**

**5.4** Define the principal molar heat capacities of a gas.

**5.5** Consider one mole of an ideal gas in a cylinder with a frictionless piston.
(a) If a quantity of heat $\Delta Q$ is given to the gas and it expands at constant pressure $p$, the volume changing from $V$ to $V + \Delta V$ and the temperature from $T$ to $T + \Delta T$, what is the work $\Delta W$ done and the relation between $\Delta Q$ and $\Delta T$ in terms of the molar heat capacity at constant pressure $C_p$?
(b) The first law of thermodynamics can be written as $\Delta Q = \Delta U + \Delta W$, where $\Delta U$ is the change in internal energy. Hence obtain the change in internal energy in terms of the molar heat capacity at constant pressure, the change in temperature and the change in volume.
(c) Now consider the situation when the heat is supplied to the gas, there is no change in the volume but the temperature change is made the same, i.e. from $T$ to $T + \Delta T$. Derive a relationship for the heat supplied in terms of the molar heat capacity at constant volume.
(d) Use the first law of thermodynamics to obtain an expression for the change in internal energy for the changes at constant volume.
(e) For the same temperature change there is always, for an ideal gas, the same change in internal energy, as the internal energy is a function only of the temperature. Hence, use your answers to (b) and (d) to obtain the relationship between the molar heat capacities at constant volume and constant pressure.
(f) The ideal-gas equation $pV = RT$ can be applied to the change at constant pressure. Hence use this to obtain the relationship $C_p - C_V = R$.

***5.6** Why is the energy needed to raise the temperature of a fixed mass of gas by a specified amount greater if the pressure is kept constant than if the volume is kept constant?

***5.7** For oxygen the molar heat capacity at constant pressure is 1.40 times that at constant volume. If oxygen has a density of 1.43 kg m$^{-3}$ at s.t.p. and the molar mass of oxygen is 32 g, what are the two principal heat capacities? (One standard atmosphere = $1.01 \times 10^5$ Pa.)

**II Read** about isothermal and adiabatic changes.
References: *D* ch. 19, *N* ch. 10, *Wh&H* ch. 25.

### Self-assessment questions

**5.8** An *isothermal* process is one at constant temperature. For an ideal gas what is the relationship between pressure and volume?

**5.9** An *adiabatic* process is one that takes place at 'constant heat', i.e. no heat enters or leaves this system during the process. For an ideal gas what is the relationship between pressure and volume?

**5.10** Use the ideal-gas equation and the pressure–volume relationship for an adiabatic change to derive a relationship between (a) the temperature and the volume, (b) the temperature and the pressure, during an adiabatic change.

***5.11** A fixed mass of gas occupies a volume of $2.0 \times 10^{-3}\ m^3$ at a temperature of 20 °C and at a pressure of $1.0 \times 10^5$ Pa. If it is expanded adiabatically to twice this volume what is the temperature change? The ratio of the specific heat capacities is 1.4.

***5.12** A fixed mass of oxygen is compressed isothermally until its volume is halved. It is then allowed to expand adiabatically until it is back to its original volume. How is the final pressure related to the initial pressure? The ratio of the specific heat capacities is 1.4.

# SELF-ASSESSMENT

## Key points

When heat $\Delta Q$ is supplied to a gas it can raise the internal energy by $\Delta U$ and do external work $\Delta W$, with $\Delta Q = \Delta U + \Delta W$. This equation is a statement of the **first law of thermodynamics** and expresses the principle of the conservation of energy in the situation.

The **principal molar heat capacity at constant volume** $C_V$ is the heat required to change the temperature of one mole of the gas by 1 K at constant volume. The **principal molar heat capacity at constant pressure** $C_p$ is the heat required to change the temperature of one mole of the gas by 1 K at constant pressure.

$$C_p - C_V = R$$

An **isothermal** process is one that takes place at constant temperature.

An **adiabatic** process is one that takes place with no heat entering or leaving the system. For an adiabatic process $pV^\gamma$ = a constant, where $\gamma = C_p/C_V$.

## Answers I

**5.1** See the key points.

**5.2** The internal energy can be considered to be the energy that the system takes from the incoming heat energy and does not devote to doing external work. In molecular terms the internal energy consists of molecular kinetic energy and potential energy due to intermolecular forces. In the case of an ideal gas there are no intermolecular forces and so the internal energy consists solely of kinetic energy. Because of this the internal energy of an ideal gas depends only on the temperature, an increase in internal energy meaning an increase in temperature.

**5.3** Work done $= F\Delta x = pA\,\Delta x = p\,\Delta V$

**5.4** See the key points.

**5.5** (a) $\Delta W = p\,\Delta V$ and $\Delta Q = C_p\,\Delta T$ (b) $C_p\,\Delta T = \Delta U + p\,\Delta V$ (c) $\Delta Q = C_V\,\Delta T$ (d) $C_V\,\Delta T = \Delta U$ as $\Delta W = 0$ (e) $C_p\,\Delta T = C_V\,\Delta T + p\,\Delta V$ (f) $p(V + \Delta V) = R(T + \Delta T)$, also initially $pV = RT$, so $p\,\Delta V = R\,\Delta T$ and hence $C_p = C_V + R$.

**5.6** With the pressure constant external work is done due to the change in volume. With constant volume no external work is done. If both have the same rise in temperature then they have the same internal energy change. Thus the difference is that there is also energy needed for the external work with constant pressure.

**5.7** $C_p = 1.4\,C_V$, also $C_p = C_V + R$ and $pV = RT$. The volume of one mole is $32 \times 10^{-3}/1.43 = 2.24 \times 10^{-2}\ m^3$. Hence $R = pV/T = 1.01 \times 10^5 \times 2.24 \times 10^{-2}/273 = 8.29\ J\ mol^{-1}\ K^{-1}$. Thus $C_p = 1.4\,C_V = C_V + 8.29$ and so $C_V = 20.7\ J\ mol^{-1}\ K^{-1}$. Hence $C_p = 29.0\ J\ mol^{-1}\ K^{-1}$.

## Answers II

**5.8** $pV$ = a constant

**5.9** $pV^\gamma$ = a constant, where $\gamma = C_p/C_V$.

**5.10** (a) $p_1V_1^\gamma = p_2V_2^\gamma$, also $p_1V_1/T_1 = p_2V_2/T_2$. Dividing the first equation by the second one gives $T_1V_1^{\gamma-1} = T_2V_2^{\gamma-1}$ and so $TV^{\gamma-1}$ = a constant.
(b) $p_1V_1/T_1 = p_2V_2/T_2$ and so $(p_1V_1/T_1)^\gamma = (p_2V_2/T_2)^\gamma$. Dividing this by $p_1V_1^\gamma = p_2V_2^\gamma$ gives $p_1^{\gamma-1}/T_1^\gamma = p_2^{\gamma-1}/T_2^\gamma$ and so $p^{\gamma-1}/T^\gamma$ = a constant.

**5.11** Using $TV^{\gamma-1}$ = a constant, from the previous question, we have $293\,V_1^{0.4} = T_2(2V_1)^{0.4}$. Hence $T_2 = 293/2^{0.4} = 222$ K and so the temperature change is $-71$ K, i.e. a temperature drop.

**5.12** For the isothermal change: $p_1V_1 = p_2V_2$ and so $p_2 = 2p_1$. For the adiabatic change $p_2V_2^\gamma = p_3V_3^\gamma$ and so $(2p_1)(\frac{1}{2}V_1)^\gamma = p_3V_1^\gamma$ and $p_3/p_1 = 2 \times (\frac{1}{2})^{1.4} = 0.76$.

# Lesson 6 Kinetic theory

## Aims

After completing this lesson you should be able to

- state the assumptions made in the kinetic theory and derive the relationship $pV = \frac{1}{3}Nm\overline{c^2}$
- identify the average kinetic energy of a molecule as being proportional to the thermodynamic temperature
- deduce from kinetic theory *Avogadro's law, Dalton's law* and *Graham's law of diffusion*
- solve problems involving the kinetic theory.

## Prerequisites

Lesson 4 in Chapter 1 on Newton's laws, Lesson 4 in this chapter on the gas laws.

# STUDY

**I Read** about the assumptions made in kinetic theory and the derivation of the relationship $pV = \frac{1}{3}Nm\overline{c^2}$. References: *B* PP ch. 9, *B* ST2 ch. 5, *D* ch. 19, *N* ch. 11, *W et al.* ch. 13(10), *Wh&H* ch. 24.

**Self-assessment questions**

**6.1** Consider a simple situation of molecules, imagined to be like ball-bearings, bouncing back and forth between the opposite walls of a box. Between their collisions with the walls they have a constant speed.
(a) The walls experience a force due to this bombardment. How is the force related to the momentum of the molecules?
(b) If the distance between the opposite walls were halved, how would the force on the walls change? Assume that there is no change in the molecular speed.
(c) Explain how the above argument leads to a law similar to Boyle's law.

**6.2** State the assumptions that are made for the kinetic theory.

**6.3** Consider a gas in a cubical box of side $L$. There are $N$ molecules in the box, each of mass $m$ and moving with a velocity $c$ (Figure 6.1).
(a) The velocity $c$ of a molecule can be resolved into three mutually perpendicular directions, parallel to the $x$, $y$ and $z$ directions. If $c_x$ is the component in the $x$ direction, what will be its momentum when it hits the wall shaded in the figure?

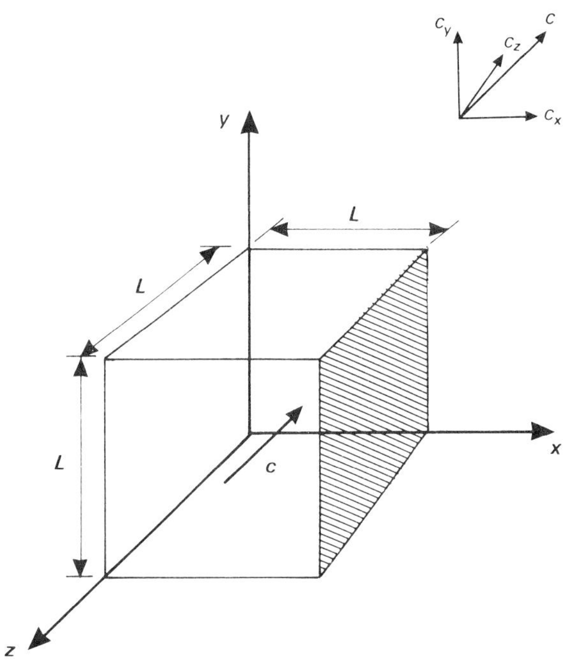


**Figure 6.1**

(b) If the molecule rebounds from its collision with the wall with the same velocity reversed, i.e. an elastic collision, what is the change in momentum?
(c) What will be the time taken for that molecule to travel to the opposite face of the box and back again to make a second collision with the shaded face?
(d) Hence, what is the rate of change of momentum at the shaded face for that molecule?
(e) Consider the $N$ molecules in the box to have velocity components in the $x$ direction of $(c_x)_1$, $(c_x)_2$, $(c_x)_3$, etc. What is the total force exerted by the $N$ molecules on the shaded face?
(f) If $\overline{c_x^2}$ represents the mean value of the squares of the velocity components in the $x$ direction, then

$$\overline{c_x^2} = \frac{(c_x^2)_1 + (c_x^2)_2 + (c_x^2)_3 + \ldots\ldots + (c_x^2)_N}{N}$$

Hence, as pressure = force/area, write an expression for the pressure on the shaded face in terms of $\overline{c_x^2}$.
(g) For any molecule $c^2 = c_x^2 + c_y^2 + c_z^2$. Because $N$ is very large and the molecules have random motion we must, on average, have as many molecules going in one direction as another. Thus $\overline{c_x^2} = \overline{c_y^2} = \overline{c_z^2}$. Hence write your answer to (f) in terms of $\overline{c^2}$.

***6.4** Hydrogen at s.t.p. has a density of 0.09 kg m$^{-3}$; what is the root-mean-square speed? 1 atmosphere pressure = $1.0 \times 10^5$ Pa. Note that the root-mean-square speed is $\sqrt{(\overline{c^2})}$ and that density $\rho = Nm/V$.

**6.5** For one mole of an ideal gas the kinetic theory equation can be written as $pV = \frac{1}{3}Lm\overline{c^2}$, where $L$ is the Avogadro number. The ideal-gas equation for one mole is $pV = RT$. Hence derive a relationship between the average translational kinetic energy of a molecule and the temperature.

***6.6** What is the total translational kinetic energy of the molecules in a mole of an ideal gas at 0 °C? ($R$ = 8.3 J mol$^{-1}$ K$^{-1}$)

***6.7** Air at s.t.p. contains $2.70 \times 10^{25}$ molecules per cubic metre. In an evacuated container the pressure of the air is $1.20 \times 10^{-1}$ Pa and the temperature 30 °C; how many molecules will it contain per cubic metre? (One standard atmosphere pressure = $1.01 \times 10^5$ Pa.)

***6.8** At s.t.p. nitrogen has a density of 1.25 kg m$^{-3}$. What is the root-mean-square velocity of nitrogen molecules at 100 °C? (One standard atmosphere pressure = $1.01 \times 10^5$ Pa.)

***6.9** A vessel of volume $4.0 \times 10^{-3}$ m$^3$ contains nitrogen at a pressure of $1.0 \times 10^2$ Pa and a temperature of 18 °C. What is (a) the number of nitrogen molecules in the vessel and (b) their root-mean-square speed? ($R$ = 8.3 J mol$^{-1}$ K$^{-1}$, Avogadro number = $6.0 \times 10^{23}$, mass of one mole of nitrogen = 28 g.)

**II Read** about the deduction from kinetic theory of Avogadro's law, Dalton's law and Graham's law. References: *D* ch. 19, *N* ch. 11, *W et al.* ch.13(10), *Wh&H* ch. 24.

**Self-assessment questions**

**6.10** *Avogadro's law* states that equal volumes of ideal gases existing under the same conditions of temperature and pressure contain equal numbers of molecules. Prove it.

**6.11** *Dalton's law* states that when two or more gases which do not react chemically are present in the same container, the total pressure is the sum of the partial pressures, i.e. the pressure that each gas would exert if it alone occupied the container. Prove it.

**6.12** *Graham's law* states that the rate of diffusion of a gas is inversely proportional to the square root of its density. Assume that the number of molecules passing through a given small hole in a barrier is proportional to the root-mean-square speed of the molecules. Hence derive Graham's law.

# SELF-ASSESSMENT

## Key points

The **assumptions of kinetic theory** are that a gas consists of particles called molecules, which are in constant random motion; the range of intermolecular forces is small compared to their average separation and consequently intermolecular forces are negligible except during collisions; the volume of the gas molecules is negligible compared with the total volume occupied by the gas; the duration of a collision is negligible compared to the time spent in motion; a molecule moves with a uniform velocity between collisions; the collisions are all elastic and Newtonian mechanics can be applied.
The equation derived on the basis of these assumptions is $pV = \frac{1}{3}Nm\overline{c^2}$, where $N$ is the number of molecules, each of mass $m$, and $\overline{c^2}$ the mean square speed.

The mean kinetic energy per molecule is $\frac{3}{2}\left(\frac{R}{L}\right)T = \frac{3}{2}kT$, where $k$ is Boltzmann's constant.

## Answers I

**6.1** (a) Force = rate of change of momentum.
(b) Twice as many molecules would hit the wall per second and hence the rate of change of momentum would be doubled. Hence the force is doubled.
(c) The pressure is proportional to the force and the volume proportional to the distance between the walls. Hence the pressure is doubled when the volume is halved. This agrees with Boyle's law.

**6.2** See the key points.

**6.3** (a) $mc_x$ (b) $mc_x - (-mc_x) = 2mc_x$ (c) $2L/c_x$ (d) $2mc_x/(2L/c_x) = mc_x^2/L$ (e) $F = (m/L)[(c_x^2)_1 + (c_x^2)_2 + (c_x^2)_3 + ..(c_x^2)_N]$ (f) Area = $L^2$ and volume $V = L^3$, hence $p = Nm\overline{c_x^2}/V$ (g) $pV = \frac{1}{3}Nm\overline{c^2}$

**6.4** $\rho = Nm/V$ and so $p = \frac{1}{3}\rho\overline{c^2}$. Therefore $c_{rms} = \sqrt{(3p/\rho)} = \sqrt{(3 \times 1.0 \times 10^5/0.09)} = 1.8 \times 10^3$ m s$^{-1}$.

**6.5** $pV = \frac{2}{3}L(\frac{1}{2}m\overline{c^2}) = RT$. Thus $\frac{1}{2}m\overline{c^2} = \frac{3}{2}(R/L)T = \frac{3}{2}kT$ where $k = R/L$ and is known as Boltzmann's constant.

**6.6** Total translational kinetic energy = $\frac{1}{2}Lm\overline{c^2} = \frac{3}{2}RT = \frac{3}{2} \times 8.3 \times 273 = 3.4 \times 10^3$ J.

**6.7** $pV = \frac{1}{3}Nm\overline{c^2}$, hence $N/V = 3p/(m\overline{c^2})$. But $\frac{1}{2}m\overline{c^2} = \frac{3}{2}kT$ and so $N/V = p/kT$. Hence $2.70 \times 10^{25} = 1.01 \times 10^5/273k$ and the required $N/V = 1.20 \times 10^{-1}/303k$. Thus eliminating $k$ gives $N/V = 2.89 \times 10^{19}$ molecules per m$^3$.

**6.8** $p = \frac{1}{3}\rho\overline{c^2}$ (see the answer to SAQ 6.4), hence $c_{rms} = \sqrt{(3p/\rho)}$. Thus at s.t.p. $c_{rms} = \sqrt{(3 \times 1.01 \times 10^5/1.25)} = 492$ m s$^{-1}$. But $\overline{c^2}$ is proportional to $T$, hence the root-mean-square velocity at 100 °C is given by $492 \times \sqrt{(373/273)} = 575$ m s$^{-1}$.

**6.9** (a) $pV = \mu RT$, hence $\mu = pV/RT = 1.0 \times 10^2 \times 4.0 \times 10^{-3}/(8.3 \times 291) = 1.66 \times 10^{-4}$. Hence as 1 mole contains $6.0 \times 10^{23}$ molecules there are $1.66 \times 10^{-4} \times 6.0 \times 10^{23} = 9.96 \times 10^{19}$ molecules.
(b) Mass of a molecule $= 28 \times 10^{-3}/6.0 \times 10^{23} = 4.67 \times 10^{-26}$ kg. $pV = \frac{1}{3}Nm\overline{c^2}$ and so $c_{rms} = \sqrt{(3pV/Nm)} = \sqrt{(3 \times 1.0 \times 10^2 \times 4.0 \times 10^{-3}/9.96 \times 10^{19} \times 4.67 \times 10^{-26})} = 508$ m s$^{-1}$.

## Answers II

**6.10** Consider two gases: $p_1V_1 = \frac{1}{3}N_1m_1\overline{c_1^2}$ and $p_2V_2 = \frac{1}{3}N_2m_2\overline{c_2^2}$. As the temperatures are the same $\frac{1}{2}m_1\overline{c_1^2} = \frac{1}{2}m_2\overline{c_2^2}$. Also $p_1 = p_2$ and $V_1 = V_2$. Hence $N_1 = N_2$.

**6.11** With ideal gases there will be no interaction between the molecules of the gases and thus the bombardment of a wall of the container will be just the sum of the bombardments of the constituent gases. Hence the pressure is the sum of the pressures due to each of the gases.

**6.12** As $p = \frac{1}{3}\rho\overline{c^2}$ then $\overline{c^2}$ is inversely proportional to the density and hence the rate of diffusion is inversely proportional to the square root of the density.

# Lesson 7 Thermal conduction

## Aims

After completing this lesson you should be able to
- define *thermal conductivity*
- solve problems involving heat conduction through single and compound objects when the lines of heat flow are parallel
- explain how the thermal conductivity can be measured for good and for bad conductors.

## Prerequisites

Lesson 2 on heat capacity and, for one problem, Lesson 3 on latent heat.

## STUDY

**I Read** about the definition of thermal conductivity and the solution of problems involving single materials when the lines of heat flow are parallel.
References: *B* ST3 ch. 2, *D* ch. 4, *N* ch. 13(11), *Wh&H* ch. 29.

### Self-assessment questions

**7.1** Sketch a graph of the way the temperature varies with distance along an unlagged bar when there is a quantity of heat flowing into the bar from one end. Also sketch the comparable graph for a lagged bar where the heat loss from the sides is negligible.

**7.2** Thermal conduction is the transfer of energy between two adjacent parts of an object due to a difference in temperature between those parts. We can therefore talk of a temperature *gradient*. In view of your answers to SAQ 7.1, what can be said about the temperature gradient along a lagged and an unlagged bar?

**7.3** Thermal conductivity is defined in terms of an equation relating the rate of flow of heat between two parts of a body and the temperature gradient between them. State the equation. Also state the units of thermal conductivity.

**7.4** In the case of a perfectly insulated bar of uniform cross-section, the lines of heat flow can be considered to be parallel to one another, passing along the length of the bar. With the unlagged bar the lines of heat flow diverge and heat passes out of the sides of the bar. Consider a perfectly lagged bar of length $L$ where heat $Q$ enters an end face in a time $t$. The temperature of that end face is $\theta_1$ and the temperature of the opposite end face through which the heat leaves the bar is $\theta_2$. If the bar has a uniform cross-sectional area of $A$, write an equation relating the heat input $Q$ and the thermal conductivity $\lambda$ of the bar material.

**7.5** A house window is made of glass 4.0 mm thick and has an area of 2.0 $m^2$. If the temperature difference between the two sides of the glass is 20 °C and the thermal conductivity of glass is 0.8 W $m^{-1}$ $K^{-1}$, what is the rate of loss of heat from the house through the window? Assume that the lines of heat flow are parallel.

**7.6** How much heat flows per second through a well lagged bar of aluminium, thermal conductivity 230 W $m^{-1}$ $K^{-1}$, if the bar has a length of 300 mm and a cross-sectional area of 100 $mm^2$, with a temperature difference of 40 °C between the ends of the bar?

**II Read** about the solution of problems involving conduction through compound objects, e.g. a cavity wall, when the lines of heat flow are parallel.
References: *B* ST3 ch. 2, *D* ch. 4, *N* ch. 13, *Wh&H* ch. 29.

**Self-assessment questions**

**7.7** Consider the flow of heat through a cavity wall. Such a wall consists of a layer of brick, then an air cavity and finally another layer of brick (as in Figure 7.1). Assume parallel lines of heat flow.

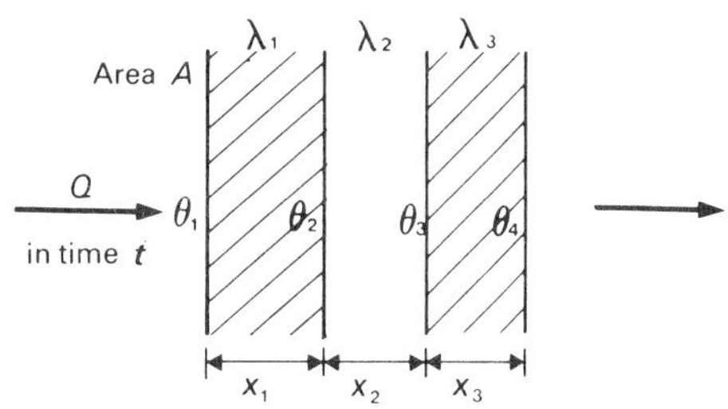


**Figure 7.1**

(a) What is the rate of heat flow through the first layer of brick if the temperatures of the surfaces of that brick are $\theta_1$ and $\theta_2$ and its thickness is $x_1$?
(b) What is the rate of heat flow through the second layer, the air, if the temperatures of the surfaces of that layer are $\theta_2$ and $\theta_3$ and its thickness is $x_2$?
(c) What is the rate of heat flow through the third layer, the outer brick wall, if the temperatures of the surfaces of that brick are $\theta_3$ and $\theta_4$ and its thickness is $x_3$?
(d) The rate of heat flow through each of the layers must be the same. Why?
(e) Rearrange the three equations to give a relationship between the rate of flow of heat through the structure and the temperature difference between the inner and outer surfaces, i.e. $\theta_1$ and $\theta_4$.

***7.8** Brick has a thermal conductivity of 1.0 W $m^{-1}$ $K^{-1}$ and air a thermal conductivity of 0.026 W $m^{-1}$ $K^{-1}$. Calculate the heat loss per square metre through a cavity wall when the temperature difference between the inside and outside is 15 °C, both brick layers being 100 mm thick and the cavity also being 100 mm across.

***7.9** A compound bar consists of a steel bar 200 mm long joined end-on to an aluminium bar of length 100 mm. The free end of the steel bar is at 100 °C and the free end of the aluminium bar at 20 °C. What is the temperature at the junction between the steel and the aluminium?
(Thermal conductivity of steel = 54 W $m^{-1}$ $K^{-1}$, thermal conductivity of aluminium = 230 W $m^{-1}$ $K^{-1}$.)

***7.10** When the heat loss through a window consisting of a single sheet of glass 3.0 mm thick and area 3.0 $m^2$ is calculated for a temperature difference of 10 °C between the air inside the room and the outside temperature, the result is found to be higher than occurs in practice. This can be accounted for if we assume there to be a layer of still air in contact with the inner glass surface of the window. What is (a) the temperature difference across this layer of air and (b) its thickness if the heat loss is 3.1 kW?
(Thermal conductivity of glass = 1.2 W $m^{-1}$ $K^{-1}$, thermal conductivity of air = 0.026 W $m^{-1}$ $K^{-1}$.)

****7.11** This question requires the use of calculus. The ice on a pond is 30 mm thick. If the air temperature above the ice is −8 °C how long will it take for the thickness of the ice to increase to 40 mm?
(Thermal conductivity of ice = 2.0 W $m^{-1}$ $K^{-1}$, specific latent heat of fusion = 3.3 × $10^5$ J $kg^{-1}$, density of ice = 0.9 × $10^3$ kg $m^{-3}$.)

**III Read** about the measurement of thermal conductivity for a good conductor by Searle's bar and a poor conductor by Lees' disc.
References: *D* ch. 4, *N* ch. 13, *Wh&H* ch. 29.

**Self-assessment questions**

***7.12** Explain how Searle's apparatus is used for the measurement of the thermal conductivity of a good conductor. Why is the bar heavily lagged?

***7.13** Explain how Lees' apparatus is used for the measurement of the thermal conductivity of a poor conductor. Why is a thin specimen used with a relatively large cross-sectional area?

# SELF-ASSESSMENT

## Key points

The **thermal conductivity** of a substance is defined by the equation

$$(dQ/dt) = -\lambda A\ (d\theta/dx)$$

where $(dQ/dt)$ is the rate at which heat passes through a section, $(d\theta/dx)$ is the temperature gradient, $A$ the area and $\lambda$ the thermal conductivity.
When the lines of heat flow are parallel the equation becomes $(Q/t) = \lambda A(\theta_2 - \theta_1)/x$, where $Q$ is the quantity of heat passing through a layer of thickness $x$ in a time $t$, $A$ is the area of the layer and $\theta_1$ and $\theta_2$ the temperatures of the surfaces of the layer.
The unit of thermal conductivity is $W\ m^{-1}\ K^{-1}$.

## Answers I

**7.1** See Figure A7.1.

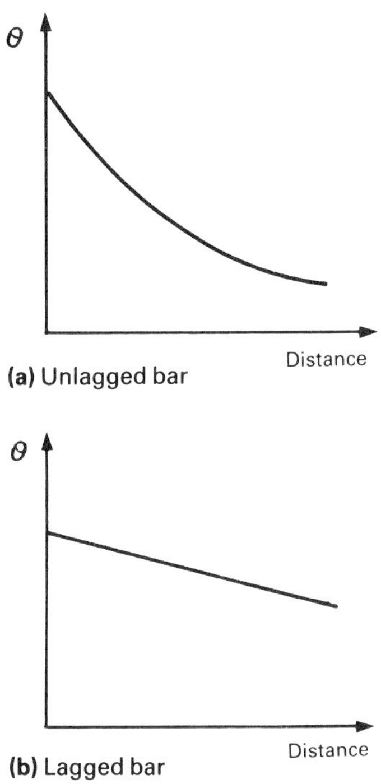


**Figure A7.1**

**7.2** With the lagged bar the temperature gradient is constant throughout the length of the bar, this however is not the case with the unlagged bar.

**7.3** See the key points.

**7.4** See the key points.

**7.5** $(Q/t) = \lambda A(\theta_2 - \theta_1)/x = 0.8 \times 2.0 \times 20/0.004 = 8000$ W

**7.6** $(Q/t) = \lambda A(\theta_2 - \theta_1)/x = 230 \times 100 \times 10^{-6} \times 40/(300 \times 10^{-3}) = 3.1$ W

## Answers II

**7.7** (a) $(Q/t) = \lambda_1 A(\theta_1 - \theta_2)/x_1$ (b) $(Q/t) = \lambda_2 A(\theta_2 - \theta_3)/x_2$ (c) $(Q/t) = \lambda_3 A(\theta_3 - \theta_4)/x_3$
(d) Because the lines of heat flow are parallel this means that the rate at which heat enters one layer must equal the rate at which it leaves and moves through the next layer, there being no loss of heat from the sides of the structure.
(e) $\theta_1 - \theta_2 = (Q/At)(x_1/\lambda_1)$, also $\theta_2 - \theta_3 = (Q/At) \times (x_2/\lambda_2)$ and $\theta_3 - \theta_4 = (Q/At)(x_3/\lambda_3)$. Adding these equations gives $\theta_1 - \theta_4 = (Q/At)(x_1/\lambda_1 + x_2/\lambda_2 + x_3/\lambda_3)$.

**7.8** Using the equation developed in 7.7(e) above, $15 = (Q/At)\ 100 \times 10^{-3}\ (1/1.0 + 1/0.026 + 1/1.0)$ and so $(Q/At) = 3.7\ W\ m^{-2}$.

**7.9** Using the equation developed in 7.7(e) above, $80 = (Q/At)\ (0.200/54 + 0.100/230)$ and so $(Q/At) = 1.93 \times 10^4\ W\ m^{-2}$. Thus using the equation in 7.7(a) we have $\theta_1 - \theta_2 = 1.94 \times 10^4 \times 200 \times 10^{-3}/54$, and as $\theta_1 = 100\ °C$ then $\theta_2 = 28.2\ °C$.

**7.10** Using the equation developed in 7.7(e) above, $10 = 3.1 \times 10^3/3.0)\ (x_1/0.026 + 0.003/1.2)$ and so $x_1 = 1.9 \times 10^{-4}$ m. The air layer is thus about 0.19 mm thick. Using the equation in 7.7(a) we have $\theta_1 - \theta_2 = (3.1 \times 10^3/3.0)\ (1.9 \times 10^{-4}/0.026) = 7.6\ °C$.

**7.11** Consider the thickness to be $x$ at time $t$. The temperature below the ice is 0 °C and so rate of flow of heat through ice $= \lambda A\theta/x$. If the ice thickness increases by $\delta x$ in a time $\delta t$ then mass of ice formed $= \rho A\ \delta x$ and so energy required in time $\delta t = l\rho A\ \delta x$. Thus rate at which energy supplied $= l\rho A\ \delta x/\delta t = \lambda A\ \theta/x$. Hence $\delta t = (l\rho/\lambda\theta)(x\ \delta x)$. Integrating this we have

$$\int_0^t dt = \frac{l\rho}{\lambda\theta}\int_{x_1}^{x_2} x\ dx$$

and so $t = \dfrac{l\rho}{2\lambda\theta}(x_2^2 - x_1^2)$. Hence $t = 6.5 \times 10^3$ s.

## Answers III

**7.12** See the references for details of the experiment. The important points are as follows. One end of the bar is heated to a constant temperature while the other end is cooled by water circulating round it. The temperature gradient along the bar is measured, using two thermocouples or thermometers inserted into holes in the bar with some mercury in the holes to ensure good thermal contact. The rate of flow of heat along the bar is measured by determining the difference in temperature between the inflow and outflow of the cooling water for a measured mass flowing in a time $t$. Then $(Q/t) = (m/t)c\,\Delta\theta$. The bar has to be well lagged, as parallel lines of heat flow are assumed.

**7.13** See the references for details of the apparatus. The temperature gradient across the thickness of the specimen is measured by determining the temperatures in copper blocks on either side of the specimen and measuring its thickness. The method used to determine the rate of flow of heat depends on the form of the apparatus used. If the heating is by an electric element then the power supplied to the element can be determined. Corrections can be made for the fact that not all the heat flows through the specimen, by considering the heat losses from the other parts of the apparatus to be proportional to their areas multiplied by their average temperature excess above the surroundings (see *Wh&H* ch. 29).
Alternatively, where the heat is supplied by steam in a chamber mounted on top of the specimen, after the temperature gradient has been determined the steam chamber is removed and the lower part of the apparatus heated to about 10 °C above the temperature it was at and then a cooling curve for it determined. Hence the rate of cooling at the appropriate temperature is determined, and from a knowledge of the mass and specific heat capacity of that part of the apparatus the rate of loss of heat is found.
The specimen is thin and its cross-sectional area relatively large in order to ensure that there is a significant rate of heat flow and the temperature of the base of the apparatus rises a measurable amount above the room temperature.

## CHAPTER 4

# Geometrical optics

## Lesson 1 Reflection at plane and spherical surfaces

### Aims

After completing this lesson you should be able to

- state the laws of reflection
- explain the formation of images in plane mirrors
- solve problems involving more than one plane mirror
- explain the formation of images in spherical mirrors.

### Prerequisites

A familiarity with simple geometry involving angles in triangles and the fact that for small angles $\tan\theta$ is approximately $\theta$, in radians.

## STUDY

**I Read** about the laws of reflection, the formation of images in plane mirrors and problems involving more than one plane mirror.
References: *B* PP ch. 6, *B* ST6 ch. 1, *D* ch. 5, *N* ch. 16, *Wh&H* ch. 31.

#### Self-assessment questions

**1.1** State the laws of reflection.

**1.2** A *ray* is the path taken by light energy in travelling from a source to the receiver. Draw a ray diagram explaining how an image is formed with a plane mirror.

**1.3** Prove that the image seen in a plane mirror has an apparent location as far behind the mirror as the object is in front.

**1.4** The image seen in a plane mirror is said to be a *virtual* image. What is meant by this term?

**1.5** How many images are formed by an object placed between two mirrors that are inclined at 60° to each other?

**1.6** Two plane mirrors are parallel to each other and 10 cm apart. An object is placed between the mirrors, 1 cm from one of the mirrors. State the position of the first two images seen in each mirror.

**II Read** about the formation of images in spherical mirrors.
References: *B* PP ch. 6, *B* ST6 ch. 1, *D* ch. 5, *N* ch. 16, *Wh&H* ch. 33.

#### Self-assessment questions

**1.7** Explain the terms *centre of curvature*, *principal axis* and *focus*.

**1.8** A concave mirror has a real focus, a convex mirror has a virtual focus. Explain this distinction.

**1.9** By applying the laws of reflection to a ray parallel to and near the principal axis, show that the focal length is half the radius of curvature.

**1.10** Ray diagrams can be used to locate the image by reflection in a spherical mirror. What happens to the following rays, drawn from the top of an object standing on the principal axis, after reflection?
(a) A ray parallel to the principal axis.
(b) A ray through the centre of curvature.
(c) A ray through the focus.

**1.11** Draw ray diagrams to locate the images in the following situations.
(a) An object placed beyond the centre of curvature, with a concave mirror.
(b) An object placed between the centre of curvature and the focus, with a concave mirror.
(c) An object placed at a distance between the focal length and the radius of curvature, with a convex mirror.

## SELF-ASSESSMENT

### Key points

The **laws of reflection**: The reflected ray is in the same plane as the incident ray and the normal to the mirror at the point of incidence. The angle of incidence equals the angle of reflection.

With a plane mirror a *virtual* image is produced as far behind the mirror as the object is in front and the line joining the object and the image is at right angles to the mirror.
The **centre of curvature** of a spherical surface is the centre of the circle forming the surface. The line joining the centre of curvature and the **pole**, the centre of the mirror surface, is the **principal axis**. A narrow beam of rays parallel to and near the principal axis is reflected to, or from, the **focus**. A concave mirror has a real focus, a convex mirror a virtual focus. The focal length is half the radius of curvature.

## Answers I

**1.1** See the key points.

**1.2** See Figure A1.1.

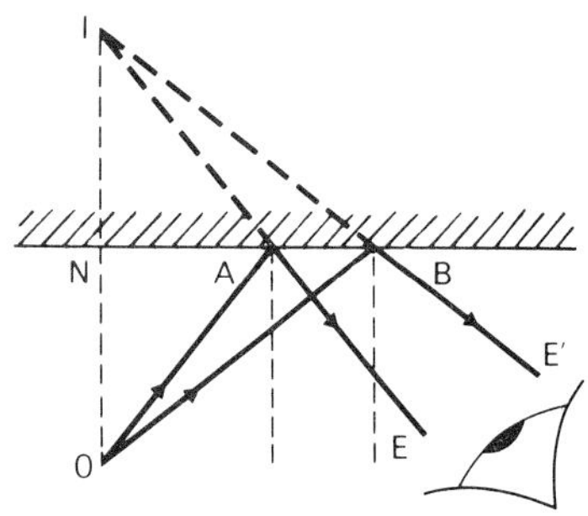


**Figure A1.1**

**1.3** Using Figure A1.1, as the angle of incidence equals the angle of reflection, angle OAN = angle IAN, also angle INA = angle ONA = 90° and NA is common, therefore the triangles INA and ONA are congruent and so IN = ON.

**1.4** The rays of light do not actually meet at the image after reflection but only appear to.

**1.5** See Figure A1.2. $I_1$ and $I_2$ are the images in the mirrors of the object P. $I_1$ then behaves as an object for $M_2$ and gives rise to an image $I_{12}$. $I_2$ acts as an object for $M_1$ and gives rise to an image $I_{21}$. Both $I_{21}$ and $I_{12}$ act as objects and give rise to a single image $I_{121}$ (or $I_{212}$). The answer is five images. See *D* ch. 5.

**1.6** For the mirror $M_1$ which is 1 cm from the object there is an image 1 cm behind $M_1$. The object is 9 cm from $M_2$ and so there is an image 9 cm behind $M_2$. This image is 19 cm from $M_1$ and so gives rise to an image 19 cm behind $M_1$. Likewise the image 1 cm behind $M_1$ gives rise to an image 11 cm behind $M_2$.

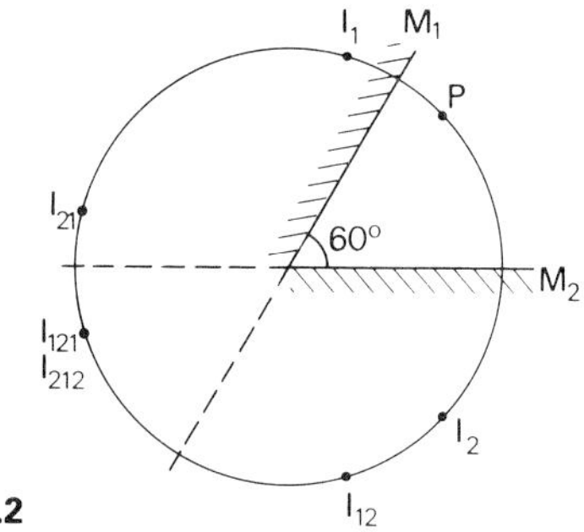


**Figure A1.2**

## Answers II

**1.7** See the key points.

**1.8** Parallel light actually does converge and pass through the real focus, whereas it only appears to diverge from the virtual focus.

**1.9** See Figure A1.3. The normal to the point of contact of the light with the mirror passes through the centre of curvature. As the angle of incidence equals the angle of reflection, angle AMC = angle CMF. But angle MCF must equal angle AMC (alternate angles with parallel lines). Thus angle CMF = angle MCF. Hence triangle FCM is isosceles and so CF = FM. To a reasonable approximation FM = FP and so $r = 2f$.

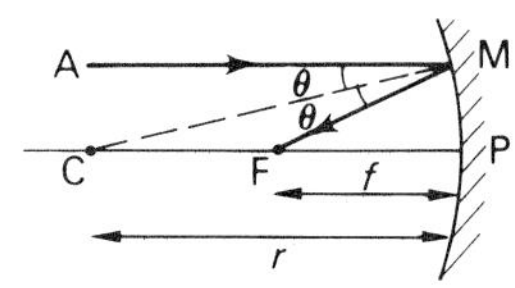


**Figure A1.3**

**1.10** (a) Passes or appears to pass through the focus. (b) Returns along the same path. (c) Returns parallel to the axis.

**1.11** See Figure A1.4.

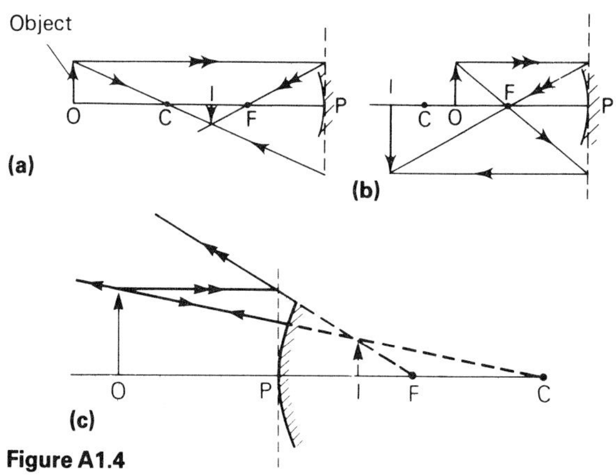


**Figure A1.4**

# Lesson 2 Refraction

## Aims

After completing this lesson you should be able to
- state the laws of refraction
- solve problems involving refraction at plane surfaces
- explain *total internal reflection*, derive the relationship between the *critical angle* and the refractive index and solve problems involving it.

## Prerequisites

Lesson 1 on reflection. Simple geometry and trigonometry is required for most of the questions.

# STUDY

**I Read** about the laws of refraction, and refraction at a plane surface.
References: *B* PP ch. 6, *B* ST6 ch. 1, *D* ch. 5, *N* ch. 17, *Wh&H* ch. 31.

### Self-assessment questions

**2.1** State the laws of refraction.

**2.2** Light rays are reversible. Explain what this statement means. Use it to explain why if the refractive index for light passing from air into water is 4/3, that from water into air is 3/4.

**2.3** Consider a ray of light travelling from medium 1, say air, to medium 2, say glass, to medium 3, say water, and then to medium 1, the air, again. If the boundaries of the media are parallel, show that ${}_1n_3 = {}_1n_2 \times {}_2n_3$, where ${}_1n_3$ is the refractive index for light passing from medium 1 to medium 3, ${}_1n_2$ is the refractive index from medium 1 to medium 2, and ${}_2n_3$ is the refractive index from medium 2 to medium 3.

**2.4** What is the refractive index from water to glass if the refractive index from air to water is 4/3 and that from air to glass is 3/2?

**2.5** What is meant by the *absolute* refractive index for a material?

**2.6** Derive the relationship $n_1 \sin i_1 = n_2 \sin i_2$, where $n_1$ is the absolute refractive index of medium 1, $n_2$ the absolute refractive index of medium 2, $i_1$ the angle between the ray and the normal in medium 1 when incident on the boundary between medium 1 and medium 2, and $i_2$ the corresponding angle in medium 2.

**2.7** Apply the relationship given in SAQ 2.6 to the solution of the following problem. A ray of light is incident at an angle of 50° on a water–glass boundary. What is the angle between the ray and the normal in the glass if the water has an absolute refractive index of 1.3 and glass an absolute refractive index of 1.5?

**2.8** One method of measuring the refractive index of a block of material is to determine the real depth of an object in the medium and then the apparent depth when viewed along the normal to the block. Derive the relationship between these measurements and the refractive index.

**2.9** When a microscope is focused on a mark on the bottom of an empty beaker and then a liquid is poured into the beaker, the microscope has to be raised through a vertical distance of 0.60 cm to bring the mark back into focus. If the depth of the liquid in the beaker is 2.00 cm, what is the refractive index of the liquid?

**II Read** about total internal reflection and the relationship between the critical angle and the refractive index.
References: *B* PP ch. 6, *B* ST6 ch. 1, *D* ch. 5, *N* ch. 17, *Wh&H* ch. 31.

### Self-assessment questions

**2.10** Explain what is meant by *total internal reflection* and define the *critical angle*.

**2.11** Show that the critical angle $C$ at a boundary between two media, when light travels from medium 1 to medium 2, is given by $\sin C = n_2/n_1$.

***2.12** Rays from a point source of light on the lower surface of a glass cube, of side 20 mm, strike the upper face and the light only emerges through a circle of radius 16 mm. What is the refractive index of the glass?

***2.13** What is the minimum refractive index the prism shown in Figure 2.1 (a) and (b) can have if it reflects light in the ways shown?

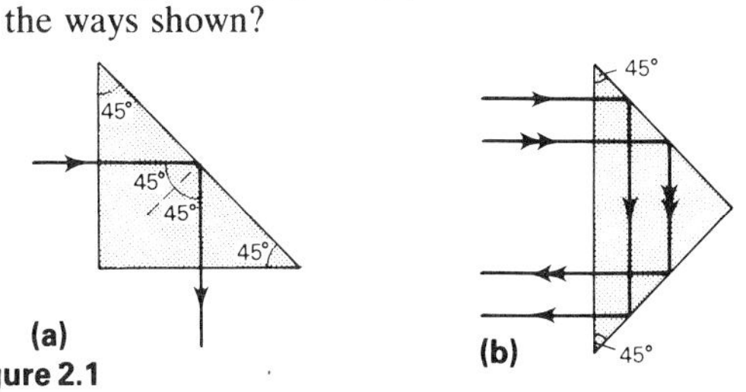


**Figure 2.1**

***2.14** For the cube shown in Figure 2.2, what is the greatest refractive index the cube can have if light incident at grazing incidence on face AB is to emerge from side BC? What would happen if the cube had a greater refractive index than this?

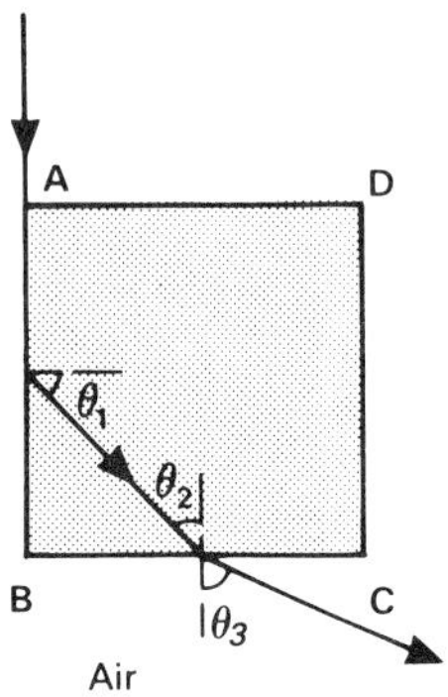


**Figure 2.2**

***2.15** In the air-cell method of determining the refractive index of a liquid, two thin plane-parallel glass plates are cemented together so as to contain a thin film of air. When this arrangement is immersed in a liquid, light will pass through the cell, up to a particular angle of incidence when total internal reflection occurs at the glass–air boundary (Figure 2.3). Derive the relationship between this angle of incidence and the refractive index of the liquid.

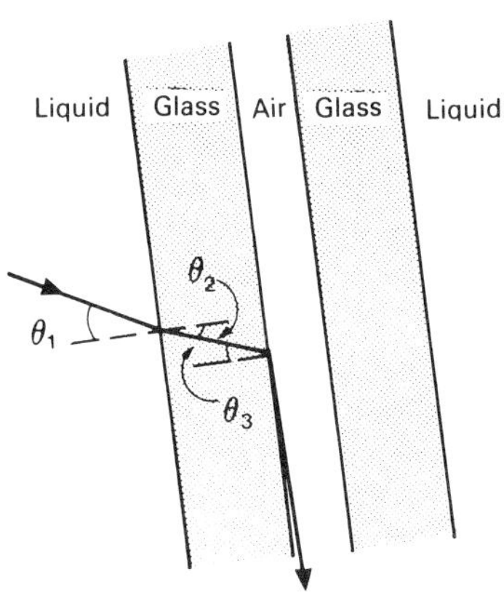


**Figure 2.3**

# SELF-ASSESSMENT

## Key points

**Laws of refraction:** The refracted ray is in the same plane as the incident ray and the normal to the surface at the point of incidence. For two particular media and a particular wavelength, the ratio of the sine of the angle of incidence to the sine of the angle of refraction is a constant, i.e. $\sin i/\sin r$ = a constant, the **refractive index** $n$ (this is **Snell's law**).

${}_1n_2 = 1/{}_2n_1$ and also ${}_1n_3 = {}_1n_2 \times {}_2n_3$.

The **absolute refractive index** of a material is the refractive index relative to a vacuum.

The symmetrical form of Snell's law is $n_1 \sin i_1 = n_2 \sin i_2$, where $i_1$ is the angle between the normal and the ray in medium 1 and $i_2$ the corresponding angle in medium 2.

The **critical angle** is the angle of incidence on a boundary between two media at which the refracted angle is 90°. At angles greater than the critical angle **total internal reflection** occurs.

## Answers I

**2.1** See the key points.

**2.2** It means that if light follows a certain path in going from, say, air to water, then light could follow the exact reverse path in going from water to air. Thus as ${}_1n_2 = \sin i_1/\sin i_2$, then ${}_2n_1 = \sin i_2/\sin i_1 = 1/{}_1n_2$. Thus ${}_an_w = 1/{}_wn_a$.

**2.3** See Figure A2.1. ${}_1n_2 = \sin i_1/\sin i_2$, also ${}_2n_3 = \sin i_2/\sin i_3$ and ${}_3n_1 = \sin i_3/\sin i_1$. The emergent ray DE, although laterally displaced, is parallel to the incident ray AB. Thus ${}_1n_3 = {}_1n_2 \times {}_2n_3$.

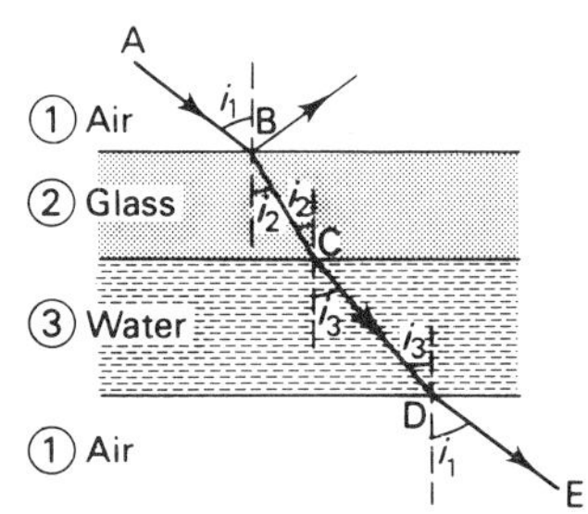


**Figure A2.1**

**2.4** ${}_wn_g = {}_wn_a \times {}_an_g = (1/{}_an_w) \times {}_an_g = (3/4) \times (3/2) = 9/8$.

**2.5** See the key points.

**2.6** For light passing from medium 1 to medium 2, ${}_1n_2 = \sin i_1/\sin i_2$. But, using the relationship derived in 2.3 above, ${}_1n_2 = {}_1n_v \times {}_vn_2$, where v stands for vacuum. Thus ${}_1n_2 = {}_vn_2/{}_vn_1$ and so ${}_vn_1 \sin i_1 = {}_vn_2 \sin i_2$. Usually the subscript v is omitted.

**2.7** $1.3 \sin 50° = 1.5 \sin i_2$ and so $i_2 = 41.6°$.

**2.8** See Figure A2.2.
${}_{a}n_{w} = \sin i_{a}/\sin i_{w}$ = (AB/IB)/(AB/OB) = OB/IB or, approximately, for near normal viewing, OA/IA = real depth/apparent depth.

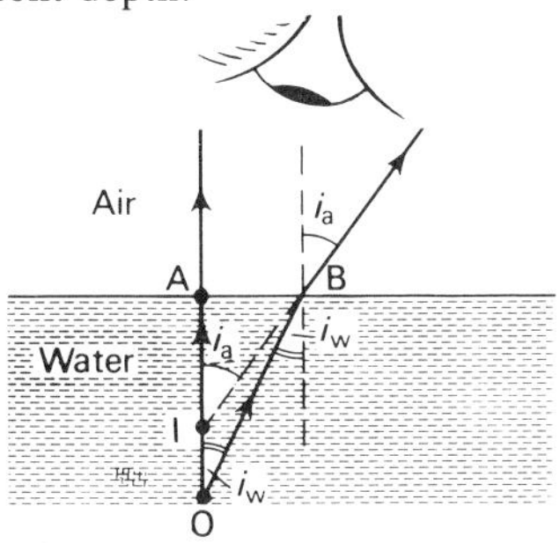


**Figure A2.2**

**2.9** ${}_{a}n_{1} = 2.00/1.40 = 1.43$

## Answers II

**2.10** When light is incident on a boundary between two media, while reflection can always occur, refraction only occurs for angles of incidence less than the critical angle. At angles greater than the critical angle all the incident energy is reflected and none is refracted. See the key points.

**2.11** $n_1 \sin C = n_2 \sin 90°$ and hence $\sin C = n_2/n_1$.

**2.12** $\sin C = 16/(20^2 + 16^2)^{\frac{1}{2}} = 0.62$ and hence ${}_{a}n_{g}$ = 1.6.

**2.13** If $C = 45°$ then ${}_{a}n_{p} = 1/\sin 45° = 1.41$.

**2.14** $\theta_1 = C$ and as angle ABC = 90° then $\theta_2 = 90° - \theta_1 = 90° - C$. For refraction at BC, $n_c \sin \theta_2 = n_a \sin \theta_3$ and so $n_c \cos C = n_a \sin \theta_3$. Thus, as $\sin C = n_a/n_c$, $\tan C = 1/\sin \theta_3$. The maximum value $\theta_3$ can have is 90°. Then $\tan C = 1$ and so $C = 45°$ and, as $n_a = 1$, $n_c = 1/\sin 45° = 1.41$. The light will be totally internally reflected at BC if the refractive index is greater than this value.

**2.15** $\theta_2 = \theta_3$. Hence $n_1 \sin \theta_1 = n_g \sin \theta_2 = n_a \sin 90°$ and so $n_1 = 1/\sin \theta_1$ as $n_a = 1$.

# Lesson 3 Refraction by prisms

## Aims

After completing this lesson you should be able to

- solve problems involving refraction through prisms, including the condition for minimum deviation
- explain the adjustments to a simple spectrometer needed for the measurement of the refractive index of a prism
- describe *dispersion* by a prism.

## Prerequisites

Lesson 2 on refraction.

## STUDY

**I Read** about refraction through prisms.
References: *B* ST6 ch. 1, *D* ch. 5, *N* ch. 18, *Wh&H* ch. 32.

### Self-assessment questions

**3.1** Consider refraction through a prism, Figure 3.1.
(a) By how much is the ray of light deviated on passing through the surface XY?

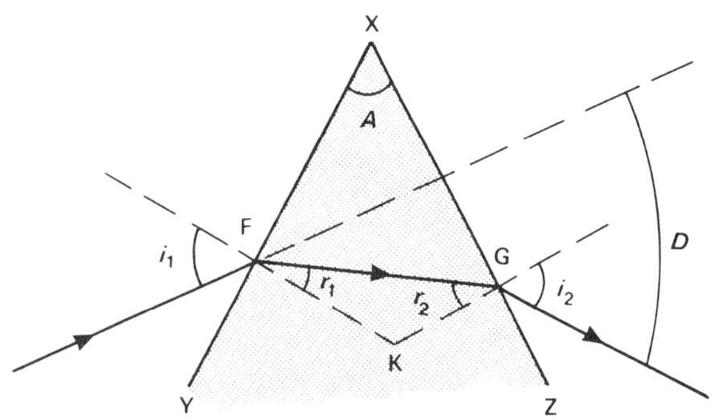


**Figure 3.1**

(b) By how much is the ray of light deviated on passing through the surface XZ?
(c) What is the total deviation *D*?
(d) For the quadrilateral XFKG, angle XFK plus angle XGK equals 180°, therefore angle *A* plus angle FKG equals 180°. Hence by considering the triangle FKG derive the relationship $A = r_1 + r_2$.
(e) The angle of deviation *D* depends on the angle of incidence $i_1$. A minimum value of angle *D* occurs when the light passes symmetrically through the prism, i.e. when $i_1 = i_2$ and $r_1 = r_2$. Hence derive the relationship

$$n = \frac{\sin[(A + D_{min})/2]}{\sin(A/2)}$$

**3.2** Calculate the angle of minimum deviation for a prism of apex angle 60° and refractive index 1.6.

***3.3** Calculate the angle of emergence ($i_2$ in Figure 3.1) and the total deviation when light is incident at an angle of 30° ($i_1$) on a prism of apex angle 60° and refractive index 1.52.

***3.4** If a prism of apex angle 60° produces a deviation of 62° for a ray of light incident at an angle of incidence of 40° on the first face of the prism, what other angle of incidence will produce the same deviation?

***3.5** Calculate the deviation produced by a prism of apex angle 60° and refractive index 1.5 when the light is incident at grazing incidence on the first face.

***3.6** Consider a prism with a very small apex angle, i.e. about 5° or less, and the light falling almost normally on the prism. When small angles are considered $\sin\theta \simeq \theta$. Making such approximations derive the relationship for the deviation, $D = (n - 1)A$.

**II Read** about the simple spectrometer and its use for the measurement of the refractive index of a prism. References: *D* ch. 5, *N* ch. 18, *Wh&H* ch. 35.

**Self-assessment questions**

**3.7** In a simple spectrometer what is the function of (a) the collimator, (b) the telescope?

**3.8** List the sequence of adjustments that have to be made before the instrument can be used.

**3.9** Explain how the simple spectrometer is used for the measurement of the refractive index of a prism.

**III Read** about the dispersion produced by a prism and the production of a pure spectrum.
References: *B* ST6 ch. 1, *D* ch. 5, *N* ch. 19, *Wh&H* ch. 35. *B* PP ch. 6 gives historical information about dispersion.

**Self-assessment questions**

**3.10** What is meant by *dispersion*?

**3.11** Violet light is deviated more by a prism than red light. For which colour of light has the prism the greater refractive index?

***3.12** The prism of a spectrometer has an apex angle of 60° and the refractive indices for red and violet are 1.515 and 1.529. What is the angular width of the spectrum produced when white light is incident on the prism and the prism is set to give minimum deviation for the red light?

**3.13** What is meant by a *pure* spectrum and how is one produced?

# SELF-ASSESSMENT

## Key points

At **minimum deviation** for a prism

$$n = \frac{\sin[(A + D_{min})/2]}{\sin(A/2)}$$

At any angle of deviation, $D = (i_1 + i_2) - (r_1 + r_2)$ and $A = (r_1 + r_2)$. For a small-angle prism with the light incident almost normally, $D = (n - 1)A$.
The separation of colours produced when white light passes through a prism is known as *dispersion*. For each of the constituent colours the prism has a a different refractive index.

## Answers I

**3.1** (a) $D_1 = i_1 - r_1$ (b) $D_2 = i_2 - r_2$ (c) $D = (i_1 - r_1) + (i_2 - r_2) = (i_1 + i_2) - (r_1 + r_2)$ (d) $r_1 + r_2$ + angle FKG = 180°, hence $r_1 + r_2 = A$ (e) $D_{min} = 2i - A$ and $2r = A$, hence $n = \sin i/\sin r = \sin[(A + D_{min})/2]/\sin(A/2)$

**3.2** $1.6 = \sin[(A + D_{min})/2]/\sin 30$ and so $\sin[(A + D_{min})/2] = 1.6 \times \sin 30 = 0.8$. Thus $(A + D_{min})/2 = 53.1°$ and hence $D_{min} = 46.3°$.

**3.3** Using the equation developed in 3.1(c) above, $D = (i_1 + i_2) - (r_1 + r_2) = (i_1 + i_2) - A$ (see 3.1(d)). Thus $D = 30° + i_2 - 60°$. But $\sin i_1/\sin r_1 = 1.52$ and so $\sin r_1 = 0.33$ and $r_1 = 19.2°$. Hence $r_2 = 60 - 19.2 = 40.8°$. But $\sin i_2/\sin r_2 = 1.52$ and so $i_2 = 83.3°$. Thus $D = 53.3°$.

**3.4** Using the equation for deviation obtained in 3.3 above, $D = (i_1 + i_2) - A$ and so $i_2 = 82°$. The two angles are $i_1$ and $i_2$, bearing in mind that the path of the light through the prism is completely reversible.

**3.5** $\sin i_1/\sin r_1 = 1.5$ and so $r_1 = 41.8°$. Thus as $A = r_1 + r_2$, then $r_2 = 18.2°$. Thus as $1.5 = \sin i_2/\sin r_2$ then $i_2 = 27.9°$. Thus $D = (i_1 + i_2) - A = 57.9°$. This is the maximum deviation for that prism.

**3.6** $i_1 = nr_1$ and $i_2 = nr_2$. Thus $D = (i_1 + i_2) - (r_1 + r_2) = (n - 1)(r_1 + r_2) = (n - 1)A$.

## Answers II

**3.7** (a) To produce parallel light. (b) To focus the parallel light onto the cross-wires and then have both in focus through the eyepiece of the telescope.

**3.8** Focus the eyepiece on the cross-wires. Then sight the telescope on a distant object and adjust the objective lens of the telescope so that the object is in focus on the cross-wires. Line up the collimator and telescope and adjust the position of the slit so that it is in focus when seen through the telescope. Level the prism table. This can be done by putting the prism on the table with one face perpendicular to the line joining two of the levelling screws. The telescope is then rotated to view the image of the slit reflected from that face. One of the screws is then adjusted until the slit image is central in the field of view. The table is then rotated so that the reflected image is seen from the next face of the prism and the third levelling screw adjusted to bring the slit image central in the field of view.

**3.9** To measure the apex angle, the apex of the prism is pointed towards the collimator. Then the reflected image of the slit from one face of the prism is found by rotating the telescope, the position of the telescope being noted. This is repeated for the image reflected from the other face and the telescope position again noted. The apex angle is half the difference between these two telescope positions.
To measure the angle of minimum deviation a monochromatic source of light is used to illuminate the collimator slit. The telescope and the prism are then moved so that the telescope is as near the 'straight-through' position as possible while still having a refracted image of the slit on the cross-wires. The position of the telescope is then noted. The prism is then removed and the position of the telescope when the slit image is on the cross-wires, i.e. the straight-through position, is noted. The difference between these two angles is the angle of minimum deviation.
The values of the apex angle and the angle of minimum deviation are then substituted in the formula developed in the answer to SAQ 3.1.

## Answers III

**3.10** See the key points.

**3.11** The refractive index for violet is greater than that for red.

**3.12** For red light $n = 1.515 = \sin[(A + D_{min})/2]/\sin(A/2)$ and so $D_{min} = 38.5°$. This means that, as $D_{min} = 2i - A$, $i = 49.25°$. This is also the angle of incidence on the first face of the violet light. Thus for the violet light $1.529 = \sin i_1/\sin r_1$ and $r_1 = 29.7°$. Thus as $A = r_1 + r_2$ then $r_2 = 30.3°$. As $1.529 = \sin i_2/\sin r_2$ then $i_2 = 50.5°$. Thus $D = (i_1 + i_2) - A = 39.7°$.

**3.13** A *pure* spectrum is one in which the different colours do not overlap. A pure spectrum can be produced by directing parallel light onto the prism and using a lens to focus the refracted light onto a screen.

# Lesson 4 Thin lenses

## Aims

After completing this lesson you should be able to
- draw ray diagrams for thin lenses
- derive and use the equation $1/f = 1/v + 1/u$
- solve problems involving thin lenses in contact
- explain methods used for the measurement of focal length.

## Prerequisites

Lesson 3, refraction by prisms, is required for one of the derivations. Lessons 1 and 2 would also be useful.

## STUDY

**I Read** about ray diagrams for thin lenses.
References: *B* PP ch. 6, *B* ST6 ch. 1, *D* ch. 5.

**Self-assessment questions**

**4.1** State what happens to the following rays after refraction by a lens. All the rays originate from the top of a small object standing on the axis of the lens.
(a) A ray parallel to the principal axis.
(b) A ray through the centre of the lens.
(c) A ray through the principal focus.

**4.2** Use ray diagrams to enable the location of the image in the following situations. State in each case whether the image is virtual or real.
(a) An object situated more than twice the focal length from a converging lens.
(b) An object situated between the focal length and twice the focal length for a converging lens.
(c) An object inside the focal length of a converging lens.

(d) An object between the focal length and twice the focal length of a diverging lens.

**II** **Read** about the derivation and use of the equation $1/f = 1/v + 1/u$ for a thin lens.
References: *B* PP ch. 6, *B* ST6 ch. 1, *D* ch. 5, *N* ch. 19, *Wh&H* ch. 32.

**Self-assessment questions**

**4.3** The following is concerned with the derivation of the relations between image and object distances for a thin lens. It is based on the relationship derived in SAQ 3.6, i.e. deviation $D = (n - 1)A$ for a prism with a small apex angle. The deviation is independent of the angle of incidence. Consider a thin lens as being made up of a number of small-angle prisms.

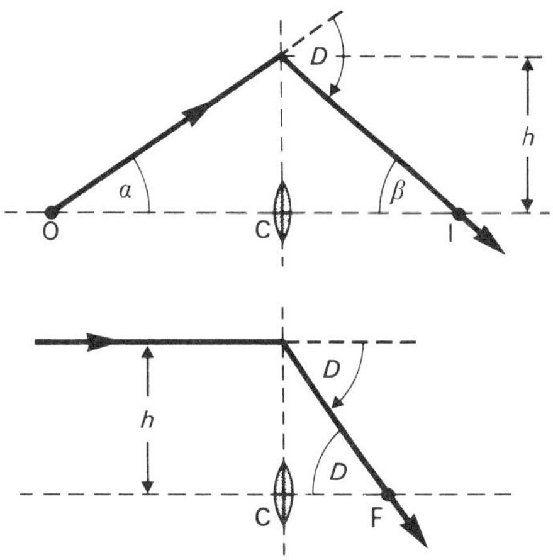


**Figure 4.1** **To relate *u*, *v* and *f*** (angles exaggerated)

(a) For the ray from the object point on the axis in Figure 4.1, what is the relationship between the deviation $D$ and the angles $\alpha$ and $\beta$?
(b) For small angles $\tan\theta$ is approximately equal to $\theta$. Express $\alpha$ and $\beta$ in terms of the object and image distances OC and CI ($u$ and $v$).
(c) For the ray parallel to the axis, express the deviation in terms of the focal length CF ($f$).
(d) Because the deviation is the same, being independent of the angle of incidence, derive the relationship between the focal length, the object and image distances.
Note that if the same derivation as above is carried through for a diverging lens a similar relationship is obtained. By adopting a sign convention, in this case that real distances are positive and virtual distances negative, the same equation can be obtained, i.e. $1/f = 1/v + 1/u$.

**4.4** *Magnification* is defined as image size/object size. Show that this also means that the magnification is image distance/object distance.

**4.5** Calculate the image distance and magnification for the following situations. In each case state whether the image is real or virtual.
(a) An object is situated 50 mm from a converging lens of focal length 120 mm.
(b) An object is situated 100 mm from a converging lens of focal length 50 mm.
(c) An object is situated 200 mm from a diverging lens of focal length 150 mm.

***4.6** Light from an object passes first through a converging lens of focal length 15 cm, placed 20 cm from the object, and then through a diverging lens of focal length 40 cm placed 40 cm away from the converging lens. Find the position of the final image. Hint: Find the intermediate image produced by the converging lens.

***4.7** A converging lens of focal length 10 cm is 15 cm from a screen and forms an image on the screen. A diverging lens of focal length 20 cm is then placed between the converging lens and the screen at a distance of 5 cm from the converging lens. By what distance must the screen be moved if the new image is to be formed on it?

***4.8** Derive a relationship for the focal length of two thin lenses in contact, in terms of the focal lengths of each lens.

**III** **Read** about methods used for the determination of the focal lengths of converging and diverging lenses.
References: *D* ch. 5, *N* ch. 19.

**Self-assessment questions**

**4.9** One method used for the measurement of the focal length of a converging lens is to place a plane mirror behind the lens and then adjust the position of an object in front of the lens so that the image is formed at the same point as the object. Why is the object distance equal to the focal length?

**4.10** With a converging lens real image distances can be determined for a number of different object distances. When a graph is plotted of $1/u$ against $1/v$, what is the form of the graph and what are the values of the intercepts?

**4.11** In the displacement method for a converging lens, an illuminated object is placed in front of the lens and the image is obtained on a screen. Keeping the object and screen positions fixed, the lens is then moved so that once again a clear image is formed on the screen. Derive an equation relating the distance $d$ between the object and the screen, the amount $L$ that the lens is moved between the two positions giving images, and the focal length $f$.

**4.12** Using an object of known size, measurements can be made, for a converging lens, of the real image size for different image distances. A graph is plotted of the magnification against the image distance. Explain how the focal length can be obtained from the graph.

**4.13** One method that is used for the measurement of the focal length of a diverging lens is to use an auxiliary converging lens to produce a real image of a real object, then use this real image as a virtual object for the diverging lens. Figure 4.2 illustrates this method. Explain what measurements are made and how the focal length can be obtained.

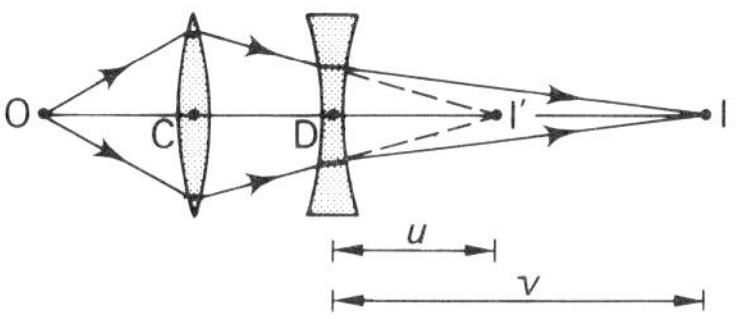


**Figure 4.2**

# SELF-ASSESSMENT

## Key points

With the 'real is positive, virtual is negative' convention for distances, $1/f = 1/v + 1/u$.
For two thin lenses in contact $1/f = 1/f_1 + 1/f_2$.
The **magnification** $m$ = image size/object size = $v/u$.

## Answers I

**4.1** (a) Passes through, or appears to pass through, the principal focus after refraction. (b) Undeviated.
(c) Refracted parallel to the principal axis.

**4.2** See Figure A4.1. (a) real (b) real (c) virtual (d) virtual

## Answers II

**4.3** (a) $D = \alpha + \beta$ (b) $\alpha = h/u$ and $\beta = h/v$, thus $D = h(1/u + 1/v)$ (c) $D = h/f$ (d) $D/h = 1/f = 1/v + 1/u$

**4.4** See Figure A4.2. $OO'/II' = u/v$ and hence $m = v/u$.

**4.5** (a) $f = +120$ mm, $u = +50$ mm, hence $1/120 = 1/v + 1/50$ and so $v = -85.7$ mm. The image is virtual. $m = v/u = 85.7/50 = 1.71$ (no sign is needed).

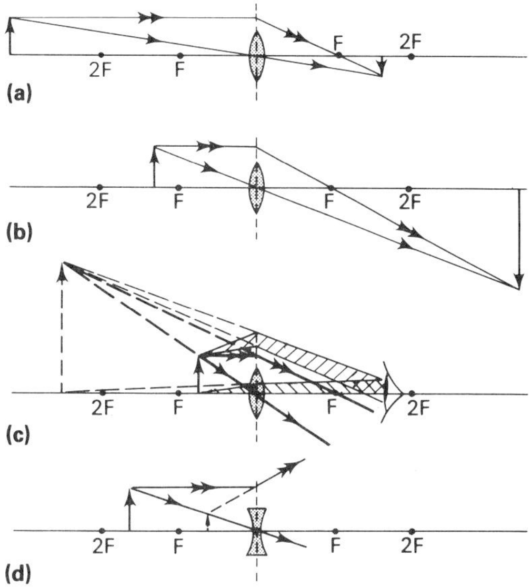


**Figure A4.1**

(b) $f = +50$ mm, $u = +100$ mm. Hence $1/50 = 1/v + 1/100$ and $v = +100$ mm. The image is real. $m = v/u = 100/100 = 1$.
(c) $u = 200$ mm, $f = -150$ mm, hence $-1/150 = 1/v + 1/200$ and $v = -85.7$ mm. The image is virtual. $m = v/u = 85.7/200 = 0.43$.
In doing such problems with the 'real is positive' convention, remember that converging lenses always have positive focal lengths, diverging lenses always have negative ones.
As a check on the answers a quick drawing of a ray diagram, as in SAQ 4.2, can be done.

**4.6** When more than one lens is involved the general procedure is to find the image produced by the first lens, in the absence of the second lens. Then consider this image as being the object for the second lens. For the converging lens $f = +15$ cm, $u = 20$ cm, thus $1/15 = 1/v + 1/20$ and so $v = +60$ cm. This means that it is $-20$ cm from the diverging lens ($f = -40$ cm). Hence $-1/40 = 1/v - 1/20$ and so $v = 40$ cm. The image is 40 cm from the diverging lens and 80 cm from the converging lens.

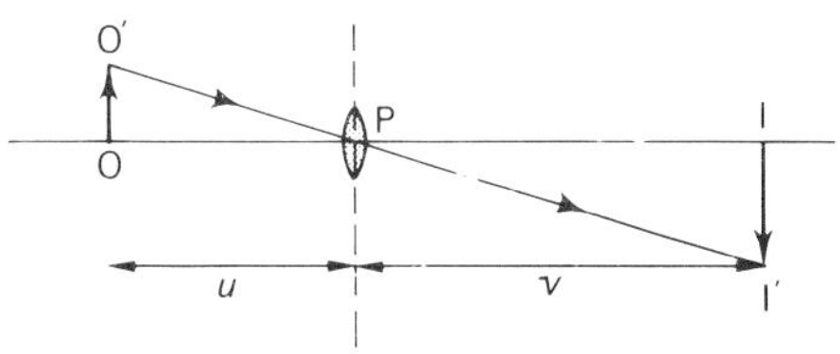


**Figure A4.2**

**4.7** This is a variation on SAQ 4.6, the intermediate image position being given. Thus the object distance for the diverging lens is $-10$ cm. For the diverging lens $f = -20$ cm. $u = -10$ cm, and so $-1/20 = 1/v - 1/10$ and $v = 20$ cm. The screen has to be moved 10 cm further from the lenses.

**4.8** For the first lens $1/f_1 = 1/v_1 + 1/u$. For the second lens the image at distance $v_1$ is a virtual object and so $1/f_2 = 1/v - 1/v_1$. Adding the two equations gives $1/f_1 + 1/f_2 = 1/v + 1/u$. The lens is thus behaving as a single lens having a focal length $f$ where $1/f = 1/v + 1/u$ and so $1/f = 1/f_1 + 1/f_2$.

## Answers III

**4.9** When the light from the lens is parallel, it meets the plane mirror along the normal to the mirror and so is reflected back along its path. This therefore means that the object must have been at the focal point.

**4.10** A straight-line graph with intercepts on both axes of $1/f$.

**4.11** This method involves object and image distances that are interchangeable. They are said to be *conjugate.* This means that in Figure A4.3 the object distance for the lens at position A is the same as the image distance when it is at B, and the image distance when it is at A is the same as the object distance when it is at B. Thus for position A, $u = \frac{1}{2}(d - L)$ and $v = \frac{1}{2}(d + L)$ and so $1/f = 2/(d + L) + 2/(d - L)$ and $f = (d^2 - L^2)/4d$.

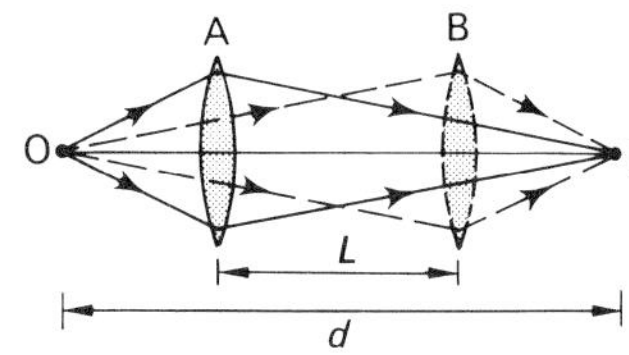


**Figure A4.3**

**4.12** $1/f = 1/v + 1/u$ and so $v/f = v/u + 1$, therefore $m = v/f - 1$. The slope of the graph is $1/f$ and the intercept on the $v$ axis when $m = 0$ is $f$.

**4.13** The converging lens is used to produce an image and its position noted. The diverging lens is then introduced and the new image position noted. Hence the distances of the object and image from the diverging lens can be obtained and so the focal length determined.

# Lesson 5 Optical instruments

## Aims

After completing this lesson you should be able to

- define the terms *near point*, *far point* and the *least distance of distinct vision*
- define *angular magnification* and *visual angle*
- derive and use relationships for the magnifying power of (a) the simple magnifying glass, (b) the compound microscope and (c) the astronomical telescope
- explain what is meant by the *eye ring*
- compare reflecting and refracting telescopes.

## Prerequisites

Lesson 4 on thin lenses.

## STUDY

**I Read** about the near point, the far point, the least distance of distinct vision, angular magnification and visual angle.

References: *B* ST6 ch. 1, *D* ch. 5, *N* ch. 20, *Wh&H* ch. 36.

### Self-assessment questions

**5.1** Define the terms *near point*, *far point* and the *least distance of distinct vision.*

**5.2** Approximately what is the least distance of distinct vision for the 'normal' eye?

**5.3** The size of the image formed by the eye is determined by the visual angle subtended at the eye by the object. Explain this.

**5.4** Define *angular magnification* (the term *magnifying power* is sometimes used).

**II Read** about the derivation of the relationships for the magnifying power of (a) the simple magnifying glass, (b) the compound microscope and (c) the astronomical telescope.

References: *B* PP ch. 6, *B* ST6 ch. 1, *D* ch. 5, *N* ch. 20, *Wh&H* ch. 36.

**Self-assessment questions**

**5.5** This question concerns the simple magnifying glass.
(a) What is the angle subtended at the eye by an object of length $h$ at the least distance of distinct vision $D$?
(b) If a converging lens is used as a magnifying glass and the final image, length $h'$, is formed at the least distance of distinct vision $D$, what is the angle subtended at the eye?
(c) Show that the angular magnification of the lens used in this way is equal to the 'linear' magnification.
(d) Hence obtain an equation for the angular magnification in terms of $D$ and $f$, the focal length.

**5.6** If a simple magnifying glass had been used to give the final image at infinity, rather than the least distance of distinct vision as in SAQ 5.5, at what distance from the lens would the object be and what would be the angular magnification?

**5.7** Figure 5.1 shows a constructional ray diagram for a compound microscope. Derive the equation for the angular magnification when the image is at the least distance of distinct vision, in terms of the linear magnifications produced by each lens.

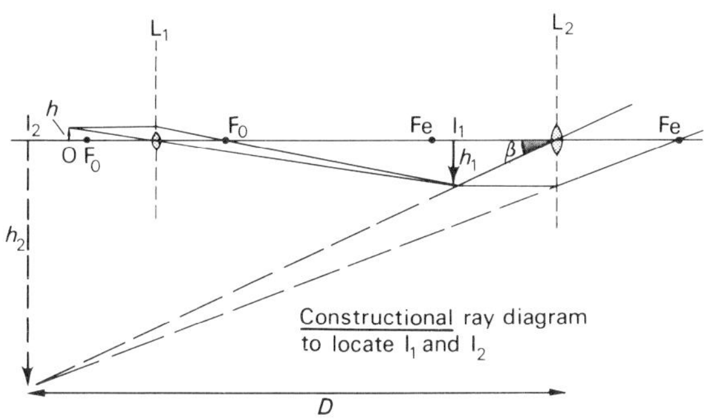


**Figure 5.1 Compound microscope**

***5.8** Calculate the angular magnification and the distance of the object from the objective lens for a compound microscope when the final image is at the least distance of distinct vision. The objective lens has a focal length of 20 mm and the eyepiece a focal length of 60 mm, the distance between them being 120 mm and the least distance of distinct vision 250 mm.

**5.9** Draw the ray diagram for a refracting astronomical telescope when the final image is formed at infinity and obtain an expression for the angular magnification.

***5.10** An astronomical telescope consists of two converging lenses, of focal lengths 100 cm and 10.0 cm. What is (a) the separation of the lenses and (b) the angular magnification if the final image is at infinity?

***5.11** A telescope consists of two converging lenses of focal lengths 40 cm and 4.0 cm, separated by 44.0 cm. If the moon subtends an angle of 0.5° at the objective, what is the angle subtended at the eye by the moon's image seen through the telescope?

**III Read** about the eye ring in relation to the microscope and the telescope.
References: *D* ch. 5, *N* ch. 20, *Wh&H* ch. 36.

**Self-assessment questions**

**5.12** What is the eye ring?

**5.13** For an astronomical telescope with the final image at infinity, what is the relationship between the diameter of the objective and the diameter of the eye ring?

***5.14** A telescope consists of two lenses, of focal lengths 5.0 cm and 100 cm. What is (a) the separation of the two lenses, (b) the angular magnification and (c) the position of the eye ring if the final image is formed at infinity?

***5.15** If the pupil of the eye has a diameter of 2.0 mm, what should be the diameter of the objective lens for the telescope in SAQ 5.14 if all the light passing through the objective is to enter the observer's eye?

**IV Read** about reflecting telescopes and how they compare with refracting telescopes.
References: *B* PP ch. 6, *B* ST6 ch. 1, *D* ch. 5, *N* ch. 20, *Wh&H* ch. 36.

**Self-assessment questions**

**5.16** The Newtonian form of reflecting telescope consists of a concave objective mirror with a small plane mirror being used to deflect the light to an eyepiece (Figure 5.2). What is the angular magnification of such an instrument if the objective mirror has a focal length of 1.0 m and the eyepiece a focal length of 5.0 cm?

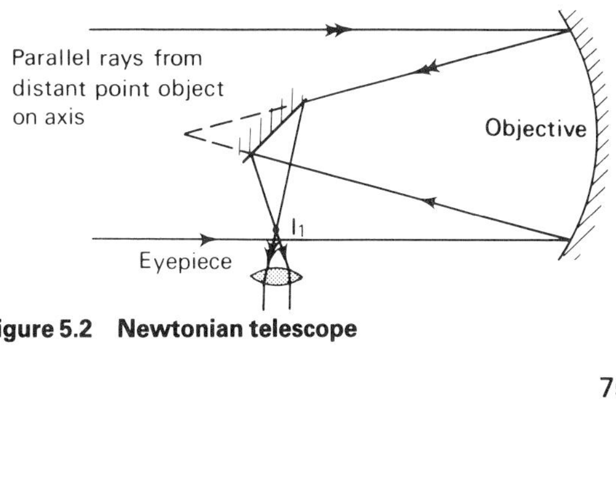


**Figure 5.2 Newtonian telescope**

***5.17** What are the advantages of reflecting telescopes over refracting telescopes?

# SELF-ASSESSMENT

## Key points

The **near point** is the position of the nearest object that can be focused clearly and without strain by the unaided eye. The **far point** is the position of the furthest object that can be focused by the unaided eye. For the normal eye the near point is about 25 cm from the eye and is called the **least distance of distinct vision**; the far point is infinity.

**Angular magnification (magnifying power)** is the angle subtended at the eye by the image seen through the instrument divided by the angle subtended by the object at the unaided eye when the object is at some stated distance. For the microscope this distance is generally taken as the least distance of distinct vision, while for a telescope the distance is infinity. For a simple magnifying glass in normal adjustment the angular magnification equals the linear magnification. For a compound microscope in normal adjustment the angular magnification is the product of the linear magnifications produced by the objective lens and the eyepiece. For a telescope in normal adjustment the angular magnification is the focal length of the objective divided by the focal length of the eyepiece, the separation of the two lenses being the sum of the focal lengths.

The **eye ring** is the circular image of the objective formed by the eyepiece and is the position at which the eye should be placed in order to gather the most light.

## Answers I

**5.1** See the key points.

**5.2** See the key points.

**5.3** A small object situated close to the eye may seem to be much bigger than a much larger object situated a long way off. The factor determining the size of the image in the eye is the visual angle subtended at the eye, see Figure A5.1.

**5.4** See the key points.

## Answers II

**5.5** (a) $h/D$ (b) $h'/D$ (c) $M = (h'/D)/(h/D) = h'/h$ = linear magnification

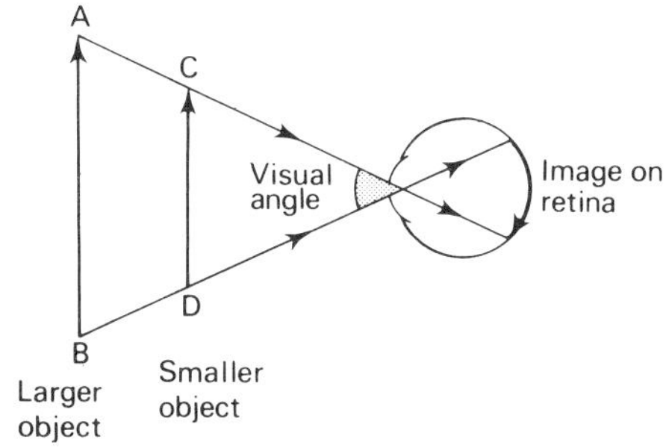


**Figure A5.1**

(d) $1/f = 1/v + 1/u$, hence multiplying throughout by $v$ gives $v/f = 1 + v/u$ and as $M = v/u$ then for $v = -D$ we have $M = -D/f - 1$. The negative sign is usually omitted. Thus $M = D/f + 1$.

**5.6** The object would have had to be at the focus. Angle subtended without instrument = $h/D$, angle with instrument = $h/f$, and so $M = D/f$.

**5.7** $M = \beta/\alpha = (h_2/D)/(h/D) = h_2/h = (h_2/h_1) \times (h_1/h)$

**5.8** For the eyepiece $v = -250$ mm, $f = +60$ mm. Thus $1/60 = -1/250 + 1/u$ and so $u = 48.4$ mm. Hence image distance for objective lens = $120 - 48.4 = 71.6$ mm. As $f = +20$ mm then $1/20 = 1/71.6 + 1/u$ and $u = 27.8$ mm. $M = (250/48.4) \times (71.6/27.8) = 13.3$.

**5.9** See Figure A5.2. $M = \beta/\alpha = (h_1/f_e)/(h_1/f_o) = f_o/f_e$.

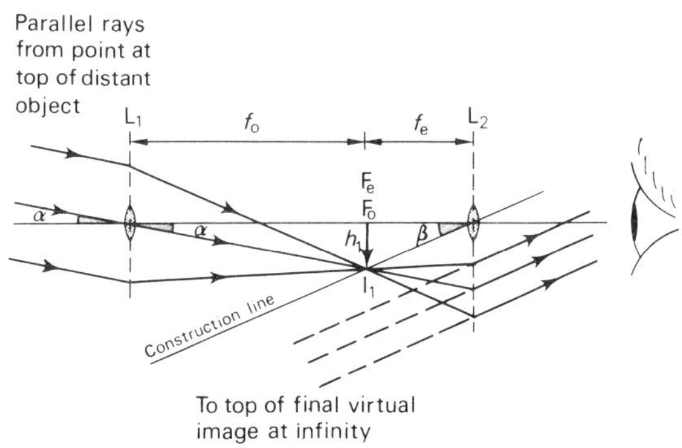


**Figure A5.2 Refracting astronomical telescope**

**5.10** (a) $f_o + f_e = 110$ cm (b) $M = f_o/f_e = 100/10 = 10$

**5.11** The separation indicates that the telescope is in normal adjustment. Thus $M = f_o/f_e = 40/4 = 10 = \beta/\alpha$ and as $\alpha = 0.5°$ then $\beta = 5°$.

## Answers III

**5.12** See the key points.

**5.13** Consider the objective lens as being the object, a distance $(f_o + f_e)$ from the eyepiece. The image, the eye ring, will be formed at an image distance given by $1/f_e = 1/(f_o + f_e) + 1/v$ and so $v = f_e\,(f_o + f_e)/f_o$. Diameter of objective/diameter of eye ring $= u/v = (f_o + f_e)/[f_e(f_o + f_e)/f_o] = f_o/f_e = M$.

**5.14** (a) $f_o + f_e = 105$ cm (b) $M = f_o/f_e = 100/5.0 = 20$ (c) Using $v = f_e(f_o + f_e)/f_o$ from 5.13 above, then $v = 5.25$ cm.

**5.15** Diameter of objective $= M \times$ diameter of eye ring $= 20 \times 2.0 = 40$ mm.

## Answers IV

**5.16** The mirror forms an intermediate image at its focal point. As the final image is at infinity the intermediate image must be at the focal point of the eyepiece. Hence the situation is comparable to that with the refracting telescope and $M = f_o/f_e = 100/5.0 = 20$.

**5.17** No chromatic aberration. A mirror can have a much larger diameter than a lens since it can be supported at the back. Only one surface needs to be accurately made. With a lens it is difficult to maintain uniform refractive index throughout the glass; this problem does not occur with a mirror.

# CHAPTER 5

# Wave motion

## Lesson 1 Describing waves

### Aims

After completing this lesson you should be able to

- define and use the terms *frequency*, *wavelength*, *amplitude* and *period*
- derive and use the relationship $v = f\lambda$
- distinguish between transverse and longitudinal waves, describing such waves graphically and algebraically
- define and use the terms *phase angle* and *phase difference*
- write and use equations for progressive waves.

### Prerequisites

A basic knowledge of linear motion, e.g. Lesson 1 in Chapter 1, and reflection and refraction, e.g. Chapter 4, Lessons 1 and 2. Lesson 1 in Chapter 1, on s.h.m., is useful for one group of questions.

## STUDY

**I Read** about frequency, wavelength, amplitude and period, the derivation and use of the relationship $v = f\lambda$.
References: *B* PP ch. 6, *B* ST6 ch. 2, *D* ch. 16, *N* ch. 21, *W et al.* ch. 17(16), *Wh&H* ch. 12.

**Self-assessment questions**

**1.1** Define the terms *frequency*, *wavelength*, *amplitude* and *period*.

**1.2** Derive the relationship between wave speed $v$, the frequency $f$ and the wavelength $\lambda$.

**1.3** What is the frequency of a wave travelling at 10 cm s$^{-1}$ with a wavelength of 2.0 cm?

**II Read** about transverse and longitudinal waves and their graphical and algebraic description.
References: *B* PP ch. 6, *B* ST6 ch. 2, *D* ch. 16, *N* ch. 21, *W et al.* (ch. 16), *Wh&H* ch. 12.

**Self-assessment questions**

**1.4** Distinctions need to be made between the displacement of the particles of a medium when a wave passes through it and the motion of the wave itself. How do transverse waves differ from longitudinal waves? How could you demonstrate both these types of wave with a stretched spring?

**1.5** Sketch a displacement–distance graph for a transverse wave motion and for a longitudinal wave motion. A longitudinal wave is said to be a sequence of compressions and rarefactions, i.e. a pressure wave. Identify the regions of compression and rarefaction in such a wave and their relation to the displacement–distance graph.

**1.6** Both the transverse and longitudinal waves referred to above are called *progressive* waves. This is because the wave profile moves along with the speed of the wave. Sketch graphs showing how the displacement varies with time for two particles half a wavelength apart and use the term *phase* in explaining the graphs.

**1.7** This follows from the previous question. The displacement of a particle a distance $x$ from the origin can be described by the equation $y = a\sin(\omega t - \phi)$.
(a) What does $\phi$ represent and why is it negative?
(b) Explain why $\phi/2\pi = x/\lambda$. Hence rewrite the equation in terms of $x$ instead of $\phi$.
(c) Another way of writing the equation is in terms of the wave speed $v$. As $\omega = 2\pi f$ (see Lesson 14 in Chapter 1) and $v = f\lambda$, rewrite the equation in terms of $v$.

***1.8** A plane wave travelling in a medium along the $x$ direction is described by the equation $y = 2.0 \times 10^{-4} \times \sin(100\pi t - 10^4\pi x/12)$, where $y$ is the displacement in metres, $t$ the time in seconds and $x$ the distance from the origin in metres. What is (a) the frequency of the wave, (b) its wavelength and (c) its speed?

***1.9** A plane wave travelling in a medium along the $x$ direction is described by the equation $y = a\sin(\omega t - kx)$,

where $y$ is the displacement at a time $t$ a distance $x$ from the origin.
(a) What would be the equation for an identical wave travelling in the opposite direction?
(b) If $a = 1.5 \times 10^{-6}$ m, $\omega = 600\,s^{-1}$ and $k = 6.0 \times 10^{-3}$ m, what is (i) the frequency of the wave, (ii) its amplitude and (iii) the wave speed?

# SELF-ASSESSMENT

## Key points

The **frequency** $f$ is the number of vibrations made per second by the wave source. The **period** $T$ is the time taken for one complete vibration: $T = 1/f$. The **wavelength** $\lambda$ is the distance between two successive similar points on the wave profile. **Wave speed** $v = f\lambda$. The **amplitude** of a wave is the maximum displacement of a particle from its mean position. With a **transverse wave** the particle displacement is at right angles to the direction of the wave motion. With a **longitudinal wave** the particle displacement is along the same direction as the wave motion.

A **progressive wave** is one where the wave profile moves along with the speed of the wave. The vibrations of the particles in a progressive wave are of the same amplitude and frequency but the **phase** of the vibration differs for different points along the wave. Progressive waves can be described by the equation $y = a\sin(\omega t - \phi)$ or $y = a\sin(\omega t - 2\pi x/\lambda)$ or $y = a\sin\frac{2\pi}{\lambda}(vt - x)$ or $y = a\sin 2\pi(t/T - x/\lambda)$, where $\omega = 2\pi f = 2\pi/T$.

## Answers I

**1.1** See the key points.

**1.2** If $f$ waves are produced per second, then the distance covered by the waves in 1 s is $f\lambda$. Hence $v = f\lambda$.

**1.3** $v = f\lambda$ and so $f = 10/2.0 = 5.0\,s^{-1} = 5.0$ Hz. (The unit of frequency is the hertz.)

## Answers II

**1.4** Transverse waves have their particle displacements at right angles to the wave motion, while longitudinal waves have their displacements along the same direction as the wave motion. See Figure A1.1.

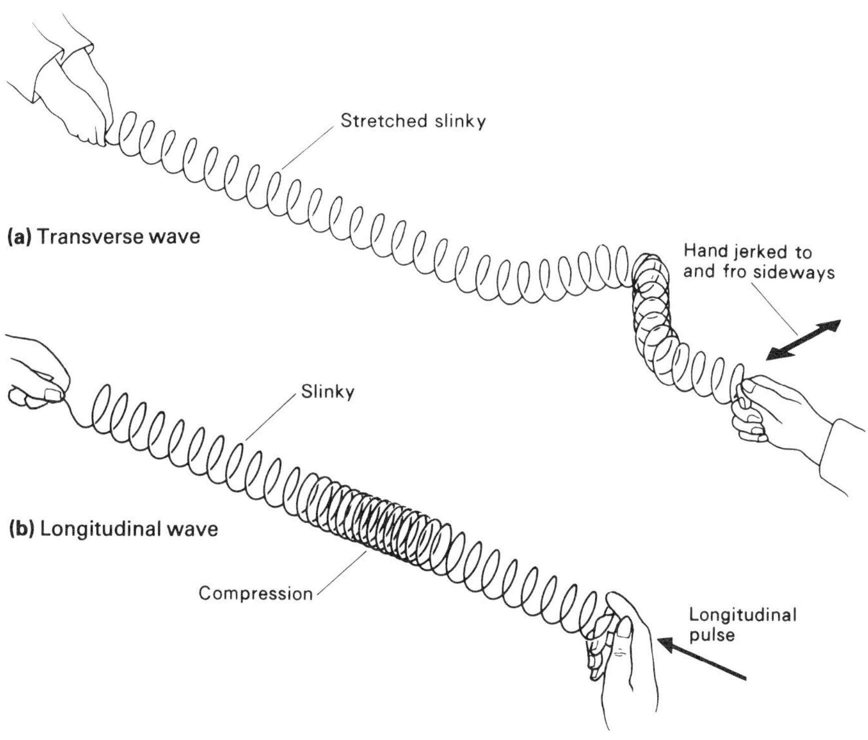


**Figure A1.1**

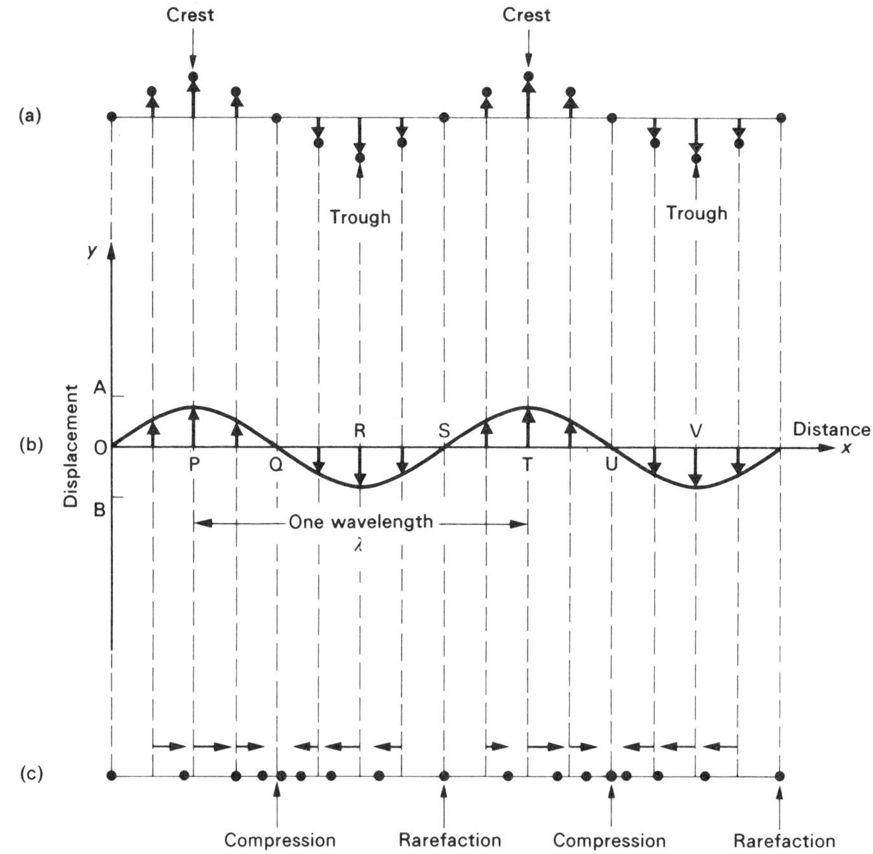


**Figure A1.2**

**1.5** See Figure A1.2. (a) is the form of the displacements for a transverse wave, (c) the form for a longitudinal wave. Both can be described by the displacement–distance graph given in (b).

**1.6** See Figure A1.3. Graph (a) could represent the motion of the particle located at point Q in Figure A1.2,

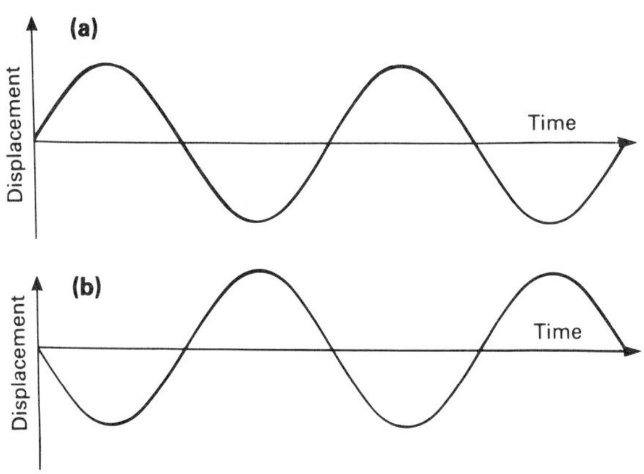


**Figure A1.3**

while graph (b) could represent the particle at point S. The particles' displacement–time graphs are out of phase by half the wave period. Though each of the points follows the same oscillation, they have the same displacements at different times.

**1.7** (a) $\phi$ is the *phase angle*. It is the amount by which the particle concerned lags behind the displacement at the origin. It is negative because it is a lag. In the previous question the phase angle would have been $\pi$, as half a complete wave is involved.

(b) $\phi/2\pi$ is the fraction of a complete wave that the particle is lagging behind the origin, there being $2\pi$ radians for one complete wave. But $x/\lambda$ also represents the same thing. Thus $\phi/2\pi = x/\lambda$. Thus $y = a\sin(\omega t - 2\pi x/\lambda)$.

(c) $\omega$ is the angular velocity of a point moving round a circle which would give as a projection on the $y$-axis the sine curve described by the equation (see Lesson 14 in Chapter 1). As $\omega = 2\pi f = 2\pi v/\lambda$ then $y = a\sin(2\pi vt/\lambda - 2\pi x/\lambda) = a\sin\frac{2\pi}{\lambda}(vt - x)$.

**1.8** (a) Compare this equation with that in 1.7(b) above, then $100\pi = \omega$ and, as $\omega = 2\pi f$, $f = 50$ Hz.
(b) With the same comparison, $2\pi x/\lambda = 10^4\pi x/12$ and so $\lambda = 2.4 \times 10^{-3}$ m.
(c) As $v = f\lambda$ then $v = 50 \times 2.4 \times 10^{-3} = 0.12\,\text{m s}^{-1}$.

**1.9** (a) $y = a\sin(\omega t + kx)$. There is a plus sign because the displacement at the point concerned would *lead* that at the origin.
(b) (i) This equation can be compared with $y = a\sin(\omega t - 2\pi x/\lambda)$. Thus $\omega = 2\pi f$ and so $f = 600/2\pi = 95.5$ Hz. (ii) Amplitude $= a = 1.5 \times 10^{-6}$ m (iii) $2\pi/\lambda = k$ and so $\lambda = 2\pi/6.0 \times 10^{-3} = 1.0 \times 10^3$ m and hence $v = f\lambda = 95.5 \times 1.0 \times 10^3 = 9.6 \times 10^4\,\text{m s}^{-1}$.

# Lesson 2 Wave properties

## Aims

After completing this lesson you should be able to:

- explain how *Huygens' construction* can be used to enable the new position of a wavefront to be found, and use it to explain reflection and refraction
- explain, qualitatively, *diffraction*
- state the *principle of superposition* and use it to explain interference patterns from two point sources of the same frequency
- explain the formation of *standing* waves.

## Prerequisites

Lesson 1 on describing waves.

## STUDY

**I Read** about Huygens' construction, paying particular attention to the terms *wavefront* and *ray*, and the application of the construction to the explanation of reflection and refraction.
References: *B* PP ch. 6, *D* ch. 16, *N* ch. 22, *W et al.* (ch. 16), *Wh&H* ch. 13.

### Self-assessment questions

**2.1** Explain the meaning of the terms *wavefront* and *ray*.

**2.2** Give a simple statement of what is meant by 'Huygens' construction', using the term *secondary wavelets* in your answer.

**2.3** Figure 2.1 shows the position of a wavefront at some instant of time and the position a time $\Delta t$ later, according to Huygens' construction. What is the distance between the new and the old wavefronts if the waves are travelling with a speed $v$?

**2.4** The wavefront in Figure 2.1 is a *plane* wavefront. What is the direction of the rays?

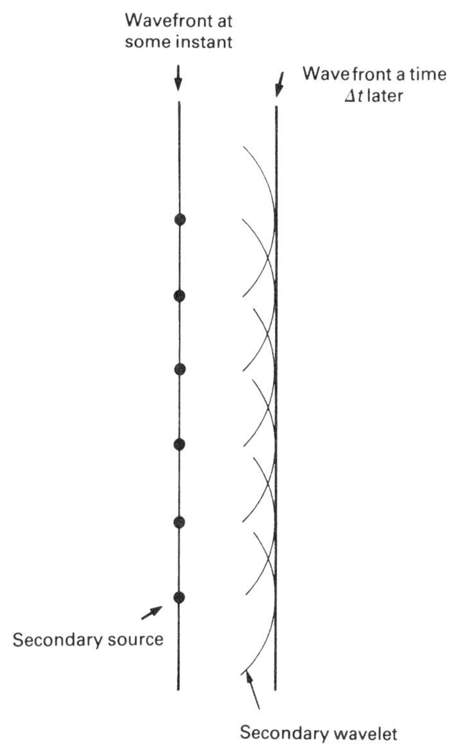


**Figure 2.1**

**2.5** Figure 2.2 shows a wavefront incident on a reflecting surface and the reflected wavefront some time later. AB is the incident wavefront and A′B′ the reflected wavefront a time $\Delta t$ later.
(a) Using Huygens' construction a secondary wavelet can be drawn with A as its centre, similarly one with B as its centre. The resulting new wavefront is A′B′. Why does the distance AA′ equal BB′?
(b) Prove that angle $i$ equals angle $r$.

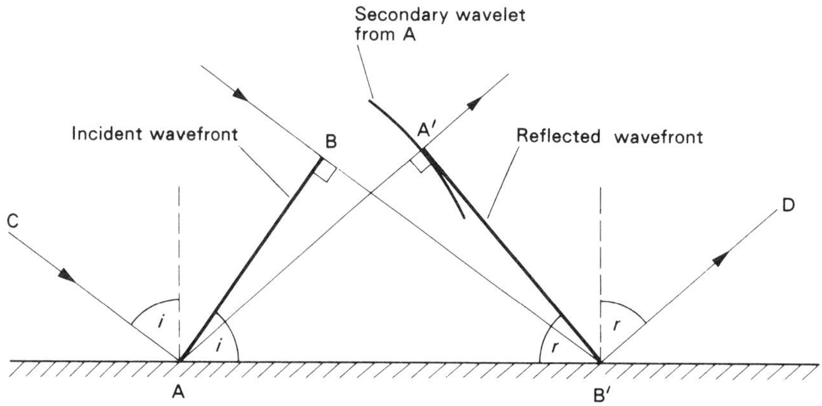


**Figure 2.2**

**2.6** Figure 2.3 shows a wavefront incident on a refracting surface and the refracted wavefront some time later. AB is the incident wavefront and A′B′ the refracted wavefront a time $\Delta t$ later.

(a) Using Huygens' construction a secondary wavelet can be drawn with A as its centre, similarly one with B as its centre. The resulting new wavefront is A′B′. If the speed of the wave motion is $v_1$ in medium 1 and $v_2$ in medium 2, what are the distances AA′ and BB′ in terms of the speeds?

(b) The refractive index for the wave motion passing from medium 1 to medium 2 is ${}_1n_2 = \sin i_1/\sin i_2$. By considering the triangles ABB′ and AA′B′, express $\sin i_1$ and $\sin i_2$ in terms of the speeds and hence obtain the refractive index in terms of the speeds.

(c) For a refracted ray bending towards the normal, i.e. $i_1 > i_2$, in which medium must the wave speed be higher?

**2.7** When white light passes from air into glass there is dispersion and the red light is refracted at a greater angle than the blue light. Which has the greater speed in glass, red light or blue light?

***2.8** Figure 2.4 shows a wavefront AB at some instant of time. If the refractive index of the glass is 1.5 and the speed of light in air is $3.0 \times 10^8$ m s$^{-1}$, draw the positions of the wavefront a time (a) $1 \times 10^{-10}$ s and (b) $2 \times 10^{-10}$ s later. Consider only the refracted wave.

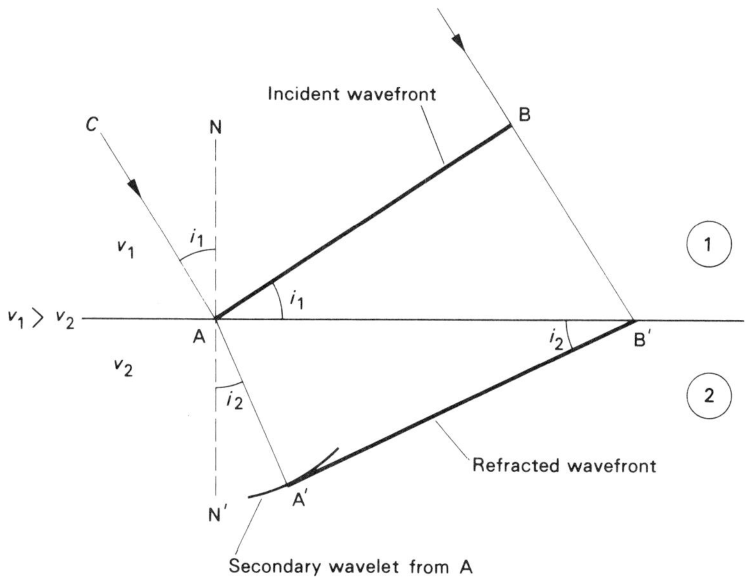


**Figure 2.3**

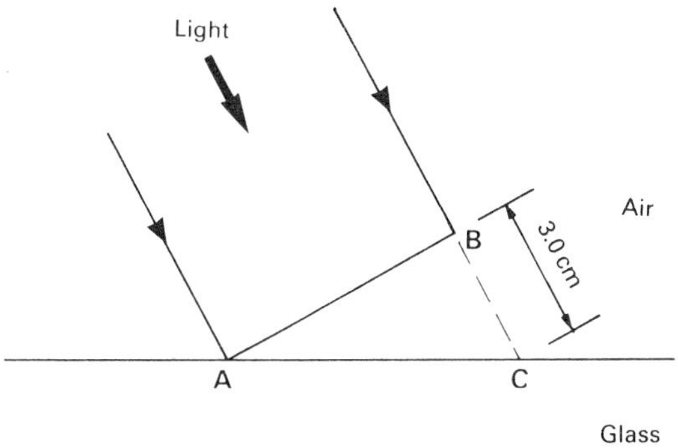


**Figure 2.4**

**II Read** about diffraction of waves, the principle of superposition and its use to explain interference patterns from two point sources of waves of the same frequency. References: *B* PP ch. 6, *B* ST6 ch. 2, *D* ch. 16, *N* ch. 21, *W et al.* ch. 23(16), *Wh&H* ch. 14.

**Self-assessment questions**

**2.9** What is *diffraction*?

**2.10** Consider diffraction of a wave motion by an aperture. Under what circumstances would the diffraction effects be (a) very noticeable, (b) barely noticeable?

**2.11** State the *principle of superposition*.

***2.12** Explain what happens when the wave trains from two point coherent sources cross.

**2.13** The point where two waves always cancel each other is called a *node*, the point where the two waves reinforce each other and produce a maximum displacement is called an *antinode*. For two point coherent sources of waves, a distance apart of a few wavelengths, sketch the nodal pattern.

**III Read** about the formation of standing waves. References: *B* PP ch. 6, *B* ST6 ch. 4, *D* ch. 16, *N* ch. 21, *W et al.* ch. 23(20), *Wh&H* ch. 14.

**Self-assessment questions**

**2.14** When a transverse wave pulse is sent along a spring and reflected from the fixed end of the spring, how is the reflected pulse related to the incident pulse?

**2.15** A spring is held taut by means of a light string tied to its far end. When a transverse wave pulse is sent along the spring and reflected from the boundary with the string, how is the reflected pulse related to the incident pulse?

**2.16** If a spring is fixed at the far end to a rigid support and a continuous transverse wave, of constant amplitude and frequency, is sent along the spring, a stationary wave pattern can be produced with the spring having fixed node positions. Explain why such a stationary wave pattern is produced.

**2.17** What are *standing* or *stationary* waves and how are they produced?

**2.18** What is the distance between nodes in a standing wave in terms of the wavelength of the wave?

***2.19** A source of electromagnetic waves is placed in front of a plane metal screen and a detector moved along the normal from the screen to the source. Successive maxima and minima are detected. Why? What is the wavelength of the waves if successive minima are 1.5 cm apart?

# SELF-ASSESSMENT

## Key points

According to **Huygens**, every point on a wavefront can be considered as a source of *secondary wavelets* which spread out with the wave speed. The new wavefront is the envelope of these secondary wavelets.
**Refraction** can be explained if light travels more slowly in the medium in which it makes a smaller angle with the normal, the refractive index ${}_1n_2$ being $v_1/v_2$.
Waves can be 'diffracted', **diffraction** being the name given to the spreading out of a wave that results from passing through an aperture or past an obstacle.
The **principle of superposition** is that the net disturbance at a given place and time caused by a number of waves which are traversing the same space is the vector sum of the disturbances which would have been produced by the individual waves separately at that place.
**Standing** or **stationary waves** result from the superposition of two wave trains which have the same amplitude, frequency and wavelength but oppositely directed, equal-size velocities. Unlike a progressive wave the waveform does not advance but remains stationary.

## Answers I

**2.1** A *wavefront* is a line or surface on which the disturbance has the same phase at all points. A *ray* is a line at right angles to a wavefront, showing the direction of travel of the wave.

**2.2** See the key points.

**2.3** Several arbitrary points have been chosen on the wavefront, these being used as centres of secondary, spherical wavelets. The radius of the secondary wavelets is chosen to be $v\Delta t$, where $v$ is the wave speed. The envelope of these secondary wavelets is the new wavefront, the envelope being the line or surface which touches all the wavelets. The distance between the new and the old wavefronts is $v\Delta t$.

**2.4** At right angles to the wavefront.

**2.5** (a) Because the secondary wavelets starting from A and B have the same wave speed and so cover equal distances in the same time interval.
(b) Consider triangles ABB′ and AA′B′. Angle ABB′ = angle AA′B′ = 90°, the rays being at right angles to the wavefronts. AA′ = BB′ and AB′ is common to both triangles. Thus the triangles are congruent and so angle $i$ equals angle $r$.

**2.6** (a) $AA' = v_2\Delta t$; $BB' = v_1\Delta t$. (b) $\sin i_1 = BB'/AB' = v_1\Delta t/AB'$. Also $\sin i_2 = AA'/AB' = v_2\Delta t/AB'$. Hence $\sin i_1/\sin i_2 = v_1/v_2 = {}_1n_2$. (c) $v_1$ must be greater than $v_2$.

**2.7** The bigger $i_2$ is, the smaller the amount of refraction, and the greater the speed in that medium. Thus red light has a higher speed in glass than blue light.

**2.8** Time taken to cover BC = $3.0 \times 10^{-2}/3.0 \times 10^{8} = 1 \times 10^{-10}$ s.
(a) In this time the wavefront at B reaches C. In the same time the wavefront at A covers a distance of $(3.0 \times 10^{8}/1.5) \times 1 \times 10^{-10} = 2.0$ cm. See Figure A2.1.

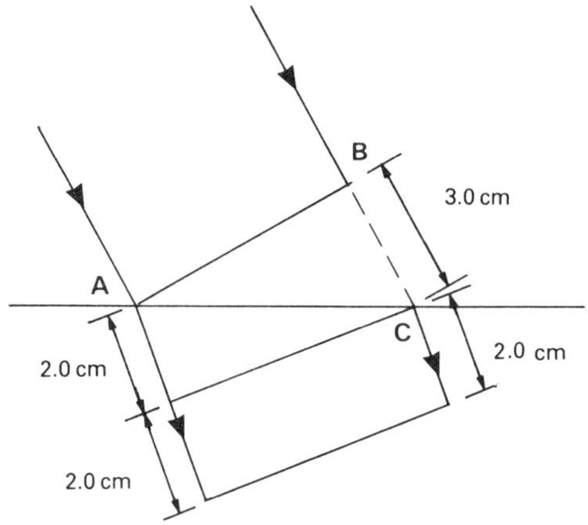


**Figure A2.1**

(b) In $2 \times 10^{-10}$ s the wavefront at A will cover a distance of 4.0 cm. The wavefront at B takes $1 \times 10^{-10}$ s to reach C and then covers a distance of 2.0 cm in the glass. See Figure A2.1.

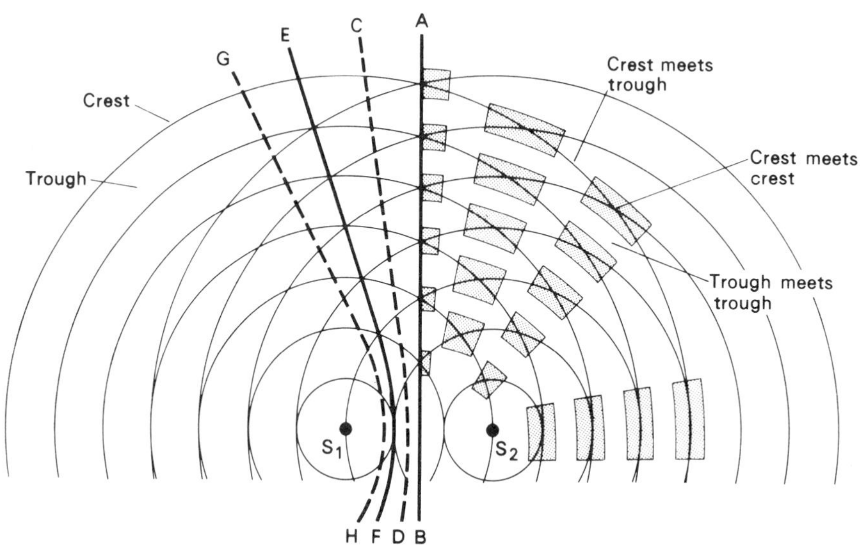


**Figure A2.2**

### Answers II

**2.9** See the key points.

**2.10** (a) When the size of the aperture is of the same order of magnitude as the wavelength. (b) When the aperture is considerably larger than the wavelength.

**2.11** See the key points.

**2.12** Superposition of the waves at the crossing points means that cancellation or reinforcement can occur. Thus at some positions there will be no resultant displacement. The result is said to be an *interference pattern*. See Figure A2.2. Along the lines AB and EF there is reinforcement when the waves meet and maximum displacement occurs. Along CD and GH there is always cancellation and no displacement.

**2.13** See Figure A2.3.

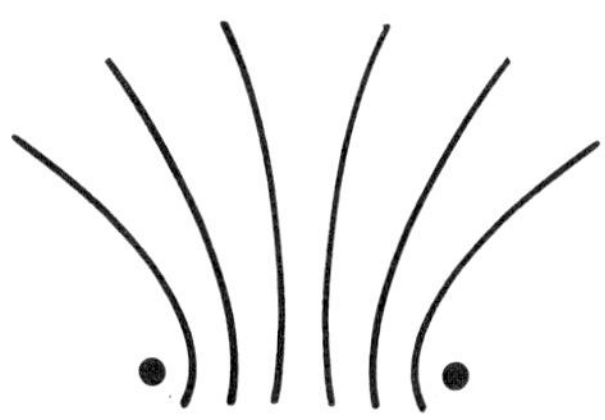

**Figure A2.3**

### Answers III

**2.14** It comes back upside down relative to the incident pulse. A phase change of $\pi$ (or $\frac{1}{2}\lambda$) has taken place.

**2.15** It is reflected the same way up as the incident pulse; no phase change has taken place.

**2.16** The incident wave and the reflected wave superimpose, the incident wave suffering a phase change of $\pi$ (or $\frac{1}{2}\lambda$) on reflection. The resulting interference shows as a pattern of nodes and antinodes. Note that in mathematical terms we have two waves represented by $y_1 = a\sin(\omega t - 2\pi x/\lambda)$ and $y_2 = a\sin(\omega t + 2\pi x/\lambda)$, see Lesson 1 and, for the sign in the sine term, SAQ 1.9. Thus $y = y_1 + y_2 = a\sin(\omega t - 2\pi x/\lambda) + a\sin(\omega t + 2\pi x/\lambda) = 2a\cos(2\pi x/\lambda)\sin\omega t$. The first term of the equation is a location-dependent term and the second a time-dependent term. At a node $y = 0$, at all times. Thus $\cos(2\pi x/\lambda) = 0$ and $2\pi x/\lambda = n\frac{1}{2}\pi$, where $n = 1, 3, 5, 7$, etc. The distance between nodes is thus $\frac{1}{2}\lambda$. At the antinodes $y$ reaches the maximum value of $2a$. This means that $\cos(2\pi x/\lambda) = \pm 1$ and so $2\pi x/\lambda = m\pi$, where $m = 0, 1, 2, 3$, etc. The distance between antinodes is $\frac{1}{2}\lambda$.

**2.17** See the key points and 2.16 above.

**2.18** $\frac{1}{2}\lambda$; see 2.16 above.

**2.19** $\frac{1}{2}\lambda = 1.5$ cm and so $\lambda = 3.0$ cm.

# Lesson 3 Sound waves

### Aims

After completing this lesson you should be able to

- recognize that sound is a longitudinal wave motion
- relate the pressure–distance and the displacement–distance graphs
- discuss the reflection, refraction, interference and diffraction of sound
- explain beats
- explain the modes of vibration of strings and solve problems involving vibrating strings
- explain the way air columns vibrate and solve problems involving such vibrations
- discuss the factors affecting the speed of sound in solids and gases.

### Prerequisites

Lessons 1 and 2 of this chapter, and for some questions a knowledge of elasticity (Lesson 1, Chapter 2) and the ideal-gas laws (Lesson 4, Chapter 3).

## STUDY

**I** **Read** about sound as a longitudinal wave and the relation between its pressure–distance graph and displacement–distance graph.
References: *B* PP ch. 6, *B* ST6 ch. 4, *D* ch. 17, *N* ch. 21, *Wh&H* ch. 41.

**Self-assessment questions**

**3.1** In what direction is the particle displacement in relation to the direction of motion of a sound wave ?

**3.2** On the same graph axes plot graphs showing how the displacement varies with distance and also how the pressure varies with distance for a sound wave.

**II** **Read** about the reflection, refraction, interference and diffraction of sound waves, also the explanation of beats.
References: *B* PP ch. 6, *B* ST6 ch. 4, *D* ch. 17, *N* ch. 21, *W et al.* (ch. 18), *Wh&H* ch. 41.

**Self-assessment questions**

**3.3** Explain echoes.

**3.4** The speed of sound in warm air is greater than that in cold air. At night the air is usually colder near the ground than it is higher up. For a sound emanating from the ground, describe the type of path the sound could follow.

**3.5** Why is the diffraction of sound through doorways a common occurrence?

***3.6** Two loudspeakers are connected in parallel to the same signal generator and emit, in phase, sound at a frequency of 4 kHz. If the speakers are arranged as shown in Figure 3.1, explain what happens when the microphone is moved along the line XY. How could the results be used to obtain a value for the speed of sound in air? (No mathematics required, just the principle.)

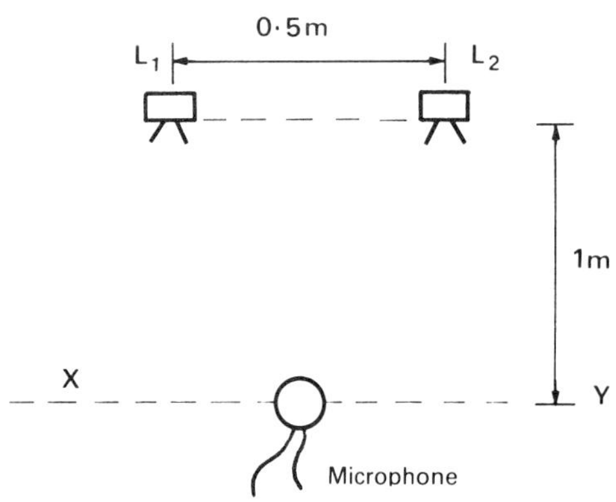


**Figure 3.1**

***3.7** Two loudspeakers are connected to the same signal generator which gives a signal of frequency 1 kHz. If the loudspeakers face each other, describe and explain the signal that would be detected by a microphone moved along the line between and joining the speakers.

**3.8** What are *beats*?

**3.9** When beats are being heard, suppose the time between two successive maxima is $T$. The beats are due to one of the sounds, frequency $f_1$, making in time $T$ one more cycle than the other sound, frequency $f_2$. Hence derive a relationship for the beat frequency.

***3.10** Two tuning forks A and B are sounded together and beats of frequency 10 Hz are heard. If A has a frequency of 456 Hz, what frequency can B be?

**III** **Read** about vibrating strings, their modes of vibration and the equation relating the fundamental frequency to the length, tension and mass.
References: *B* ST6 ch. 4, *D* ch. 17, *N* ch. 27, *Wh&H* ch. 41.

**Self-assessment questions**

**3.11** For a standing wave on a string fixed at both ends and of length $L$, what are (a) the positions of the nodes and antinodes when the string is vibrating with the maximum wavelength possible, (b) the length of this maximum wavelength, (c) the length of the next-longest wavelength?

**3.12** The speed with which a transverse wave travels along a string of mass $\mu$ per unit length and under tension $T$ is given by $v = \sqrt{(T/\mu)}$. How is the fundamental frequency related to the tension, mass per unit length and length $L$ of the string?

***3.13** A sonometer wire of length 50 cm and mass per unit length $1.2 \times 10^{-3}$ kg m$^{-1}$ is stretched by a load of 3.0 kg. What will be the frequency of the note emitted if the string is plucked at its mid-point? ($g = 9.8$ m s$^{-2}$)

***3.14** A piano wire has a diameter of $7.0 \times 10^{-4}$ m. If it is replaced by another wire, of the same material but with a diameter of $7.2 \times 10^{-4}$ m, what percentage change in the tension is needed to restore the original fundamental frequency?

***3.15** A horseshoe magnet is so placed over a sonometer wire that when an alternating current of frequency 50 Hz passes along the wire it is set into resonant vibration. If the magnet is at the mid-point of the wire, and the wire has a length of 80 cm and is under a tension of 50 N, what is the mass per unit length of the wire?

**IV** **Read** about the vibrations of air columns.
References: *B* ST6 ch. 4, *D* ch. 17, *N* ch. 27, *Wh&H* ch. 41.

**Self-assessment questions**

**3.16** For an organ pipe, closed at one end and open at the other, vibrating in its fundamental mode, what are the positions of (a) the displacement node and antinode, (b) the pressure node and antinode?

**3.17** What is the longest wavelength that the air in a pipe of length $L$ can oscillate at if (a) the pipe is open at both ends, (b) the pipe is closed at one end and open at the other?

***3.18** An organ pipe closed at one end and open at the other has a length of 30 cm. What are the wavelengths of (a) the fundamental note, (b) the first overtone and (c) the second overtone?

**3.19** Because the air at the end of an open pipe is free to move, the antinode for the displacement is not exactly at the end of the pipe but a little distance $c$ from the open end. This end correction means the effective length of a pipe open at just one end is $L + c$, if open at both ends $L + 2c$. Explain how, by using a resonance tube, i.e. a tube in which the level of water and hence the length of the air column in the tube can be varied, with a series of tuning forks of known frequencies, the velocity of sound (and the end correction) can be obtained from a graph of length of air column above the water against the reciprocal of the frequency.

***3.20** An open pipe of length 15 cm and a pipe closed at just one end of length 11.5 cm are each sounding their first overtones. If the notes are of the same frequency what is the end correction, if it is assumed to be the same for the open ends of both pipes?

***3.21** A uniform tube of length 80 cm is vertical with its lower end dipping into water. It can be moved into the water to vary the length of air column. Neglecting the end correction, what will be the lengths of air column above the water at which the tube will resonate when a tuning fork of frequency 512 Hz is sounded above the open end of the tube? Take the speed of sound to be 340 m s$^{-1}$.

**V** **Read** about the factors affecting the speed of sound in solids and gases, the equations $v = \sqrt{(E/\rho)}$ and $v = \sqrt{(\gamma r T)}$ or $\sqrt{(\gamma RT/M)}$.
References: *B* ST6 ch. 4, *D* ch. 17, *N* ch. 27, *Wh&H* ch. 41.

**Self-assessment questions**

**3.22** The speed of sound in any medium is given by

$$\text{speed} = \sqrt{\left(\frac{\text{elastic modulus of medium}}{\text{density of medium}}\right)}$$

In a simple qualitative manner give an explanation of the relevance of the elastic modulus in determining the speed of sound in a medium.

**3.23** What is the speed of sound in steel if it has a density of $7.8 \times 10^3$ kg m$^{-3}$ and Young's modulus is $2.0 \times 10^{11}$ Pa?

**3.24** For gases the elastic modulus relevant to the speed of sound equation in SAQ 3.22 is the bulk modulus. Under the conditions in which sound travels through a gas the bulk modulus is $\gamma p$, where $\gamma$ is the ratio of the principal specific heats of the gas and $p$ the pressure.
(a) For a mass $m$ of an ideal gas we can write $pV = mrT$ (see Lesson 4, Chapter 3). Hence write an expression for the speed of sound in terms of the temperature.
(b) Does the speed of sound in an ideal gas depend on the pressure? Explain your answer.
(c) What are the factors which affect the speed of sound in an ideal gas?

**3.25** If the velocity of sound in air at 0 °C is 331 m s$^{-1}$, what is its value at 20 °C?

***3.26** What is the length of the air column in a pipe, open at one end and closed at the other, that gives the fundamental note if the air in the pipe is at 18 °C and a tuning fork of frequency 256 Hz is sounded above the open end? Speed of sound in air at 0 °C = 330 m s$^{-1}$.

# SELF-ASSESSMENT

## Key points

Sound is a **longitudinal** wave motion, transmitted through materials as a **pressure wave**. Like other waves, sound can be reflected, refracted, diffracted and show interference effects.
**Beats** occur when two notes of slightly different frequencies but similar amplitudes are sounded together: the loudness increases and decreases periodically. The beat frequency is the difference between the two frequencies of the notes.

The speed at which a transverse pulse or wave travels along a stretched string is $\sqrt{(T/\mu)}$ where $T$ is the tension and $\mu$ the mass per unit length. The **fundamental mode** of oscillation for a string of length $L$ is when $L = \frac{1}{2}\lambda$, the **first overtone** is $L = \lambda$, the **second overtone** is $L = 1\frac{1}{2}\lambda$. Stationary longitudinal waves can be set up in pipes when the wave 'fits' the length of the tube and gives a **displacement node** at a closed end and a **displacement antinode** at an open end.
The **speed of sound** in any medium is $\sqrt{(\text{elastic modulus of medium/density of medium})}$. In the case of a solid this is $\sqrt{(E/\rho)}$. For a gas the speed is $\sqrt{(\gamma p/\rho)}$. This is, for an ideal gas, $\sqrt{(\gamma rT)}$.

## Answers I

**3.1** Along the direction of the wave motion as it is a longitudinal wave.

**3.2** See Figure A3.1, also Figure A1.2 (page 80).

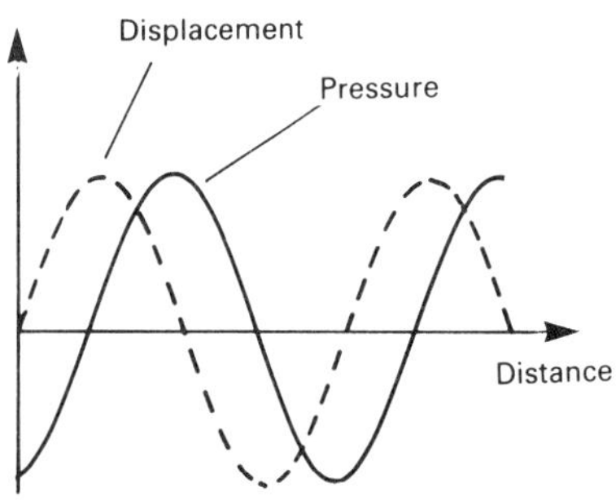


**Figure A3.1**

## Answers II

**3.3** An *echo* is a reflection of a pulse of sound from some distant reflector such that a noticeable interval occurs between the initial sound and the reflected sound.

**3.4** See Figure A3.2.

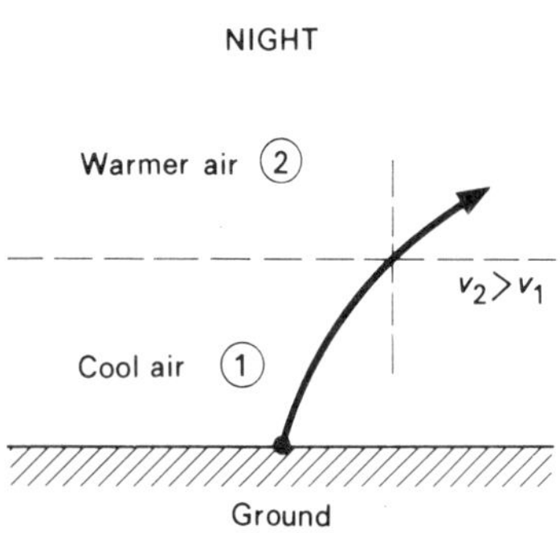


**Figure A3.2**

**3.5** Because the size of the aperture is comparable to the wavelength.

**3.6** A series of maxima and minima will be detected. If the microphone is initially equidistant from the loudspeakers then a maximum is detected because it receives waves in phase. When it is moved to the first minima, the waves are out of phase by half a wavelength. If the distances from the microphone to $L_1$ and $L_2$ are measured then the wavelength can be found by subtracting them. As the frequency of the sound is known, the speed can then be determined from $v = f\lambda$. Further results for the wavelength can be obtained from measurements made on further minima, or maxima.

**3.7** A series of maxima and minima will be detected due to superposition of the two waves. A standing wave pattern will be produced.

**3.8** See the key points.

**3.9** Number of cycles of frequency $f_1$ in time $T$ is $f_1T$. Number of cycles of frequency $f_2$ in time $T$ is $f_2T$. But $f_2T - f_1T = 1$. Thus $f_2 - f_1 = 1/T$ = beat frequency.

**3.10** B can be 446 Hz or 466 Hz.

## Answers III

**3.11** (a) A node at each end and an antinode in the middle. (b) $2L$. This is called the *fundamental wavelength*. (c) $\lambda = L$

**3.12** $\lambda = 2L$ and so $v = 2Lf$. Thus $f = (1/2L)\sqrt{(T/\mu)}$.

**3.13** Plucking the string at the mid-point means an antinode at that point. Hence we can assume that it is the fundamental mode. Thus $\lambda = 2L = 1.0$ m. $v = f\lambda = 1.0f = \sqrt{(T/\mu)} = \sqrt{(3.0 \times 9.8/(1.2 \times 10^{-3}))}$ and so $f = 157$ Hz.

**3.14** The mass per unit length is proportional to $d^2$, thus as $f$ is proportional to $\sqrt{(T/\mu)}$ then $f$ is proportional to $\sqrt{(T/d^2)}$. Hence $T/d^2$ must be constant for constant frequency. Thus $T_2/T_1 = 7.2^2/7.0^2 = 1.058$. Hence the tension must be increased by 5.8%.

**3.15** $v = f\lambda = \sqrt{(T/\mu)}$, hence $\mu = T/f^2\lambda^2$. Natural frequency = driving frequency at resonance, and $\lambda = 2L$, so $\mu = 50/(4 \times 50^2 \times 0.8^2) = 7.8 \times 10^{-3}$ kg m$^{-1}$.

## Answers IV

**3.16** (a) Displacement node is at the closed end and antinode at the open end. (b) Pressure node at the open end and antinode at the fixed end.

**3.17** (a) $\lambda = 2L$ (b) $\lambda = 4L$ These are known as the *fundamental* wavelengths.

**3.18** (a) $\frac{1}{4}\lambda = 30$ cm and so $\lambda = 120$ cm (b) $L = \frac{3}{4}\lambda$ and so $\lambda = 40$ cm (c) $L = \frac{5}{4}\lambda$ and so $\lambda = 24$ cm

**3.19** The length of the air column is adjusted so that resonance occurs when a tuning fork is sounded above the open end of the tube. This is done for the fundamental note only. Then $L + c = \frac{1}{4}\lambda = \frac{1}{4}v/f$. The experiment is repeated for a number of different frequencies and a graph plotted of $L$ against $1/f$. As $L = (v/4)f - c$ then the slope of the graph is $v/4$ and so the velocity can be obtained. The end correction is the intercept with the $L$ axis when $1/f = 0$.

**3.20** $15 + 2c = \lambda$ and $11.5 + c = \frac{3}{4}\lambda$. Thus $c = 0.5$ cm.

**3.21** $\lambda = v/f = 340/512 = 0.664$ m $= 66.4$ cm. Resonance occurs at $\frac{1}{4}\lambda = 16.6$ cm, at $\frac{3}{4}\lambda = 49.8$ cm, at $\frac{5}{4}\lambda = 83.0$ cm. This last one cannot be achieved and so only the fundamental and the first overtone are possible.

## Answers V

**3.22** The pressure wave travels through the medium by pushing molecules back and forth. This means stretching and compressing intermolecular bonds. If the molecules are thought of as balls tethered to each other by springs then pushing the balls back and forth means compressing and extending the springs. Hence the importance of the elasticity of the material which is a measure of the difficulty of 'stretching' the intermolecular bonds.

**3.23** $v = \sqrt{(E/\rho)} = \sqrt{(2.0 \times 10^{11}/(7.8 \times 10^3))} = 5064$ m s$^{-1}$

**3.24** (a) $v = \sqrt{(\gamma p/\rho)} = \sqrt{(\gamma rT)}$, as $p/\rho = rT$.
(b) No. When the pressure changes the density also changes and $p/\rho$ remains constant.
(c) Temperature; humidity, this affecting both $\gamma$ and $\rho$; wind direction.

**3.25** $v \propto \sqrt{T}$ and so $v_{20}/v_0 = \sqrt{(293/273)}$ and $v_{20} = 342.9$ m s$^{-1}$.

**3.26** $v_{18}/v_0 = \sqrt{(291/273)}$ and so $v_{18} = 341.7$ m s$^{-1}$. Hence $\lambda = v/f = 341.7/256 = 1.335$ m. Thus $L = \frac{1}{4}\lambda = 0.334$ m.

---

# Lesson 4 Interference of light

## Aims

After completing this lesson you should be able to

- state the conditions necessary for the interference of light to be shown
- explain the principle of, and carry out calculations involving, Young's double-slit experiment
- explain Lloyd's mirror experiment
- explain the principle of, and carry out calculations involving, interference fringes produced by thin wedge-shaped films, including Newton's rings.

## Prerequisites

Lesson 2 on wave properties. Also required is a basic knowledge of refraction, as in Lesson 2, Chapter 4.

## STUDY

**I Read** about the conditions necessary for the interference of light to be shown, Young's double-slit experiment and Lloyd's mirror.
References: *B* PP ch. 6, *B* ST6 ch. 2, *D* ch. 18, *N* ch. 23, *W et al.* ch. 23(18), *Wh&H*, ch. 38.

### Self-assessment questions

**4.1** What are the conditions necessary for interference effects to be shown? If you use the word 'coherent' explain what it means.

**4.2** Explain the principle of Young's double-slit experiment and state the measurements that have to be made in order to calculate the wavelength of the light used.

**4.3** Figure 4.1 shows the measurements involved in Young's double-slit experiment. $S_1$ and $S_2$ are the slits, and a screen on which the interference fringes are formed is a distance $D$ away.
(a) Why will there be a central bright fringe, i.e. a maximum, at Q?
(b) For the waves from the two slits arriving and superposing at P the path difference between the two wave motions is $p$. Consider the sine of $\theta$ in triangle $S_1S_2N$ and the tangent of the same angle in triangle PQR and, by making the assumption that because the angle $\theta$ is very small the tangent and sine can be equated, arrive at a relationship for $p$ in terms of $d$, $s$ and $D$. Also, you

ought to prove for your own satisfaction that angle $S_2S_1N$ and angle PRQ *are* the same angle $\theta$.
(c) What condition must be met by this path difference $p$ if there is to be a bright fringe at P?
(d) Hence obtain a relationship for the distance between bright fringes.
(e) What is the condition to be met if there is a dark fringe at P? Hence what is the distance between dark fringes?
(f) Which would give the greater number of fringes per centimetre on the screen, red light or violet light?
(g) What would be seen on the screen if white light were used?

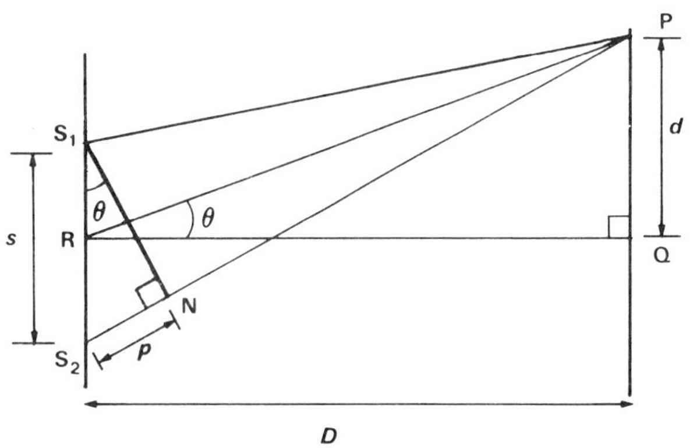


**Figure 4.1**

***4.4** Interference fringes were produced with a Young's double-slit arrangement using light of wavelength $6 \times 10^{-7}$ m. If the separation of the two slits was 0.70 mm and the screen on which the fringes were observed was 1.0 m from the slits, what was the fringe separation?

***4.5** Show that if a layer of transparent material of thickness $d$ and refractive index $n$ is introduced into the path of the light from one of the slits in the Young's double-slit experiment it is equivalent to increasing the path length by $(n - 1)d$. Hint: consider the number of wavelengths in the material, remembering that the wavelength in the material is different to that in air.

***4.6** State and explain what happens to the fringe pattern when a thin sheet of transparent material is used to cover one of the slits in a Young's double-slit experiment.

**4.7** Figure 4.2 shows the arrangement of Lloyd's mirror experiment. Explain how interference fringes are produced on the screen when there is only a single slit and why the central fringe at Q is dark and not bright.

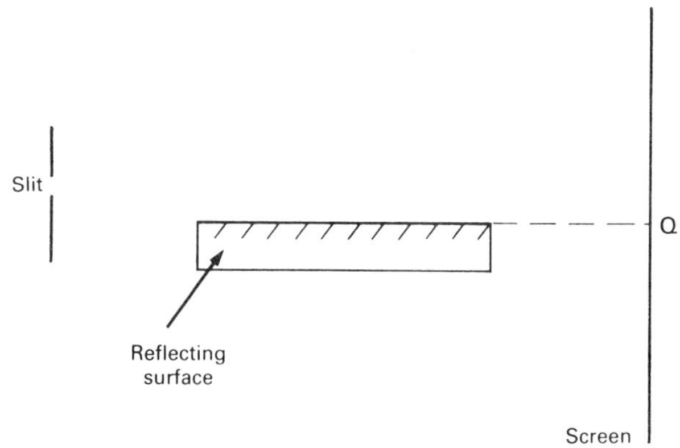


**Figure 4.2**

**II** **Read** about the production of interference fringes by thin wedge-shaped films.
References: *B* PP ch. 6, *B* ST6 ch. 2, *D* ch. 18, *N* ch. 23, *W et al.* ch. 23, *Wh&H* ch. 38.

**Self-assessment questions**

**4.8** Interference fringes can be seen when a thin wedge-shaped film is illuminated by monochromatic light. Explain how the fringes are formed.

**4.9** Figure 4.3 shows a wedge-shaped air film illuminated by monochromatic light of wavelength $\lambda$ at near-normal incidence.

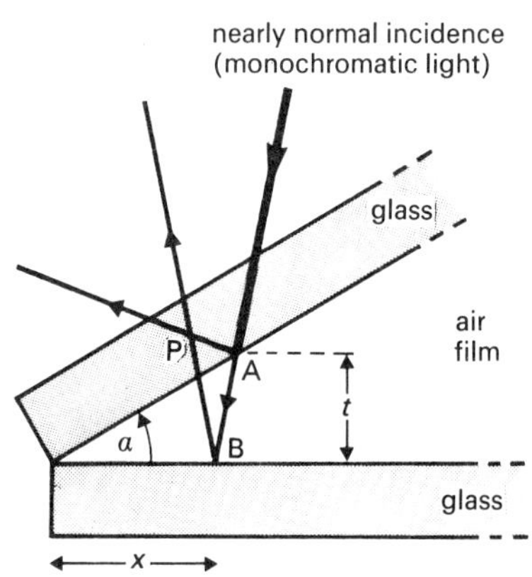


**Figure 4.3**

(a) Interference occurs between the light reflected from A and from B when the two wave trains superpose at P. The physical path difference between these wave trains after separating at A is effectively $2t$, the angle $\alpha$ in the figure having been considerably exaggerated. However the *optical* path difference, i.e. the path difference in terms of the number of waves that can be fitted into it, is $2t + \frac{1}{2}\lambda$. Why the $\frac{1}{2}\lambda$?

(b) Hence arrive at the condition for a bright fringe to be formed at P.
(c) Because of the smallness of $\alpha$ we can make the approximation $\alpha = t/x$. Derive an expression for the fringe separation in terms of $\alpha$ and $\lambda$.

**4.10** With a wedge-shaped air film illuminated by monochromatic light, at the position where there is zero film thickness a dark fringe is observed. Why dark and not bright, in view of there apparently being zero path difference?

***4.11** When light of wavelength $5.9 \times 10^{-7}$ m falls normally on a wedge-shaped air film between two glass plates, interference fringes are formed and 10 fringes are found to occupy a distance of 2.3 mm. What is the angle of the film?

***4.12** Explain the colours seen in soap films.

**III** **Read** about Newton's rings.
References: *B* ST6 ch. 2, *D* ch. 18, *N* ch. 23, *Wh&H* ch. 38.

**Self-assessment questions**

**4.13** Explain how Newton's rings can be observed.

**4.14** Newton's rings are an example of interference with a wedge-shaped air film. For a dark ring, number $m$ from the centre, the radius is $r_m$ and the film thickness is $t_m$. The theorem of intersecting chords gives $(2R - t_m)t_m = r_m^2$, where $R$ is the radius of curvature of the lens. Hence derive a relationship between the radius of the dark fringes and the wavelength of the light used.

**4.15** When Newton's rings are used to determine the wavelength of the light, the square of the diameter of a ring is plotted against its number when counted out from the centre.
(a) How is the wavelength obtained from such a graph?
(b) There is always difficulty in being sure of the number of a ring, since the rings near the centre of the lens are difficult to count. How would this affect the result given by the graph?

***4.16** A lens was placed with a convex surface of radius of curvature 50.0 cm in contact with a plane surface. Newton's rings were observed when the lens was illuminated with monochromatic light. If the radius of the 15th ring was 2.13 mm what was the wavelength?

***4.17** Newton's rings were obtained when a biconvex lens resting on a flat glass plate was illuminated by light of wavelength $5.9 \times 10^{-7}$ m. When the lens was lifted vertically through a small distance, 30 fringes were counted moving through the centre spot. How far was the lens moved?

****4.18** When Newton's rings are being observed with air between the spherical surface of a lens and a plane glass sheet, the diameter of the 10th dark fringe is 12.0 mm. When a liquid is introduced between the spherical surface and the plane surface the diameter of the 10th ring becomes 10.2 mm. What is the refractive index of the liquid?

# SELF-ASSESSMENT

## Key points

The **conditions for interference** of light are that the sources must have roughly the same amplitude, they must be *coherent*, i.e. have a constant phase relationship and the same frequency, and must be unpolarized or have significant plane-polarized components in the same direction.
With Young's double-slit experiment two coherent sources are obtained by division of the wavefront. With wedge-shaped films the sources are obtained by division of the wave amplitude by partial reflection and partial transmission.
For Young's double slits, the fringe separation is $\lambda D/s$, where $s$ is the slit separation and $D$ the distance between the slits and the screen on which the fringes are observed. With wedge-shaped films the fringe separation is $\lambda/2\alpha$, where $\alpha$ is the angle of the wedge.
With Newton's rings the diameter of the $m$th ring equals $\sqrt{(4R\lambda m)}$, where $R$ is the radius of curvature of the surface in contact with the plane surface.

## Answers I

**4.1** See the key points.

**4.2** See Figure A4.1. Light from a single slit is divided into two parts by double slits. These slits are narrow enough for appreciable diffraction to occur. Where the diffracted waves from each slit overlap superposition occurs and interference effects are found. The wavelength of the light can be calculated from the fringe separation = $\lambda D/s$, where $D$ is the screen-to-slit distance and $s$ the separation of the two slits. Thus $D$, $s$ and the fringe separation have to be measured.

**4.3** (a) Because the path length followed by the light from each slit is the same there is no path difference and the waves must arrive in phase.

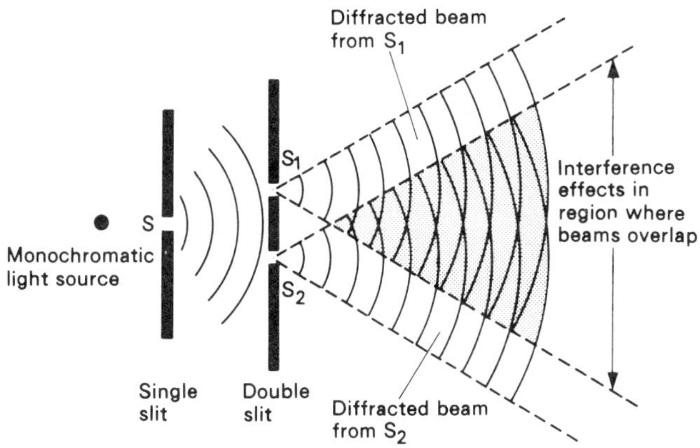


**Figure A4.1 Young's double-slit experiment**

(b) $\sin\theta = p/s$ and $\tan\theta = d/D$. Thus $p = ds/D$.
(c) $p = m\lambda$, where $m$ has the value 0, 1, 2, 3, 4, etc.
(d) $d = \lambda D/s$
(e) $p = (m + \frac{1}{2})\lambda$ and so again $d = \lambda D/s$. The bright and dark fringes show the same spacing.
(f) Red light has a longer wavelength than violet light and thus red light will have a greater fringe separation. Hence there will be more violet fringes per centimetre than red fringes.
(g) A central white fringe because the zero path difference applies to all wavelengths. Either side of this central fringe coloured fringes are found, having a violet edge nearest to the centre and a red edge furthest from it. Fewer fringes are visible than with monochromatic light because after a few fringes on either side of the centre the various coloured fringes overlap to such an extent that no clear fringes can be seen.

**4.4** $d = \lambda D/s = 6 \times 10^{-7} \times 1.0/(0.70 \times 10^{-3}) = 8.57 \times 10^{-4}$ m

**4.5** $n = c/v$, where $c$ is the velocity in a vacuum or, to a reasonable approximation, air, and $v$ the velocity in the material. But $c = f\lambda$ and $v = f\lambda_m$, where $\lambda$ is the wavelength in a vacuum and $\lambda_m$ the wavelength in the material. Thus $n = \lambda/\lambda_m$. In a vacuum the number of waves that would have fitted into a distance $d$ is $d/\lambda$, in the case of the material this is $d/\lambda_m = nd/\lambda$. Thus the path length is greater by $(nd/\lambda - d/\lambda)\lambda = (n - 1)d$.

**4.6** The path length of one of the wavefronts is increased and thus the fringe pattern becomes displaced. For example, the position of zero path difference is no longer central.

**4.7** Interference occurs between the light received on the screen direct from the source and that reflected from the reflector. The situation can be considered to be effectively two-source interference, the two sources being the slit and its image. The central fringe is dark because a phase change of $\frac{1}{2}\lambda$ takes place on reflection.

## Answers II

**4.8** Interference takes place between the light reflected from the upper surface of the film and that reflected from the lower surface of the film, see Figure 4.3 (page 90).

**4.9** (a) Because a phase change of $\frac{1}{2}\lambda$ takes place for the reflection at B, but not at A. (b) $2t + \frac{1}{2}\lambda = m\lambda$ (c) $x = (m - \frac{1}{2})\lambda/2\alpha$ and hence fringe separation $x_{m+1} - x_m = \lambda/2\alpha$.

**4.10** Because there is a phase change of $\frac{1}{2}\lambda$ for the reflection at the lower surface but not at the upper surface of the air film.

**4.11** Using the equation developed in 4.9(c) above, fringe separation $= 2.3 \times 10^{-3}/10 = 5.9 \times 10^{-7}/2\alpha$ and hence $\alpha = 1.28 \times 10^{-3}$ rad.

**4.12** Light is reflected from the front and the rear surfaces of the soap film, as in Figure A4.2. Because there is a path difference between these reflections interference occurs when they are brought together by the eye. The condition for a bright fringe to be produced for a particular wavelength is that $2nt\cos\theta + \frac{1}{2}\lambda = m\lambda$. When $t$ is small, reinforcement at a particular angle may occur for one or two wavelengths while others show destructive interference. The result is that a colour is seen. At different angles different colours will be seen. Very thin soap films appear to be black, i.e. there is no reflected light, because of the phase change that occurs at the upper surface of the film, but not at the lower surface.

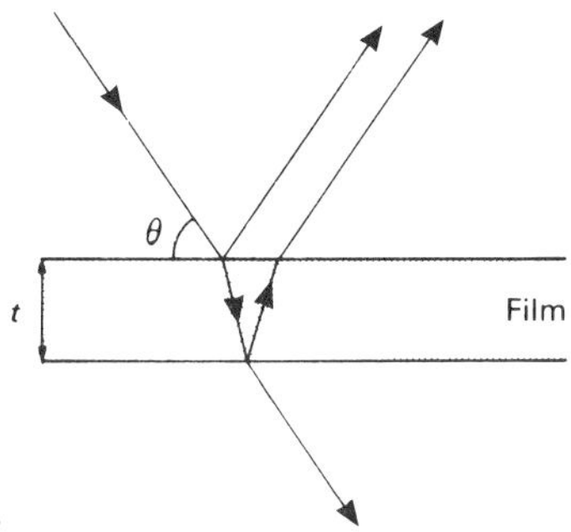


Figure A4.2

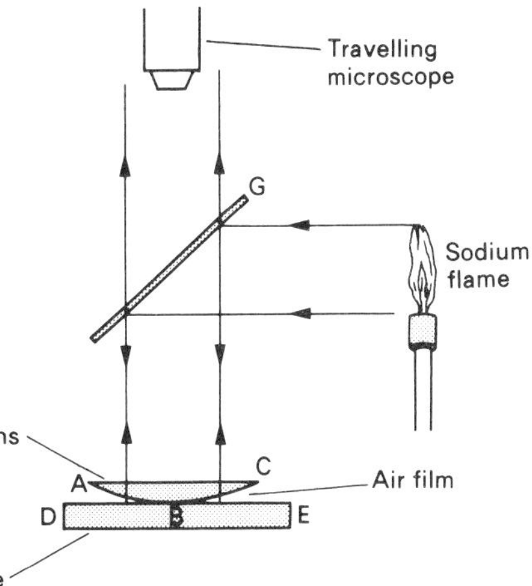


Figure A4.3 Observation of Newton's rings

## Answers III

**4.13** See Figure A4.3.

**4.14** Path difference = $2t_m + \frac{1}{2}\lambda$ and hence for the $m$th dark fringe $2t_m + \frac{1}{2}\lambda = (m + \frac{1}{2})\lambda$ which simplifies to $2t_m = m\lambda$. As $t_m$ is considerably less than $R$, $(2R - t_m)t_m = r_m^2$ simplifies to $r_m^2 = 2Rt_m$. Thus $r_m^2 = Rm\lambda$.

**4.15** (a) Using the relationship derived in 4.14 above, $d_m^2 = 4r_m^2 = (4R\lambda)m$. Thus a graph of $d_m^2$ against $m$ is a straight line through the origin with a slope of $4R\lambda$.
(b) If there is an error in the counting of $m$ the graph still has the same slope but does not pass through the origin. Thus if the observed number is $m'$ when it should be $m$ we have $m = m' + c$, where $c$ is the error. Thus $d_m^2 = 4R\lambda\, m' + 4R\lambda\, c$. When $d^2 = 0$ then $c = m'$ and thus can be obtained from the intercept on the axis.

**4.16** $r_m^2 = Rm\lambda$ and so $\lambda = (2.13 \times 10^{-3})^2/(50.0 \times 10^{-2} \times 15) = 6.05 \times 10^{-7}$ m.

**4.17** Lifting the lens through a distance $h$ increased the path length by $2h$. Thus $2h = 30 \times 5.9 \times 10^{-7} = 1.77 \times 10^{-5}$ m.

**4.18** Using the equation developed in the answer to SAQ 4.14, we have for air $r_m^2 = Rm\lambda$, where $\lambda$ is the wavelength in air. Hence $\lambda = (12.0 \times 10^{-3})^2/4Rm$. When the liquid is introduced we have $\lambda_l = (10.2 \times 10^{-3})^2/4Rm$ where $\lambda_l$ is the wavelength in the liquid. Thus as $n = \lambda/\lambda_l$ (see the answer to SAQ 4.5), $n = (12.0 \times 10^{-3})^2/(10.2 \times 10^{-3})^2 = 1.38$.

# Lesson 5 Diffraction of light

## Aims

After completing this lesson you should be able to

- explain the action of, derive the equation for, and solve problems involving the diffraction grating
- explain diffraction at a single slit, deriving an equation for it
- explain what is meant by *resolving power* of an aperture and apply *Rayleigh's criterion* to optical instruments.

## Prerequisites

Lesson 2 on wave properties and Lesson 4 on the interference of light. For those questions concerned with resolving power, Lesson 5 in Chapter 4 is useful. The grating question involving the simple spectrometer is based on the discussion of that instrument given in Lesson 3 of Chapter 4.

## STUDY

**I Read** about the diffraction grating and the derivation of the equation for it.
References: *B* PP ch. 6, *B* ST6 ch. 2, *D* ch. 18, *N* ch. 24, *W et al.* ch. 23(21), *Wh&H* ch. 39.

**Self-assessment questions**

**5.1** What is a diffraction grating?

**5.2** Each slit in a diffraction grating acts as a source of secondary waves and the resulting effects can be determined by considering the superposition of all these waves. Figure 5.1 illustrates this situation. Explain why the grating only produces wavefronts moving in a few specified directions.

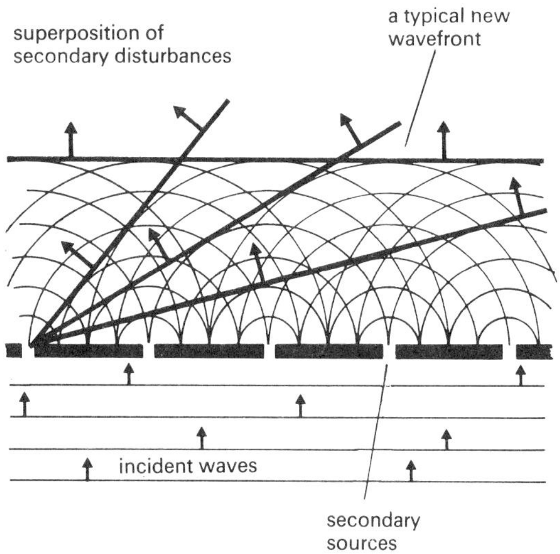


**Figure 5.1**

**5.3** Figure 5.2 shows a diffraction grating with the rays of light all with the same path difference $p$ being brought to a focus by a lens.
(a) What is the path difference $p$ in terms of $\theta$ and $s$?
(b) What is the condition that when these waves superpose at P a maximum will be produced?
(c) Hence obtain the relationship for maximum light intensity in terms of $\theta$.

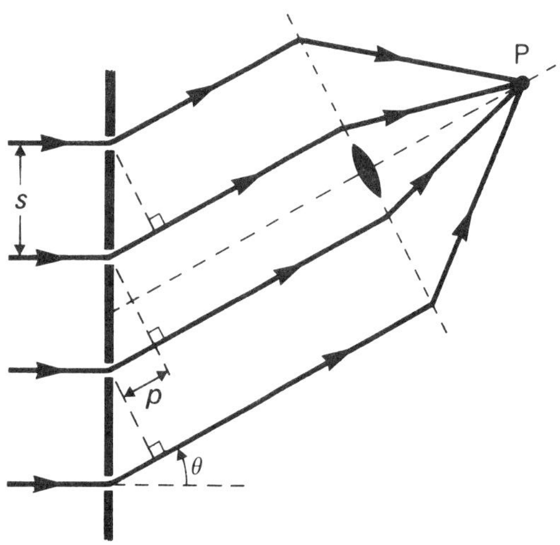


**Figure 5.2**

***5.4** When a source of white light is viewed through a diffraction grating a central white image is seen with a number of spectra on either side of the centre. Explain why (a) there is a central white image, (b) dispersion occurs and (c) a number of spectra are produced.

***5.5** A diffraction grating has 500 lines per millimetre. When used with monochromatic light of wavelength $6 \times 10^{-7}$ m at normal incidence, at what angles will bright diffraction images be observed?

***5.6** Explain how a spectrometer can be used with a diffraction grating to determine the wavelength of the light emitted from a sodium flame.

***5.7** A spectrometer used with a diffraction grating at normal incidence to determine the wavelength of the light emitted by a source gives a telescope setting of $16° 23'$ from the straight-through position for the first-order image. If the grating has 5000 lines per centimetre, what is the wavelength of the light?

***5.8** A lamp emits two wavelengths, $4.20 \times 10^{-7}$ m and $6.00 \times 10^{-7}$ m. What will be the angular separation of these two wavelengths in the third order of a diffraction pattern produced by a diffraction grating having 4000 lines per centimetre, when the light is at normal incidence on the grating?

**II Read** about diffraction at a single slit and the equation for it.
References: *B* PP ch. 6, *B* ST6 ch. 2, *D* ch. 18, *N* ch. 24, *W et al.* (ch. 21), *Wh&H* ch. 39.

**Self-assessment questions**

***5.9** Describe an arrangement by which diffraction at a single slit could be demonstrated.

**5.10** Figure 5.3 shows the situation where a plane wave is incident normally on a single slit of width $a$.
(a) Consider all the light from the slit that is emerging from it at an angle $\theta$. If we divide the slit into half and calculate the path difference between a ray from the top of the top half of the slit, point A, and a ray from the top of the bottom half of the slit, point C, we can state that there will be comparable pairs of points with the same path difference for all points between A and B. What is this path difference, CD, in terms of the slit width and $\theta$?
(b) If this path difference leads to a minimum at this angle, what is the relationship between $\theta$ and $\lambda$?
(c) Sketch a graph showing how the intensity of light detected on a distant screen varies with angle (use $\sin \theta$ since for small angles $\sin \theta$ is approximately $\theta$) and indicate the angles of the various minima.

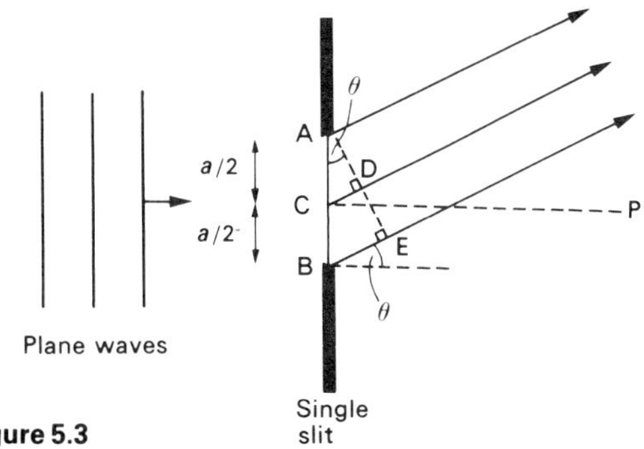


**Figure 5.3**

***5.11** If light of wavelength $6.0 \times 10^{-7}$ m is incident normally on a single slit of width 0.20 mm, what is the angular width of the central maximum? What would be the width of this central maximum on a screen placed 1.2 m from the slit?

**III** **Read** about resolving power, Rayleigh's criterion and the problems of resolution with optical instruments. References: *B* PP ch. 6, *B* ST6 ch. 2, *D* ch. 18, *N* ch. 24, *W et al.* (ch. 21), *Wh&H* ch. 39.

**Self-assessment questions**

**5.12** When you look at an object you are looking at it through an aperture, the pupil of your eye. When you use an optical instrument you are seeing an object by light that has passed through an aperture, the lens of the instrument. When light passes through an aperture diffraction occurs. What effect has this on what is seen?

**5.13** State *Rayleigh's criterion* for resolution.

**5.14** When small angles are involved $\sin\theta$ approximates to $\theta$. Hence explain how the resolving power of a slit is given by $\theta = \lambda/a$, where $a$ is the width of the slit.

***5.15** The resolving power of a circular aperture is given by $\theta = 1.22\lambda/d$, where $d$ is the diameter of the aperture.
(a) What is the resolving power of the eye if the pupil has a diameter of 2 mm and the light has an average wavelength of $5 \times 10^{-7}$ m?
(b) The objective of the Mount Palomar telescope has a diameter of 5 m. What is the resolving power for light of average wavelength $5 \times 10^{-7}$ m?

# SELF-ASSESSMENT

## Key points

An optical transmission **diffraction grating** consists of a large number of fine equidistant parallel lines ruled on a transparent plate to effectively give a series of equidistant parallel slits through which light can pass. For such a grating, when a plane wavefront is incident at normal incidence on it constructive interference only occurs for the light diffracted through certain angles, these angles being given by $s \sin\theta = m\lambda$, with $s$ being the distance between successive slits and $m$ having the value 0, 1, 2, 3, etc. and being known as the **order number**. When a plane wavefront is incident on a single slit of width $a$ then the angle at which the diffracted light gives constructive interference is given by $\sin\theta = m\lambda/a$. Two objects can be said to be **resolved**, when viewed through an aperture, if the maximum intensity of the central diffraction pattern of one object falls on the first minimum of the other. This is known as **Rayleigh's criterion**. The **resolving power**, i.e. the minimum angle that can be resolved, for a slit is $\lambda/a$ and for a circular aperture $1.22\lambda/D$, where $D$ is the diameter of the aperture.

## Answers I

**5.1** See the key points.

**5.2** Only in certain specified directions do the secondary wavelets combine to produce a continuous wavefront. If you look at the figure, only for certain directions can tangents be drawn to a number of the secondary wavelets.

**5.3** (a) $p = s \sin\theta$ (b) $p = m\lambda$ (c) $s \sin\theta = m\lambda$

**5.4** (a) Because only for $\sin\theta = 0$ will the equation developed in 5.3(c) be satisfied by all wavelengths.
(b) Dispersion occurs at each order number, other than $m = 0$, because $\sin\theta$ depends on the value of $\lambda$.
(c) A number of spectra are produced, one for each order number, other than $m = 0$.

**5.5** $s = 1/\text{number of lines per unit length} = 10^{-3}/500 = 2.0 \times 10^{-6}$ m. Hence, as $s \sin\theta = m\lambda$ we have $\sin\theta = m \times 6 \times 10^{-7}/(2.0 \times 10^{-6}) = 0.3m$. Thus for $m = 1$, $\theta = 17.5°$, for $m = 2$, $\theta = 36.9°$, for $m = 3$, $\theta = 64.2°$. No other orders are possible since the value of $\sin\theta$ cannot exceed 1.

**5.6** The collimator and the telescope are adjusted in the same way as when used with the prism in Chapter 4, SAQ 3.8. The telescope and collimator are then placed at 90° to each other. The grating is then mounted on the table at right angles to a line joining two of the levelling screws, say A and B. The table is rotated until the reflected image of the slit is on the cross-wires, the screws being adjusted to ensure this. The grating is then rotated through 135° so that it is at right angles to the

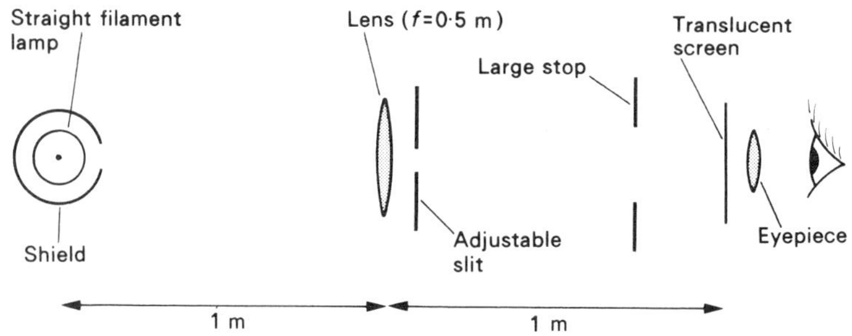


Figure A5.1

light coming from the collimator. The telescope is then rotated to bring a first-order image onto the cross-wires. If the image is not central the levelling screw C may need adjustment. The angle of the telescope is noted. The telescope is then rotated to bring the other first-order image onto the cross-wires and the angle is again noted. The difference between these two angles is $2\theta$. The experiment can be repeated for other orders and the grating equation used to obtain $\lambda$.

**5.7** $s = 10^{-2}/5000$ m and so as $s \sin\theta = 1\lambda$ we have $\lambda = 5.6 \times 10^{-7}$ m.

**5.8** $s = 10^{-2}/4000$ m and so as $s \sin\theta = 3\lambda$ we have for the $4.20 \times 10^{-7}$ m wavelength $\theta = 30.26°$. For the $6.00 \times 10^{-7}$ m wavelength $\theta = 46.05°$ and so the angular separation is 15.79°.

## Answers II

**5.9** See Figure A5.1.

**5.10** (a) $\frac{1}{2}a \sin\theta$ (b) $\frac{1}{2}a \sin\theta = \frac{1}{2}m\lambda$ and so $\sin\theta = m\lambda/a$ (c) See Figure A5.2.

**5.11** $\sin\theta = 1 \times \lambda/a = 6.0 \times 10^{-7}/(0.20 \times 10^{-3}) = 3.0 \times 10^{-3}$. For such a small angle $\sin\theta$ is approximately $\theta$ and thus angular width $= 2 \times 3.0 \times 10^{-3} = 6.0 \times 10^{-3}$ rad. $x/D = \tan\theta$ or approximately $\theta$. Hence $x = 1.2 \times 3.0 \times 10^{-3}$ m and so width $= 2 \times 1.2 \times 3.0 \times 10^{-3} = 7.2 \times 10^{-3}$ m $= 7.2$ mm.

## Answers III

**5.12** Diffraction occurs and a pattern rather like that shown in Figure A5.2 is produced. This means the image is not as sharply defined as might have been expected. Not only is the central area more diffuse but subsidiary maxima appear.

**5.13** See the key points.

**5.14** As derived in the answer to SAQ 5.10(b), $\sin\theta = m\lambda/a$ and thus, as $\sin\theta$ approximates to $\theta$, $\theta = m\lambda/a$. Thus resolving power = minimum angle that can be resolved = angle between centre and first minimum = $\lambda/a$.

**5.15** (a) resolving power $= 1.22 \times 5 \times 10^{-7}/(2 \times 10^{-3}) = 3.1 \times 10^{-4}$ rad (b) resolving power $= 1.22 \times 5 \times 10^{-7}/5 = 1.22 \times 10^{-7}$ rad

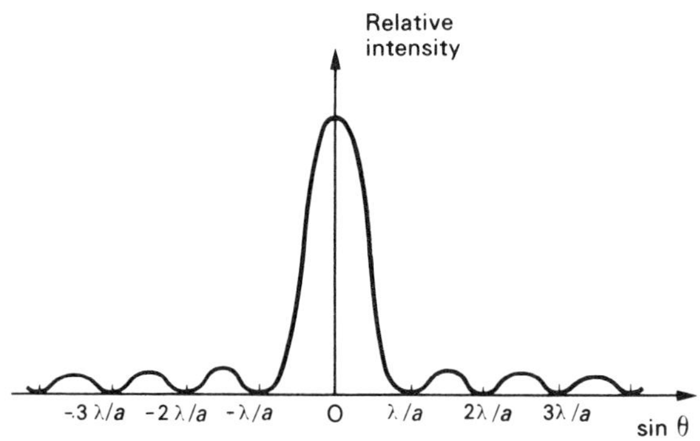


Figure A5.2

# Lesson 6 Polarization

## Aims

After completing this lesson you should be able to

- explain how polarization experiments demonstrate the transverse nature of light waves
- explain what is meant by *plane-polarized* light
- explain how plane-polarized light can be produced by (a) reflection, (b) double refraction, (c) selective absorption
- explain the applications of polarized light in sunglasses, photoelasticity and optical rotation.

## Prerequisites

Lesson 1, describing waves, and Lessons 1 and 2 of Chapter 4 on reflection and refraction.

# STUDY

**I Read** about the polarization of light and how experiments give evidence for the transverse nature of light waves, also how plane-polarized light can be produced.
References: *B* PP ch. 6, *B* ST6 ch. 2, *D* ch. 18, *N* ch. 25, *W et al.* (ch. 18), *Wh&H* ch. 40.

### Self-assessment questions

**6.1** How can two pieces of Polaroid be used to demonstrate the transverse nature of light?

**6.2** What is meant by *plane-polarized* light?

**6.3** When unpolarized light is reflected from glass or water or certain other materials, the reflected ray is found to be partially plane-polarized. When the reflected ray and the refracted ray are at right angles to each other the reflected ray is completely plane-polarized, while the refracted ray is partially plane-polarized in a plane at right angles to the plane of polarization of the reflected ray. Derive a relationship between the angle of incidence for this condition, known as the *Brewster angle*, and the refractive index.

***6.4** What is the angle of incidence for light to be completely plane-polarized on reflection from a surface having a refractive index of 1.5?

**6.5** Many crystals show *double refraction*, an unpolarized incident ray being split up on refraction into two rays. One ray is called the *ordinary ray*, because it obeys the laws of refraction, and the other is called the *extraordinary ray*, because it does not obey the laws of refraction. The two rays are plane-polarized in planes at right angles to each other. Describe how this effect is used in the Nicol prism to produce a beam of plane-polarized light.

**6.6** Polaroid is made from tiny crystals of quinine iodosulphate, all lined up in the same direction in a sheet of nitrocellulose. When unpolarized light is incident on such crystals the vibrations in one plane are absorbed while no absorption occurs in the plane at right angles. Explain what happens when a single sheet of Polaroid is used to view unpolarized light and the sheet is rotated, and what happens when a sheet of Polaroid is used to view plane-polarized light and the sheet is rotated.

**II Read** about the applications of polarized light in sunglasses, photoelasticity and optical rotation.
References: *B* PP ch. 12, *B* ST6 ch. 2, *D* ch. 18, *N* ch. 25, *Wh&H* ch. 40.

### Self-assessment questions

***6.7** Explain how Polaroid is used in sunglasses to reduce reflected light, i.e. glare, and also to reduce the brightness of the sky.

**6.8** In photoelasticity, when Perspex, or other plastic, models are viewed between crossed Polaroids in white light, coloured fringes are seen round the regions of highest stress. Why can only certain materials be used: what property must they have?

**6.9** Certain crystals have the property of rotating the plane of polarization when plane-polarized light is passed through them. They are said to be *optically active*. Some liquids also show the same effect, the molecules in the liquid being responsible for the rotation. For such liquids, the angle rotated depends on the concentration and the length of the liquid through which the light passes. Explain how optical rotation is used to determine the concentration of sugar solutions.

****6.10** Two sheets of Polaroid are lined up so that their polarization directions are initially parallel. When one of the sheets is rotated, how does the transmitted light intensity vary with the angle between the polarization directions of the Polaroids? Through what angle must the Polaroid be rotated to reduce the light intensity by 50%?

# SELF-ASSESSMENT

## Key points

Light is a transverse wave motion. Light is **unpolarized** if the plane of vibration is continually and randomly changing. **Plane-polarized** light contains only one, fixed, plane of vibration. The plane of vibration for an electromagnetic wave is the plane that contains the electric field direction.
When light is reflected from certain materials the reflected light is partially plane-polarized, though at a particular angle it is completely plane-polarized. This angle is when the reflected and refracted rays are at right angles to each other.
Some materials show **double refraction**, an incident unpolarized ray being split into two refracted beams, each plane-polarized. Some materials exhibit **selective absorption**, preferentially absorbing one plane of vibration.

## Answers I

**6.1** When light is viewed through just one piece of Polaroid, though there is some general reduction in intensity there is no apparent effect on rotating the sheet. However, if the light is viewed through two pieces of Polaroid and one is rotated relative to the other, the light intensity transmitted varies and is virtually zero at one particular orientation. These effects can only be explained on the basis of light being a transverse wave motion. With a transverse wave motion more than one plane of vibration is possible, this not being the case with a longitudinal wave. Thus if Polaroid only transmits in one plane then the first Polaroid isolates one plane of vibration and the second only transmits the component of that vibration which is parallel to its plane of transmission.

**6.2** See the key points.

**6.3** $i + r = 90°$, hence $n = \sin i/\sin r = \sin i/\sin(90 - i) = \sin i/\cos i = \tan i$.

**6.4** $\tan i = 1.5$ and so $i = 56°$.

**6.5** See Figure A6.1. The refractive index of the calcite for the ordinary ray is 1.66 and for the extraordinary ray 1.49. The Canada balsam has a refractive index of about 1.55 for both rays. If the angles of the crystal are cut carefully the light hits the Canada balsam layer at such an angle that the ordinary ray is totally internally reflected and the extraordinary ray transmitted.

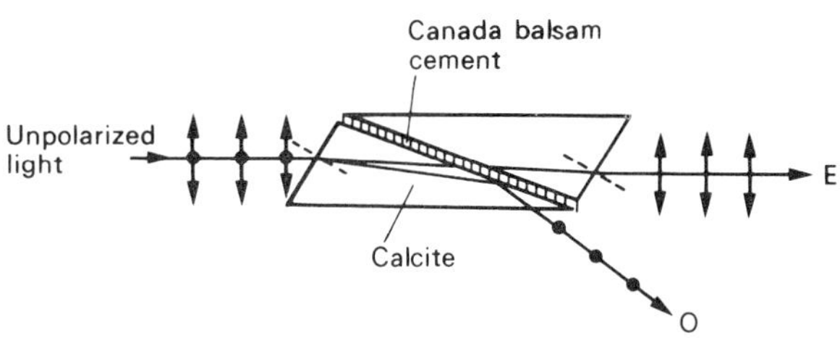


**Figure A6.1 Nicol prism**

**6.6** When the single sheet is used to view unpolarized light and rotated, the plane transmitted by the Polaroid changes according to the angle of the sheet of Polaroid. However, because the incident light contains all planes in equal amounts, the intensity is the same. If the sheet of Polaroid is used to view plane-polarized light this is not the case. Only in one direction, when the transmission plane of the Polaroid is parallel to the plane of polarization of the light, will the light be transmitted unaffected. At all other angles only the component of the light parallel to the transmission plane of the Polaroid will be transmitted. The light intensity will thus decrease, becoming zero when the plane of transmission of the Polaroid is at right angles to the plane of the light.

## Answers II

**6.7** Light reflected from a surface such as that of the sea is partially plane-polarized. Thus Polaroid sunglasses can reduce this reflected light by selective absorption of the light in a particular plane. Light from a blue sky is also partially plane-polarized and thus can be reduced by Polaroid sunglasses. See *B* PP ch. 12 for an explanation of polarization by scattering from particles in the sky, also ch. 8 for another aspect of such scattering.

**6.8** They must be double-refracting when stressed.

**6.9** The instrument used is called a saccharimeter, see *N* ch. 25. The basic principle of the equipment is to direct monochromatic light through a Nicol prism to give plane-polarized light. This is then passed through a tube, usually 100 mm long, containing the solution. A second Nicol prism is then used to view the transmitted light. With no solution present the second Nicol prism is rotated to find the angle for no light transmission. Then with the solution in the tube the new angle for no transmission is found. The difference between the two angles is the angle by which the liquid has rotated the plane of polarization. Tables can be used, or calibration experiments carried out, to determine how this angle relates to the concentration.

**6.10** If $\theta$ is the angle between the transmission planes of the Polaroid sheets then the transmitted component is $I\cos\theta$. To reduce the light by 50% means that $\cos\theta$ must have a value of 0.5 and so $\theta = 60°$

# Lesson 7 Electromagnetic waves

## Aims

After completing this lesson you should be able to

- describe the electromagnetic spectrum, indicating approximate wavelengths and frequencies for the various types of wave
- solve problems involving the interference, diffraction and polarization of electromagnetic waves
- describe how the speed of light can be determined.

## Prerequisites

Lessons 1, 2, 4, 5 and 6 in this chapter, i.e. all the lessons concerned with basic wave properties and the interference, diffraction and polarization of light.

## STUDY

**I Read** about the electromagnetic spectrum and the general properties of electromagnetic waves.
References: *B* PP ch. 12, *B* ST6 ch. 3, *D* ch. 18, *N* ch. 21, *W et al.* (ch. 18), *Wh&H* ch. 37.

### Self-assessment questions

**7.1** Sketch a diagram showing the types of wave that constitute the electromagnetic spectrum and some indication of their wavelengths in a vacuum.

**7.2** Calculate the frequencies of the following.
(a) Radio waves with a wavelength of 1.0 m.
(b) Infra-red waves with a wavelength of $2 \times 10^{-5}$ m.
(c) X-rays with a wavelength of $1 \times 10^{-10}$ m.
(Velocity of the radiation = $3 \times 10^8$ m s$^{-1}$)

***7.3** A microwave transmitter and receiver are lined up and a strong signal is detected by the receiver. A grid of parallel wires is then placed between the transmitter and receiver. When this is rotated, about a direction which always maintains the grid at right angles to the transmitter-to-receiver axis, the signal detected varies and falls to zero at a particular angle. Explain these results.

***7.4** A light meter is placed a distance *d* from a small light source. When the distance is increased to 2*d* the light intensity detected falls to one quarter of its original value. Give a reason for this and state what reading you would expect when the distance is increased to 3*d*.

***7.5** A source of radio waves and a receiving dipole aerial are lined up. Along the same line and beyond the receiver a vertical metal rod is placed, parallel to the dipole. When the receiving dipole is moved along this line the detected signal rises and falls. Why?
One form of receiving aerial consists of a dipole placed $\frac{1}{4}\lambda$ in front of a metal rod. Explain the purpose of the metal rod and why the distance is $\frac{1}{4}\lambda$.

***7.6** Figure 7.1 shows a microwave transmitter ($\lambda$ = 3 cm) in front of an aperture with a probe for detecting the microwaves on the other side of the aperture. Explain what will be detected by the probe as the aperture is gradually made smaller.

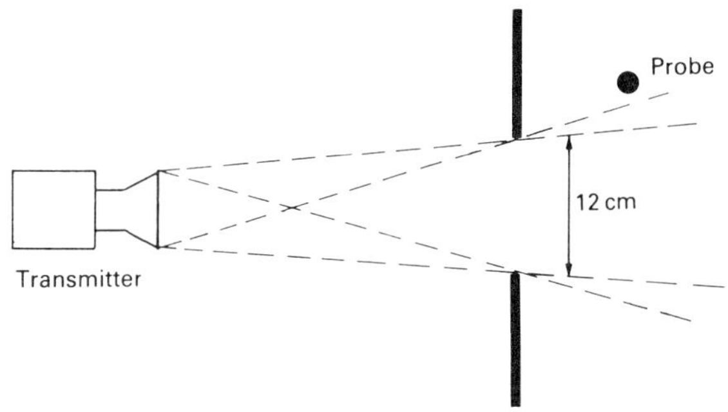


**Figure 7.1**

**II Read** about the measurement of the speed of light.
References: *B* PP ch. 12, *B* ST6 ch. 3, *D* ch. 18, *N* ch. 22, *W et al.* (ch. 18), *Wh&H* ch. 37.

### Self-assessment questions

**7.7** The first terrestrial method for the determination of the speed of light was by Fizeau in 1849. Essentially his method consisted of sending a pulse of light to a distant mirror and back again, then determining the time taken for the round trip.
(a) Fizeau used a mirror 8.6 km away. What time would he need to determine? Speed of light = $3.0 \times 10^8$ m s$^{-1}$.
(b) A rotating toothed wheel was used to produce brief pulses of light. The light passed through a gap between teeth in the wheel, passed to the distant mirror and back again to the toothed wheel. If the wheel had rotated sufficiently the pulse would arrive back in time to hit the next tooth on the wheel. When this happened no light passed through the wheel to the observer. Fizeau used a wheel with 720 teeth and found that when the wheel revolved at 12.6 revolutions per second no light passed through to the observer. What value do these results give for the speed of light?

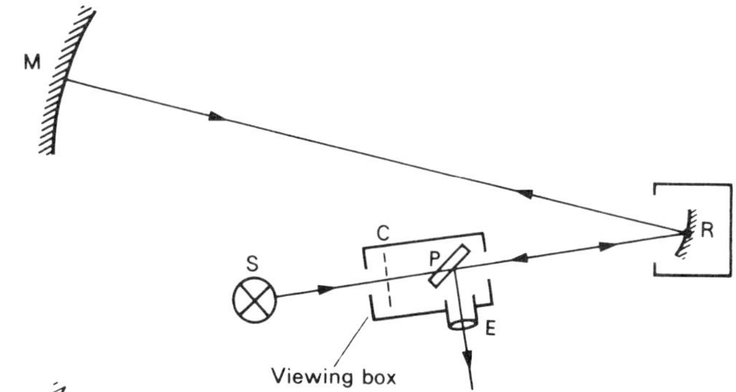


Figure 7.2

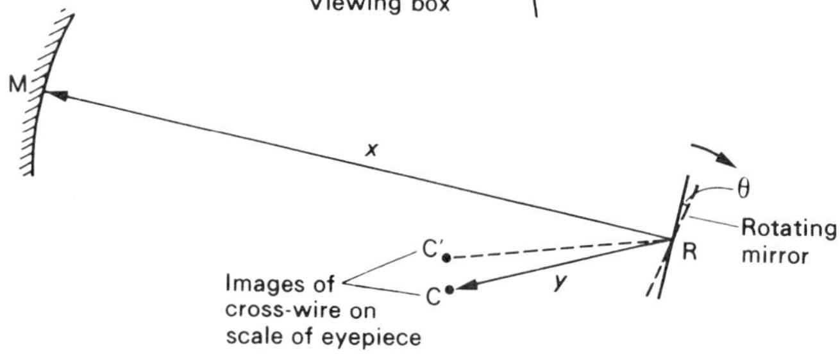


Figure 7.3

**7.8** Figure 7.2 shows the basic principles of a version of the Foucault method of measuring the speed of light. Light from the source S illuminates cross-wires at C and then travels on through a glass plate P to the rotating mirror R. This reflects the light to a distant mirror M. This mirror reflects the light back again to R and then to the plate P. Here the light is reflected to the eyepiece E where an image of C is formed on the eyepiece scale. When R is rotated at high speed the image of C in the eyepiece is displaced by a very small distance to C′.

(a) Figure 7.3 shows a simplified diagram of the appropriate parts of the ray paths for the light. If the mirror R makes $n$ revolutions per second what is the time taken for it to rotate through an angle $\theta$?

(b) If during the time taken for the mirror to rotate through $\theta$ the light travels to the mirror M and back again, a distance of $2x$, how is the speed of light related to $\theta$?

(c) The reflected ray from the mirror R is deflected through $2\theta$ when the mirror turns through $\theta$. How then is the distance CC′ related to $\theta$?

(d) Hence arrive at a relationship for the speed of light involving $n$, $x$, $y$ and $d$.

***7.9** In the Foucault method for the determination of the speed of light a plane mirror rotating at 70 revolutions per second is used to reflect a beam of light to a mirror 100 m away. This stationary mirror reflects the light back to the rotating mirror. If during the time taken to traverse this path the mirror has rotated through 1.0 second of arc, what is the speed of light?

# SELF-ASSESSMENT

## Key points

Electromagnetic waves are transverse waves which all have the same speed in a vacuum, $3 \times 10^8\ \mathrm{m\,s^{-1}}$, and for which the disturbance is a variation with time of the magnitude of the electric and magnetic fields at a point.

## Answers I

**7.1** See Figure A7.1.

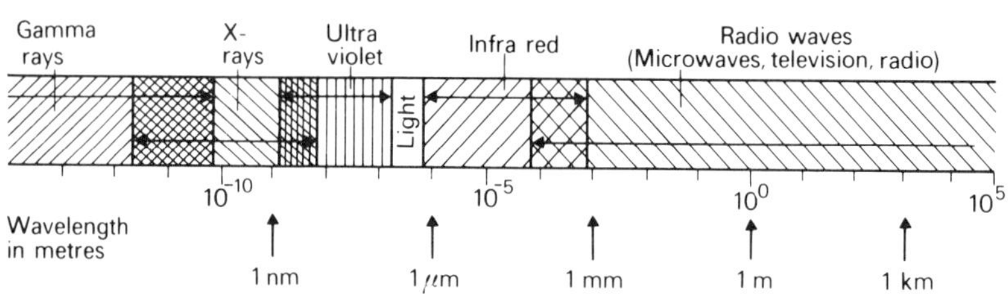


Figure A7.1

**7.2** (a) $f = v/\lambda = 3 \times 10^8/1.0 = 3 \times 10^8$ Hz (b) $f = v/\lambda = 3 \times 10^8/(2 \times 10^{-5}) = 1.5 \times 10^{13}$ Hz (c) $f = v/\lambda = 3 \times 10^8/(1 \times 10^{-10}) = 3 \times 10^{18}$ Hz

**7.3** This indicates that radio waves are transverse waves, the grid of wires acting like a sheet of Polaroid, see SAQ 6.6. The transmitter is emitting plane-polarized waves.

**7.4** Intensity $\propto 1/(\text{distance})^2$, hence at $3d$ an intensity 1/9th of that at distance $d$ would be expected.

**7.5** Standing waves are set up (see Lesson 2). The purpose of the metal rod is to reflect the radio waves and so give a bigger signal at the receiving dipole. The $\frac{1}{4}\lambda$ is because there is a phase change of $\frac{1}{2}\lambda$ on reflection and so this is the position of the first maximum in the standing wave pattern.

**7.6** As the aperture is made smaller more diffraction occurs and so a greater percentage of the signal passing through the aperture can be detected by the probe.

## Answers II

**7.7** (a) Time $= 2 \times 8.6 \times 10^3/(3.0 \times 10^8) = 5.8 \times 10^{-5}$ s
(b) Time from a gap to a tooth $= 1/(2 \times 720 \times 12.6) = 5.51 \times 10^{-5}$ s. Hence speed $= 2 \times 8.6 \times 10^3/(5.51 \times 10^{-5}) = 3.12 \times 10^8$ m s$^{-1}$.

**7.8** (a) $\theta/2\pi n$ (b) $c = 2x/(\theta/2\pi n) = 4\pi nx/\theta$
(c) arc CC′ = radius × angle subtended = $2\theta y$
(d) $c = 8\pi nxy/d$, where $d$ is the distance CC′ which for such a small angle is reasonable approximation to the arc.

**7.9** Using the equation developed in 7.8(b) above, $c = 4\pi nx/\theta = 4\pi \times 70 \times 100/(1/60 \times 60) = 3.10 \times 10^8$ m s$^{-1}$.

## CHAPTER 6

# Electromagnetic fields

## Lesson 1 Electric forces and fields

### Aims

After completing this lesson you should be able to

- state how the force between two particles due to their charges depends on the magnitude of the charges and their distance apart, i.e. *Coulomb's law*
- explain the terms *permittivity* and *permittivity of free space*
- solve problems involving Coulomb's law and two or more charged particles
- explain what is meant by an electric field and its *strength*
- describe the general shape of the fields produced by point charges and charged parallel plates
- derive and use the equation for the field strength due to a point charge
- calculate the field strength due to a number of point charges and also that due to a charged sphere.

### Prerequisites

A basic knowledge of electrostatics, the principle that like charges repel and unlike charges attract, and electrostatic induction. Lesson 1, Chapter 1, on the resolution of forces is also necessary.

## STUDY

**I Revise** the basic principles of electrostatics and then read about Coulomb's law.
References: *B* PP ch. 2, *B* ST5 ch. 2, *D* ch. 11, *N* ch. 28, *W et al.* ch. 35(30), *Wh&H* ch. 42.

**Self-assessment questions**

**1.1** If you rub a strip of polythene with a cloth the polythene will pick up small pieces of paper.
(a) What is happening when the polythene is being rubbed with the cloth?
(b) Explain how the polythene is able to pick up the pieces of paper.

**1.2** State *Coulomb's law.*

**1.3** Explain the terms *permittivity* and *permittivity of free space.*

**1.4** Calculate the force between two charged particles in a vacuum if each carries a charge of $+2.0 \times 10^{-6}$ C and they are 50 mm apart. Take the permittivity of free space to be $8.85 \times 10^{-12}\ \mathrm{C^2\,N^{-1}\,m^{-2}}$ or use the fact that $1/4\pi\epsilon_0$ is about $9.0 \times 10^9\ \mathrm{N\,m^2\,C^{-2}}$.

***1.5** Two point charges A and B are situated 60 mm apart. If A has a charge of $+q$ and B a charge of $-2q$, where should a point charge of $-q$ be placed so that it experiences no resultant electrostatic force?

****1.6** Equal charges of $+q$ are placed at the corners A, B and C of a square ABCD having sides of length $L$. What is the force acting on a charge of $-q$ placed at the corner D if the arrangement is in a vacuum?

**II Read** about the concept of an electric field, the definition of field strength, lines of force and the field line patterns for a point charge and a pair of charged parallel plates.
References: *B* PP ch. 11, *B* ST5 ch. 2, *D* ch. 11, *N* ch. 28, *W et al.* ch. 33(30), *Wh&H* ch. 43.

**Self-assessment questions**

**1.7** Define *electric field strength* and state its units. How is the direction of the field strength specified?

**1.8** What is presented by a line of force (or electric field line)?

**1.9** Sketch the electric field line patterns for (a) a point positive charge, (b) a point negative charge, (c) a pair of parallel conducting plates when one plate has a positive charge and the other an equal negative charge.

**III Read** about the derivation of the field strength a distance $r$ from a point charge using Coulomb's law, the solution of problems using the relationship, and the calculation of the field strength due to a charged conducting sphere.

References: *B* PP ch. 11, *B* ST5 ch. 2, *D* ch. 11, *N* ch. 28, *W et al.* ch. 35(30), *Wh&H* ch. 43.

**Self-assessment questions**

**1.10** (a) What is the force acting on a test point charge $+Q_0$ placed a distance $r$ from a point charge $Q$?
(b) What is the force per unit charge at the test charge, i.e. the electric field strength a distance $r$ from a point charge $Q$?

**1.11** If the distance from a point charge is doubled how is the field strength changed?

**1.12** A point charge of $+2.0 \times 10^{-7}$ C is situated 40 mm from a point charge of $-3.0 \times 10^{-7}$ C in a vacuum. What is the electric field strength midway between the two charges? ($\epsilon_0 = 8.85 \times 10^{-12}\,\mathrm{C^2\,N^{-1}\,m^{-2}}$)

***1.13** Equal point charges of $+q$ are placed at the corners A, B and C of a square ABCD having sides $L$. At corner D there is a point charge of $-2q$. What is the electric field strength at the centre of the square?

***1.14** A point charge of $+q$ is placed 20 mm from another point charge of $-2q$ in a vacuum. At what position is the electric field strength zero?

**1.15** The electric field strength outside an isolated charged conducting sphere is exactly the same as if the entire charge were concentrated as a point charge at the centre of the sphere. Thus for a charge $Q$ on the sphere, of radius $a$, how does the field strength vary with distance from the sphere centre for distances greater than $a$?

**1.16** Inside a hollow charged conducting sphere, or inside a solid conducting sphere with all the charge located at the surface, there is no electric field. With hollow conductors of any shape the field inside is zero. Coaxial cables have an inner wire surrounded by an outer network of meshed wires, insulated from the inner wire. Why the outer mesh?

***1.17** Sketch a graph illustrating the variation of electric field strength with distance from the centre of a uniformly charged solid metal sphere.

***1.18** What is the electric field strength at the surface of a charged metal sphere of radius 100 mm if it carries a charge of $+2.0 \times 10^{-7}$ C in a vacuum? ($\epsilon_0 = 8.85 \times 10^{-12}\,\mathrm{C^2\,N^{-1}\,m^{-2}}$)

**1.19** (a) What is the electric field strength at the surface of a sphere of radius $a$?
(b) If we define charge density $\sigma$ to be charge per unit area of surface, what is the relationship between the electric field at the surface and the charge density?

# SELF-ASSESSMENT

## Key points

**Electrostatic induction** is when a charged body A brought near to an uncharged body B causes a redistribution of charge within B so that the part of it nearest to A has a net charge opposite to the charge on A and is therefore attracted to it.

**Coulomb's law:** the force $F$ between two point charges $Q_1$ and $Q_2$ is directly proportional to the product of their charges and inversely proportional to the square of their distance $r$ apart:

$$F = \frac{Q_1 Q_2}{4\pi\epsilon r^2}$$

where $\epsilon$ is the **permittivity** of the medium in which the charges are located. The permittivity of a vacuum is called the **permittivity of free space** ($\epsilon_0$).

**Electric field strength** $E = F/Q$, where $F$ is the force experienced by a test charge of $+Q$ placed at the point concerned, the unit of field strength being $\mathrm{N\,C^{-1}}$ or $\mathrm{V\,m^{-1}}$. An electric field can be described pictorially by **lines of force** (or **field lines**), the tangent to a line of force at a point on the line giving the direction of the field at that point. Point charges and charged spheres have radial lines of force. The field strength a distance $r$ from a point charge $Q$ is

$$E = \frac{Q}{4\pi\epsilon r^2}$$

For a charged sphere the field at a distance $r$, greater than the radius of the sphere, is the same as if the entire charge were concentrated at the centre of the sphere. The field strength at the surface of a charged sphere is therefore

$$E = \frac{Q}{4\pi\epsilon a^2} = \frac{\sigma}{\epsilon}$$

where $a$ is the radius, $Q$ the charge, and $\sigma$ the charge density, i.e. charge per unit area.

Within *any* closed conductor there is always zero electric field.

## Answers I

**1.1** A movement of charge occurs between the two bodies and one becomes positively charged and the

other (the polythene) negatively charged: electrons are moving from the cloth to the polythene.
(b) Electrostatic induction occurs. The paper has at all times no net charge but the presence of the negatively charged polythene causes a redistribution of charge within the paper such that the end nearest the polythene has a net positive charge, and so is attracted to the polythene, and the end furthest from the polythene has a net negative charge.

**1.2** See the key points.

**1.3** See the key points.

**1.4** $F = Q_1Q_2/4\pi\epsilon r^2 = 2.0 \times 10^{-6} \times 2.0 \times 10^{-6}/(4\pi \times 8.85 \times 10^{-12} \times 0.050^2) = 14.4\,\text{N}$

**1.5** The force on the point charge due to A is $-q^2/(4\pi\epsilon r^2)$, where $r$ is the distance from the point charge to A. The force due to B is $+2q^2/(4\pi\epsilon R^2)$, where $R$ is the distance from the point charge to B. For no resultant force the sum of these forces must be zero. This can only happen if the forces are both along the same line. Thus $-(1/r^2) + (2/R^2) = 0$ and so $R^2 = 2r^2$. The position of the point charge must be along the line through A and B. If you doubt this, draw diagrams showing the directions of the forces on the charge at non-axial points; at all such points there is a resultant force. So $R = r + 60$ and hence $(r + 60)^2 = 2r^2$ and $r = 146\,\text{mm}$.

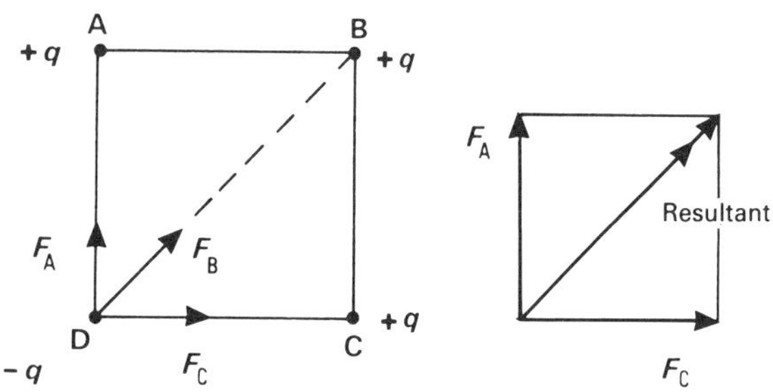


**Figure A1.1**

**1.6** Figure A1.1 shows the arrangement. Distance AD $= L$, CD $= L$ and BD $= L\sqrt{2}$ (using Pythagoras). Force due to the charge at A $= F_A = -q^2/(4\pi\epsilon_0 L^2)$, force due to charge at C $= F_C = -q^2/(4\pi\epsilon_0 L^2)$. The resultant of these two forces is $\sqrt{(F_A{}^2 + F_C{}^2)} = -q^2\,\sqrt{2}/(4\pi\epsilon_0 L^2)$. This is in the same direction as the force due to the charge at B, $F_B = -q^2/(4\pi\epsilon_0 2L^2)$. Hence the force on the charge $= -1.9\,q^2/(4\pi\epsilon_0 L^2)$.

## Answers II

**1.7** See the key points. The direction of the field at a point is the direction of the force that would act on a positive charge placed at that point.

**1.8** The tangent to a line of force at a point gives the direction of the field at that point. Note that lines of force can never cross, as this would imply that at the cross-over point the field had simultaneously two directions.

**1.9** See Figure A1.2.

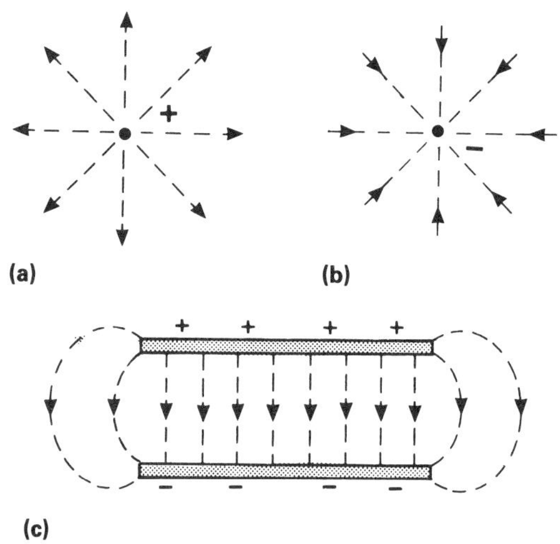


**Figure A1.2**

## Answers III

**1.10** (a) Using Coulomb's law, $F = QQ_0/(4\pi\epsilon r^2)$.
(b) $E = F/Q_0 = Q/(4\pi\epsilon r^2)$. The electric field follows an inverse-square law for distance.

**1.11** Reduced to a quarter of its initial value.

**1.12** $E = 2.0 \times 10^{-7}/(4\pi \times 8.85 \times 10^{-12} \times 0.020^2) - 3.0 \times 10^{-7}/(4\pi \times 8.85 \times 10^{-12} \times 0.020^2) = -2.25 \times 10^6\,\text{N}\,\text{C}^{-1}$. The direction of the field is towards the negative charge.

**1.13** The fields due to the opposite corners A and C will cancel out, thus only the fields due to B and D have to be considered. Both these will produce a field at the centre in the same direction, towards D. The resultant electric field is thus $E = q/(4\pi\epsilon\,\frac{1}{2}L^2) + 2q/(4\pi\epsilon\,\frac{1}{2}L^2) = 6q/(4\pi\epsilon L^2)$.

**1.14** Zero field strength is when $q/(4\pi\epsilon_0 r^2) - 2q/(4\pi\epsilon_0 R^2) = 0$, i.e. when $R^2 = 2r^2$. But $R = r + 20$, hence $r = 48\,\text{mm}$. The point is 48 mm from the $+q$ charge and 72 mm from the $-2q$ charge, along the line through the two charges (see the answer to SAQ 1.5).

**1.15** $E = Q/(4\pi\epsilon r^2)$ when $r \geqslant a$.

**1.16** To stop external fields having any effect on the current in the central wire.

**1.17** See Figure A1.3.

**1.18** $E = 2.0 \times 10^{-7}/(4\pi \times 8.85 \times 10^{-12} \times 0.100^2) = 1.8 \times 10^5\,\mathrm{N\,C^{-1}}$

**1.19** (a) $E = Q/4\pi\epsilon a^2$ (b) $\sigma = Q/4\pi a^2$ and so $E = \sigma/\epsilon$. This is a general relationship that applies to other surfaces as well as spheres. For example, a plane surface in the above derivation can be considered to be a surface of infinite radius.

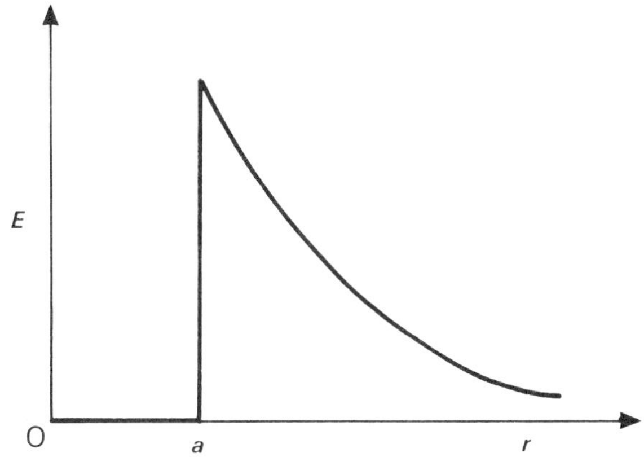


**Figure A1.3** $E \propto r^{-2}$ when $r \geqslant a$, the sphere radius

# Lesson 2 Electric potential

## Aims

After completing this lesson you should be able to

- explain what is meant by electric potential and define the term
- explain what is meant by an *equipotential surface* and sketch the form of such surfaces for point charges, charged spheres and between parallel charged plates
- derive the equation for the electric potential at a point due to an isolated point charge
- derive the relationship between electric field strength and electric potential
- solve problems involving electric potential, electric fields and point charges, charged spheres or charged parallel plates.

## Prerequisites

Lesson 1 on electric forces and fields, also Lesson 5 in Chapter 1 on work, energy and power. Lesson 10 in Chapter 1, on gravitational fields, could usefully precede this lesson. Calculus is needed for one question.

## STUDY

**I** **Read** about the concept of electric potential, its definition and the form of equipotential surfaces for point charges, charged spheres and between charged parallel plates.

References: *B* ST5 ch. 2, *D* ch. 11, *W et al.* (ch. 30), *Wh&H* ch. 45. *B* PP ch. 11 and *W et al.* ch. 38(29) discuss the concept of potential with regard to gravitational fields.

**Self-assessment questions**

**2.1** Why is work done when a charge is moved in an electric field in such a direction that a component of its motion is in the direction of the field?

**2.2** Consider a uniform electric field of strength $E$. What is the force acting on a charge $Q$ in that field? What is the work done in moving the charge a distance $d$ against the electric field? Energy is conserved, so what happens to the energy transferred in doing the work?

**2.3** Define *electric potential.*

**2.4** What is the work done in moving a point charge of $2.0 \times 10^{-10}\,\mathrm{C}$ through a distance of 10 mm in the opposite direction to an electric field of $50\,\mathrm{N\,C^{-1}}$?

**2.5** What is the potential difference between two points if 6.0 J of work has to be done in moving a charge of 0.5 C between them?

**2.6** Is *potential* a vector or a scalar quantity?

**2.7** What is an *equipotential surface*?

**2.8** Has work to be done to move a charge along an equipotential surface?

**2.9** Why must equipotentials be at right angles to the direction of the electric field?

**2.10** Sketch the form of the equipotentials, and the electric lines of force, for (a) a point charge, (b) a charged conducting sphere, (c) a pair of parallel conducting plates when one plate has a positive charge and the other an equal negative charge.

**II Read** about the derivation of the equation for the electric potential at a point due to an isolated point charge, and about the potential both within and external to a charged conducting sphere.
References: *B* ST5 ch. 2, *D* ch. 11, *N* ch. 28, *Wh&H* ch. 45. *B* PP ch. 11 and *W et al.* ch. 38(29) discuss the similar equation for gravitational fields.

**Self-assessment questions**

**2.11** Consider the situation shown in Figure 2.1, where initially a point charge of $+Q$ is at O and a point charge $+Q_0$ at point C.

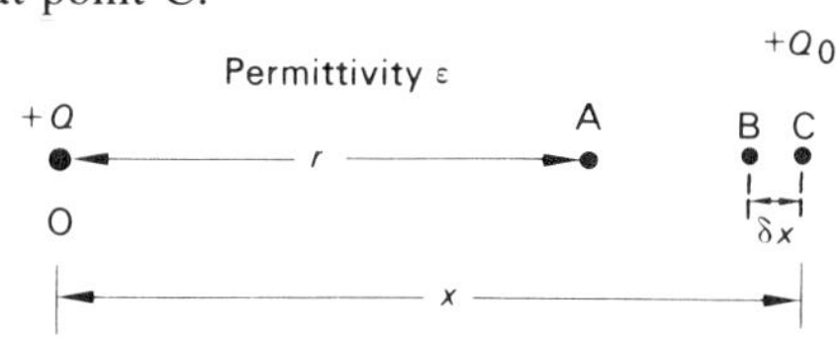


**Figure 2.1**

(a) What is the direction of the force acting on the charge at C?
(b) If the charge $Q_0$ is now moved to point B, through a distance $\delta x$, what is the work done by an external agent against the force $F$ due to the field?
(c) The answer to (b) is $\delta W = -F\delta x$. Why the minus sign?
(d) But $F = QQ_0/(4\pi\epsilon x^2)$, hence determine by integration the work done in fetching the charge $Q_0$ from infinity to the point A.
(e) What is the potential at the point A?

**2.12** Figure 2.2 shows the force–distance graph for a point charge $+Q_0$ at different distances from a point charge $+Q$.
(a) What does the shaded area under the graph represent when the charge $+Q_0$ is moved from C to B through a distance $\delta x$?
(b) How could the work done in fetching the charge $+Q_0$ from infinity to the position A be determined from the graph?
Note: you might like to refer to *B* ST5 ch. 2, where the equation obtained in the answer to SAQ 2.11(d) is derived without the use of calculus by considering the area under a force–distance graph.

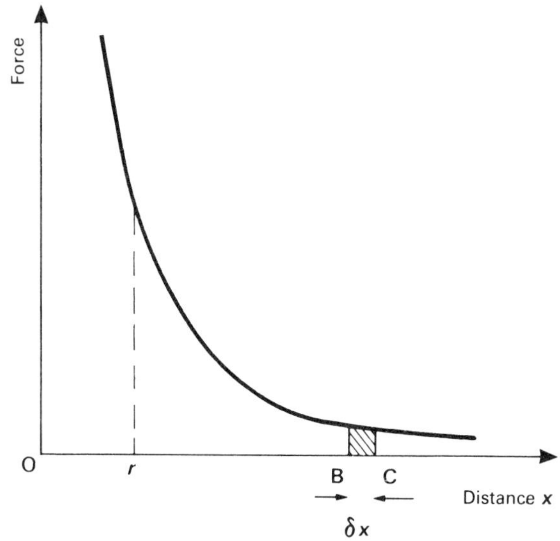


**Figure 2.2**

**2.13** Calculate the potential at a distance of 6.0 mm from a point charge of $+2.0 \times 10^{-9}$ C in a vacuum. ($\epsilon_0 = 8.85 \times 10^{-12}\,C^2\,N^{-1}\,m^{-2}$)

**2.14** If the distance from a point charge is doubled, how does the potential change?

**2.15** As the charge of an isolated conducting sphere can be considered to be effectively located as a point charge at the centre of the sphere for the calculation of field strengths for distances greater than or equal to the sphere radius, so potentials can be similarly calculated. What is the potential at the surface of a sphere of radius 100 mm in a vacuum when the sphere carries a charge of $+3.0 \times 10^{-5}$ C? ($\epsilon_0 = 8.85 \times 10^{-12}\,C^2\,N^{-1}\,m^{-2}$)

**2.16** Inside a charged conducting sphere the electric field strength is zero. This means that no work has to be done in moving charges within the sphere because there are no electric forces acting on them. As a charged sphere has a potential at its surface, what can be said about the potential inside the sphere?

***2.17** Sketch graphs showing how (a) the electric field strength and (b) the potential varies within an isolated charged conducting sphere and external to it.

**III Read** about the derivation of the relationship between electric field strength and electric potential, i.e. $E = -$(potential gradient).
References: *B* PP ch. 11, *B* ST5 ch. 2, *D* ch. 11, *W et al.* ch. 39(29), *Wh&H* ch. 45.

**Self-assessment questions**

**2.18** (a) What is the force acting on a positive charge $Q_0$ when in an electric field of strength $E$?
(b) Suppose the charge is moved a distance $\delta x$ in the opposite direction to that of the field; what is the work done by an external agent to produce that movement?
(c) The answer to (b) is $\delta W = -EQ_0\,\delta x$. If $\delta V$ is the change in potential, i.e. the work done per unit charge in moving it through that distance, write the relationship in terms of $\delta V$ and $E$.
(d) As $\delta x \to 0$ so the answer to (c) becomes $E = -\mathrm{d}V/\mathrm{d}x$. But $\mathrm{d}V/\mathrm{d}x$ is the potential gradient. What therefore can be said about the potential gradient in a uniform electric field?

***2.19** A potential difference of 1000 V is maintained between two parallel conducting plates a distance 10 mm apart. What is the electric field strength between the plates?

***2.20** What is the maximum potential to which a spherical conducting sphere can be raised in air if it has a radius of 0.20 m and electrical breakdown of the air occurs when the electric field strength exceeds $3 \times 10^6\,\mathrm{V\,m^{-1}}$?

# SELF-ASSESSMENT

## Key points

If an external agent does work $W$ in bringing from infinity a test charge $+Q_0$ to a particular point in an electric field, then that point is said to have an **electric potential** equal to $W/Q_0$. The unit of potential is $\mathrm{J\,C^{-1}}$ or volt, these being identical.
The **potential difference** between two points is the work done per unit charge in moving a charge between the points.
All the points in an electric field which are at the same potential can be considered to be lying on a surface, such a surface being known as an **equipotential surface**.
The potential at a distance $r$ from a point charge $Q$ is given by $V = Q/(4\pi\epsilon r)$.
Equipotential surfaces and electric field lines are perpendicular to each other because there is no component of the electric field parallel to an equipotential surface. The **electric field strength** $E = -\mathrm{d}V/\mathrm{d}x$, where $\mathrm{d}V/\mathrm{d}x$ is the potential gradient.

## Answers I

**2.1** In an electric field a force acts on a charge, and so moving the charge in such a direction that a component of its motion is in the direction of the field means work is done. No work is done if the charge is moved at right angles to the electric field because then there is no component of the motion in the direction of the force.

**2.2** $F = EQ$, hence work $= EQd$. The potential energy of the charge changes.

**2.3** See the key points: $V = W/Q$ or alternatively $V =$ potential energy$/Q$.

**2.4** $W = EQd$, see 2.2 above, hence $W = 50 \times 2.0 \times 10^{-10} \times 0.010 = 1 \times 10^{-10}\,\mathrm{J}$.

**2.5** Potential difference $= 6.0/0.5 = 12\,\mathrm{V}$.

**2.6** A scalar quantity.

**2.7** See the key points.

**2.8** No, because there is no change in potential.

**2.9** Because no work is done in moving along an equipotential surface there can be no force component in that direction acting on the charge. Hence the electric field must be at right angles to the equipotential surface.

**2.10** See Figure A2.1.

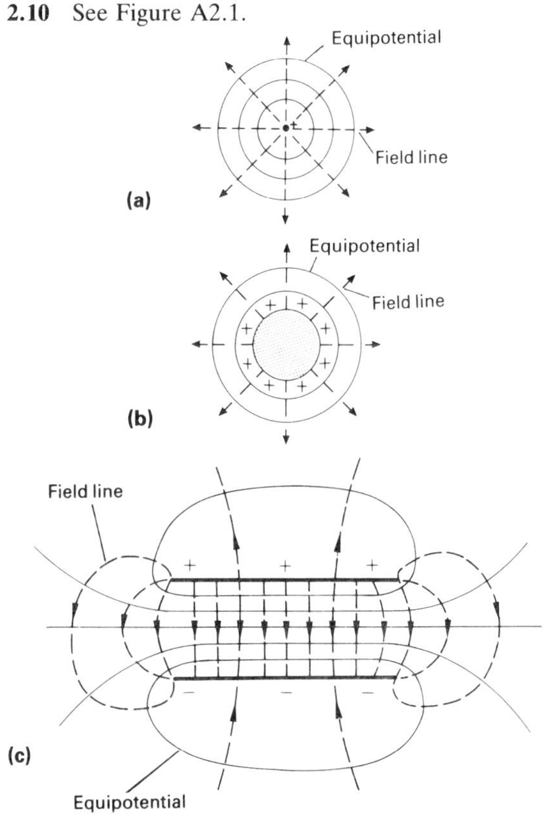


**Figure A2.1**

## Answers II

**2.11** (a) In the direction OC, i.e. away from $Q$.
(b) $\delta W = -F\delta x$
(c) Because the distance $x$ is being reduced by $\delta x$.

(d) $\delta W = \dfrac{QQ_0}{4\pi\epsilon}\left(-\dfrac{\delta x}{x^2}\right)$

$$W = -\frac{QQ_0}{4\pi\epsilon}\int_{\infty}^{r}\frac{dx}{x^2} = \frac{QQ_0}{4\pi\epsilon r}$$

(e) $V = \dfrac{W}{Q_0} = \dfrac{Q}{4\pi\epsilon r}$

**2.12** (a) The work done in moving charge $+Q_0$ from C to B.
(b) Determine the area under the graph from infinity to position A.

**2.13** $V = Q/4\pi\epsilon r = 2.0 \times 10^{-9}/(4\pi \times 8.85 \times 10^{-12} \times 0.006) = 3.0 \times 10^3$ V

**2.14** Halved.

**2.15** $V = Q/4\pi\epsilon_0 r = 3.0 \times 10^{-5}/(4\pi \times 8.85 \times 10^{-12} \times 0.100) = 2.7 \times 10^6$ V

**2.16** The potential will be constant throughout the entire sphere, being the same as the value at the surface.

**2.17** See Figure A2.2.

## Answers III

**2.18** (a) $EQ_0$ (b) $\delta W = -F\delta x = -EQ_0\,\delta x$ (c) $\delta V = -E\,\delta x$ (d) $E = -dV/dx$. In a uniform field there is a constant potential gradient.

**2.19** $E = -dV/dx = -1000/0.010 = -1.0 \times 10^5$ V m$^{-1}$

**2.20** $E = Q/4\pi\epsilon_0 r^2$ and hence $Q/4\pi\epsilon_0 = Er^2 = 3 \times 10^6 \times 0.20^2 = 1.2 \times 10^5$ C$^{-1}$ N m. $V = Q/4\pi\epsilon_0 r = 1.2 \times 10^5/0.2 = 6.0 \times 10^5$ V.

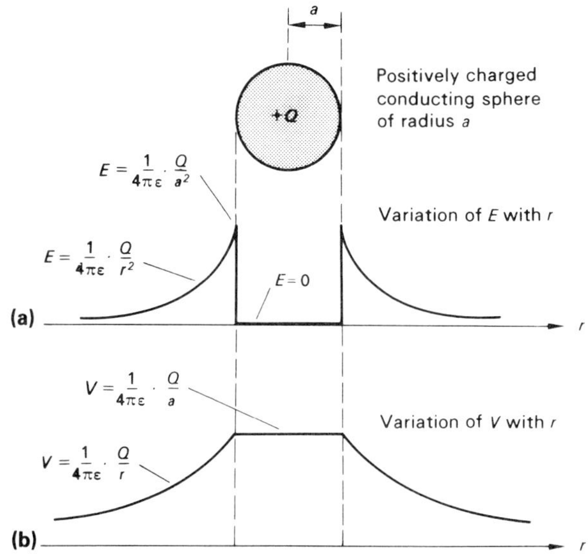


**Figure A2.2**

# Lesson 3 Capacitance

## Aims

After completing this lesson you should be able to
- define capacitance
- derive relationships for the capacitance of (a) an isolated conducting sphere and (b) a pair of parallel conducting plates
- explain the terms *permittivity*, *relative permittivity* or *dielectric constant*, and *permittivity of free space*
- explain the form of practical capacitors
- explain the action of a capacitor when being charged and discharged
- explain the reed-switch method of measuring capacitance
- derive and use the relationship for energy stored by a capacitor.

## Prerequisites

Lessons 1 and 2 on electric forces and fields and electric potential. Lesson 5 of Chapter 1, on work, energy and power, is required for a group of questions.

# STUDY

**I Read** about the definition of capacitance and the derivation of the relationships for capacitance for (a) an isolated conducting sphere and (b) a pair of parallel conducting plates, and also the definitions of the terms permittivity, relative permittivity, dielectric constant and permittivity of free space.

References: *B* PP ch. 11, *B* ST5 ch. 2, *D* ch. 12, *N* ch. 29, *W et al.* ch. 28(26), *Wh&H* ch. 46, ch. 47.

**Self-assessment questions**

**3.1** Define *capacitance* and state its unit.

**3.2** Consider an isolated sphere, of radius $a$ and having a charge $Q$, in a medium of permittivity $\epsilon$.
(a) What is the potential at the surface of the sphere?
(b) As $C = Q/V$, what is the capacitance of the sphere?

***3.3** If the earth has a radius of $6.4 \times 10^6$ m, what is its capacitance? ($\epsilon_0 = 8.85 \times 10^{-12}\,C^2\,N^{-1}\,m^{-2}$)

**3.4** Consider a pair of conducting parallel plates, each of common area $A$ and separated by a medium of thickness $d$ and permittivity $\epsilon$ (Figure 3.1). One plate has a charge $+Q$ and the other a charge $-Q$.

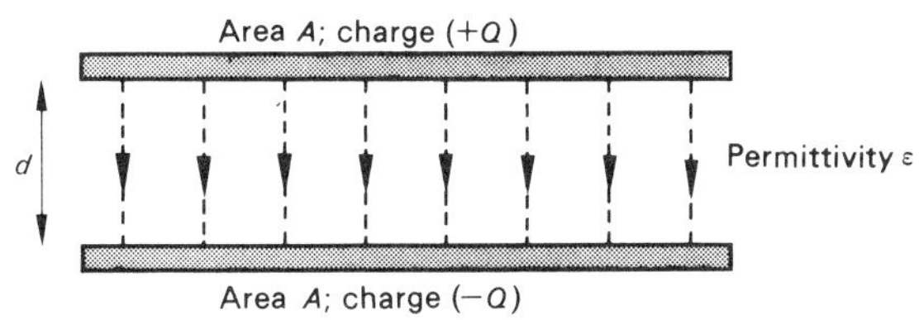


**Figure 3.1**

(a) What is the charge density $\sigma$ on the plates? (See Lesson 1, SAQ 1.19.)
(b) $E = \sigma/\epsilon$, hence what is the electric field between the plates? The field between the plates is uniform, apart from near the edges.
(c) If $V$ is the potential difference between the plates, what is the potential gradient between the plates? Hence what is the electric field in terms of this potential difference? (See Lesson 2, SAQ 2.18).
(d) Hence as $C = Q/V$, derive a relationship for the capacitance of the parallel plates.

***3.5** What is the capacitance of a parallel-plate capacitor if it consists of a sheet of mica 0.10 mm thick and 1.5 cm square, silvered on both sides? The permittivity of the mica is 6.0 times that of a vacuum. ($\epsilon_0 = 8.85 \times 10^{-12}\,C^2\,N^{-1}\,m^{-2}$)

**3.6** Define *relative permittivity* (or *dielectric constant*).

**II Read** about the form of practical capacitors: variable air, paper, ceramic, mica and electrolytic.
References: *B* ST4 ch. 2, *D* ch. 12, *N* ch. 29.

**Self-assessment questions**

**3.7** Describe the basic form and materials of (a) a paper capacitor, (b) an electrolytic capacitor.

**3.8** Describe how the capacitance is varied with a variable air capacitor.

***3.9** A paper capacitor consists of a sheet of paper 35 mm wide, 10 m long and $2.0 \times 10^{-2}$ mm thick between sheets of metal foil. If the relative permittivity of the paper is 2.7, what is the capacitance? ($\epsilon_0 = 8.85 \times 10^{-12}\,C^2\,N^{-1}\,m^{-2}$)

**III Read** about the action of a capacitor in being charged and discharged and the measurement of capacitance by the use of a vibrating reed switch.
References: *B* ST5 ch. 2, *D* ch. 12, *N* ch. 29, *W et al.* ch. 28(26), *Wh&H* ch. 46.

**Self-assessment questions**

**3.10** For the circuit shown in Figure 3.2, what might the meters be expected to indicate when the switch is closed?

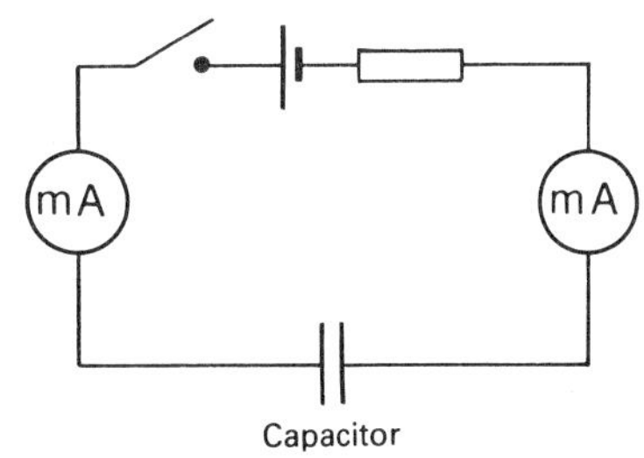


**Figure 3.2**

**3.11** For the circuit shown in Figure 3.3, what might the meters be expected to indicate when the switch is closed, for an initially charged capacitor?

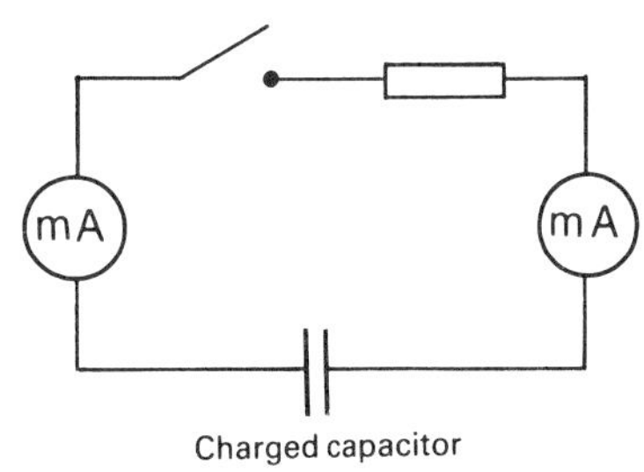


**Figure 3.3**

**3.12** Figure 3.4 shows a method by which capacitance can be measured. It involves a vibrating reed switch. This enables the capacitor to be alternately charged and discharged through the galvanometer, the number of charge–discharge actions per second equalling the frequency of the a.c. supply to the coil of the switch. Because the current pulses from the discharging follow

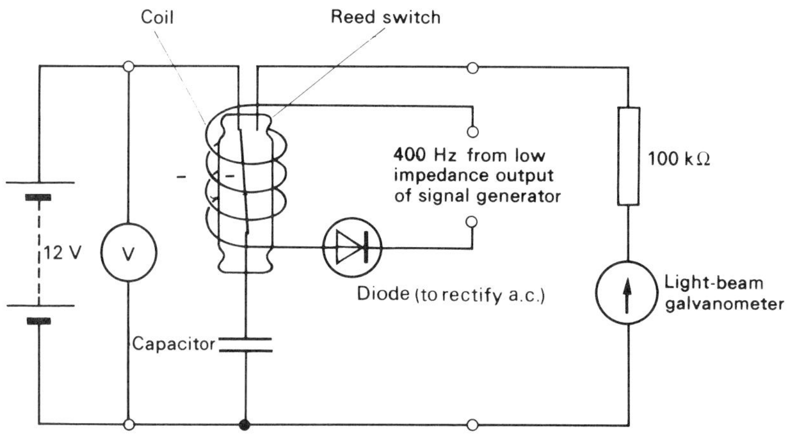


**Figure 3.4 Measurement of capacitance**

each other so rapidly the galvanometer gives a steady reading. This steady current is the charge passing through the instrument per second, current being rate of movement of charge. Hence derive a relationship between the current and the charge on the capacitor when fully charged. As the charging potential difference is known, the capacitance can be obtained. If the experiment were repeated for a number of different potential differences, with the reed-switch frequency constant, how could the capacitance be obtained from a graph of the current plotted against the potential difference?

**3.13** Explain how the capacitance measurement method described in SAQ 3.12 could be used in order to obtain a value of $\epsilon_0$.

***3.14** In a vibrating reed-switch experiment the frequency applied to the reed switch is 100 Hz. If the current indicated is 50 $\mu$A when the charging potential difference is 12 V, what is the capacitance? If the frequency is changed to 200 Hz at the same potential difference, what is the value of the new current?

**IV Read** about the derivation of the equation for the energy stored by a charged capacitor.
References: *B* PP ch. 11, *B* ST5 ch. 2, *D* ch. 12, *N* ch. 29, *W et al.* (ch. 26), *Wh&H* ch. 46.

**Self-assessment questions**

**3.15** Figure 3.5 shows how the charge on a capacitor varies with the potential difference, the charge being directly proportional to the potential difference. Consider a capacitor with a charge $Q$.

(a) If the capacitor begins to discharge and loses a small amount of charge $\delta Q$, and the potential difference $V$ can be considered to be constant for this small change, what is the energy loss?
(b) How is this loss related to the area marked on the graph?
(c) If the capacitor discharges completely, what is the total energy loss? Use your answer to (b) and the graph.
(d) What therefore is the energy stored in a capacitor when it has been charged to a charge $Q$ by a potential difference $V$?

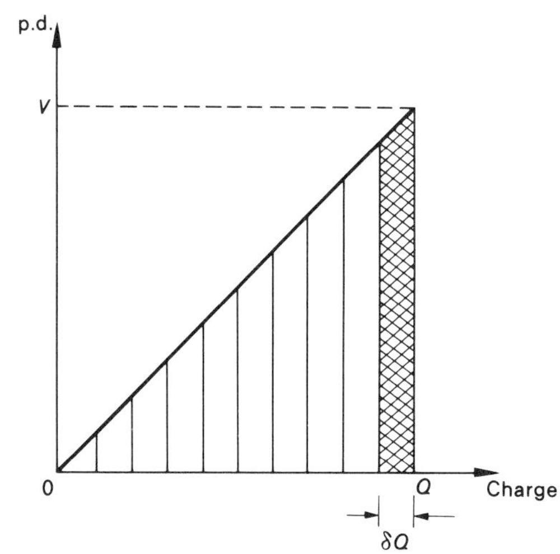


**Figure 3.5**

***3.16** What is the energy stored in a 16 $\mu$F capacitor charged to 12 V?

# SELF-ASSESSMENT

## Key points

If a charge $Q$ is put on an insulated object and its potential changes by $V$ then it is said to have a **capacitance** of $C$, where $C = Q/V$. The unit of capacitance is the farad. The capacitance of an isolated sphere of radius $a$ is $4\pi\epsilon a$. The capacitance of a parallel-plate capacitor is $\epsilon A/d$, where $A$ is the plate area and $d$ their separation.

The **relative permittivity** or **dielectric constant** is $\epsilon/\epsilon_0$ where $\epsilon$ is the permittivity and $\epsilon_0$ the permittivity of free space, i.e. a vacuum.

When a parallel-plate capacitor is charged a current flows to one plate and away from the other, the net result being that one plate acquires a positive charge and the other an equal-sized but negative charge. As the charge builds up on the plates so the potential difference between the capacitor plates builds up; at any instant the ratio $Q/V$ is equal to the capacitance. When a capacitor is discharged the charges move to equalize the potentials and thus the final result is no net charge on either plate.

The **energy stored** by a charged capacitor is $\frac{1}{2}QV$.

## Answers I

**3.1** See the key points.

**3.2** (a) $V = Q/4\pi\epsilon a$ (See Lesson 2.) (b) $C = 4\pi\epsilon a$

**3.3** $C = 4\pi\epsilon_0 a = 4\pi \times 8.85 \times 10^{-12} \times 6.4 \times 10^6 = 7.1 \times 10^{-4}$ F

**3.4** (a) $\sigma = Q/A$ (b) $E = \sigma/\epsilon = Q/A\epsilon$ (c) $E = V/d$ (d) $C = A\epsilon/d$

**3.5** $C = A\epsilon/d = 0.015^2 \times 6.0 \times 8.85 \times 10^{-12}/(0.10 \times 10^{-3}) = 1.19 \times 10^{-10}$ F

**3.6** See the key points.

## Answers II

**3.7** (a) See Figure A3.1. (b) See Figure A3.2.

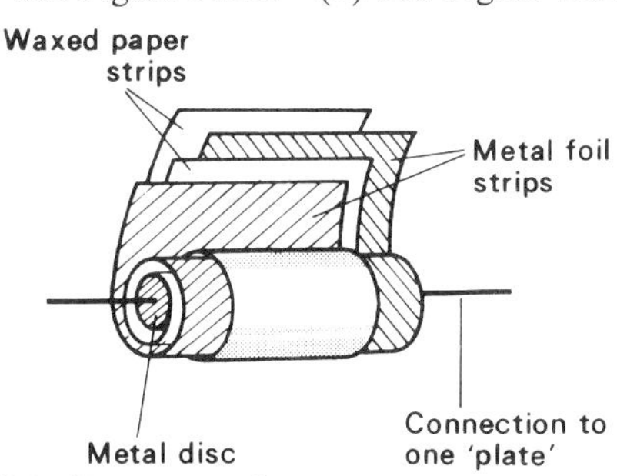


**Figure A3.1 Paper capacitor**

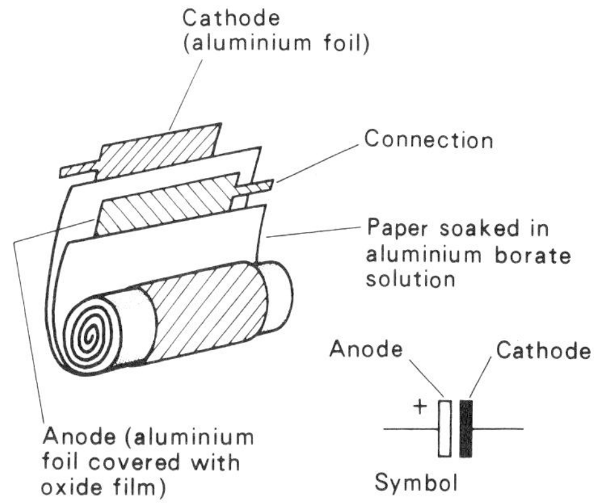


**Figure A3.2 Electrolytic capacitor**

**3.8** By varying the area by which the plates overlap; see Figure A3.3.

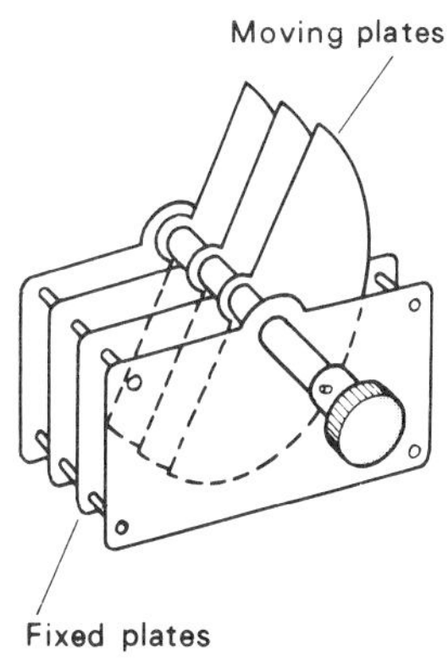


**Figure A3.3 Variable air capacitor**

**3.9** $C = A\epsilon/d = 10 \times 35 \times 10^{-3} \times 2.7 \times 8.85 \times 10^{-12}/(2.0 \times 10^{-5}) = 4.2 \times 10^{-7}$ F

## Answers III

**3.10** Current flowing in the same direction for each meter but decaying with time to zero. The current in the same direction indicates charge flowing onto one plate and off the other. The currents have the same values because the plates have, at any instant, equal-sized charges.

**3.11** Current flowing in the same direction for each meter but decaying with time to zero. Charge is flowing off one plate and onto the other so that the final effect is zero net charge on each plate.

The currents are in the opposite directions to those which occurred during the charging.

**3.12** Current is rate of movement of charge. Thus if charge $Q$ is moved every $1/f$ s then $I = Q/(1/f) = Qf$. As $Q = CV$ then $I/f = CV$. Thus a graph of current against $V$ is a straight line through the origin with a slope of $fC$. The capacitance can thus be obtained from the slope of the graph.

**3.13** The capacitance of an air capacitor can be measured and then, using the equation arrived at in the answer to SAQ 3.4, $\epsilon$ can be calculated if the plate area and separation are known. For air $\epsilon$ is effectively $\epsilon_0$.

**3.14** Using the equation developed in 3.12, $I = fCV$ and thus $C = 50 \times 10^{-6}/(12 \times 100) = 4.2 \times 10^{-8}$ F. If the frequency is doubled to 200 Hz, the current is doubled.

## Answers IV

**3.15** (a) Energy loss = work done = $V\delta Q$ (b) It equals the hatched area. (c) Total energy loss = total area under graph = $\frac{1}{2}QV$ (d) $\frac{1}{2}QV$

**3.16** Energy = $\frac{1}{2}QV = \frac{1}{2}CV^2 = \frac{1}{2} \times 16 \times 10^{-6} \times (12)^2 = 1.2 \times 10^{-3}$ J

# Lesson 4 Capacitor networks

## Aims

After completing this lesson you should be able to

- calculate the capacitance of capacitors in (a) series and (b) parallel connections
- solve problems with combinations of capacitors, involving their charges, potential differences and stored energy.

## Prerequisites

Lesson 3 on capacitance.

## STUDY

**Read** about capacitors in series and in parallel.
References: *B* ST5 ch. 2, *D* ch. 12, *N* ch. 29, *W et al.* (ch. 26), *Wh&H* ch. 46.

### Self-assessment questions

**4.1** Figure 4.1 shows three capacitors in parallel.
(a) What is the potential difference across each capacitor?
(b) The total charge that would have to be on a single capacitor which could replace the three capacitors is $Q = Q_1 + Q_2 + Q_3$. Hence derive a relationship between the capacitance of the single replacement capacitor and the capacitances of the three capacitors.

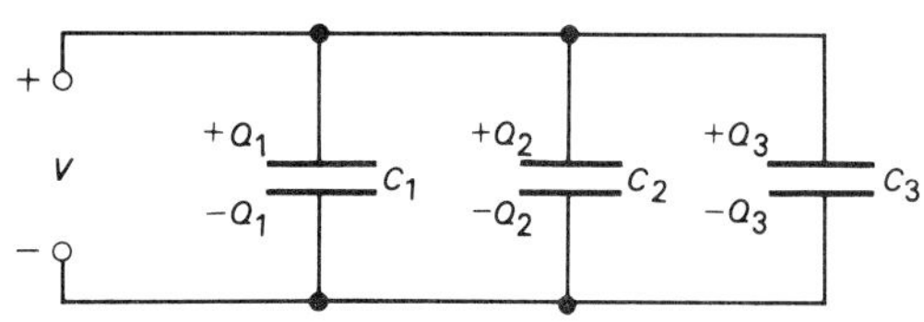


**Figure 4.1**

**4.2** An 8μF capacitor and a 16μF capacitor are connected in parallel. What is the combined capacitance of this arrangement?

**4.3** Figure 4.2 shows three capacitors in series. The charges are the same on each capacitor, even though they have different capacitances. How is the potential difference across the three related to the potential differences across each capacitor? Hence derive a relationship between the capacitance of a single replacement capacitor and the capacitances of the three capacitors.

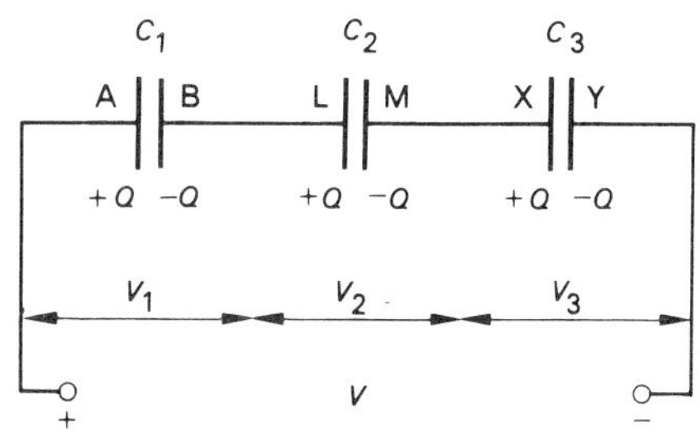


**Figure 4.2**

**4.4** An 8μF capacitor and a 16μF capacitor are connected in series. What is the combined capacitance of this arrangement?

***4.5** A 1μF capacitor and a 4μF capacitor are connected in parallel and then charged to a potential difference of 12 V. What is (a) the charge on each capacitor, (b) the combined capacitance?

***4.6** A 4μF capacitor is charged to a potential difference of 10 V. The charged capacitor is then removed from the voltage supply and connected across an uncharged 8μF capacitor. What is (a) the charge on

each capacitor, (b) the total capacitance of the two capacitors?
Hint: Consider what the charge is initially on the 4$\mu$F capacitor and then consider what happens to this charge.

*4.7 Two capacitors of capacitances 2$\mu$F and 4$\mu$F are connected in series and the combination is connected across a 12 V battery until the capacitors become fully charged. The charged capacitors are then disconnected and reconnected so that they are in parallel with each other without the battery being present. If they are connected so that similarly charged plates are connected together, what is the resulting potential difference across the combination?

*4.8 A capacitor is charged by a battery. It is then disconnected from the battery and connected across another capacitor of the same capacitance but previously uncharged. How is (a) the charge on the capacitors, (b) the potential difference across the combination and (c) the total electrical energy stored related to that of the single charged capacitor?

*4.9 Two capacitors of capacitances 2$\mu$F and 3$\mu$F are connected in series and the combination is connected across a 6 V battery until the capacitors are fully charged. What is (a) the energy stored by each capacitor, (b) the energy stored by each capacitor when they are disconnected from the battery and connected in parallel, similarly charged plates being connected together, (c) the difference in the total energy stored by the system when in series and when in parallel, (d) the reason for their being a difference in the energy stored?

*4.10 A charged metallized polystyrene ball comes into contact with an uncharged identical ball. (a) What happens to the charge and (b) why do the two balls then fly apart?

# SELF-ASSESSMENT

## Key points

The capacitance of a system of capacitors in parallel is given by $C = C_1 + C_2 + C_3 +$ etc. For capacitors in series $1/C = 1/C_1 + 1/C_2 + 1/C_3 +$ etc. When capacitors are connected in parallel the potential difference is the same across each but the charges can be different. When capacitors are connected in series the charges are the same on each capacitor but the potential differences can be different.

## Answers

**4.1** (a) The same, $V$. (b) As $Q_1 = C_1V$ and $Q_2 = C_2V$ and $Q_3 = C_3V$ we can write $CV = C_1V + C_2V + C_3V$ and so $C = C_1 + C_2 + C_3$.

**4.2** $C = 8 + 16 = 24\mu$F

**4.3** $V = V_1 + V_2 + V_3$ and so $(Q/C) = (Q/C_1) + (Q/C_2) + (Q/C_3)$ and thus $1/C = 1/C_1 + 1/C_2 + 1/C_3$.

**4.4** $1/C = 1/8 + 1/16 = 3/16$ and so $C = 5.3\mu$F.

**4.5** (a) $Q = CV$ and thus $Q_1 = 1 \times 10^{-6} \times 12 = 1.2 \times 10^{-5}$ C. $Q_2 = 4 \times 10^{-6} \times 12 = 4.8 \times 10^{-5}$ C. (b) $C = C_1 + C_2 = 1 + 4 = 5\mu$F

**4.6** (a) Initially, charge on 4$\mu$F capacitor $Q = CV = 4 \times 10^{-6} \times 10 = 4 \times 10^{-5}$ C. This charge is then shared between the two capacitors. $Q = Q_1 + Q_2 = C_1V + C_2V$, where $V$ is the new potential difference across the capacitors. Thus $4 \times 10^{-5} = 4 \times 10^{-6}\,V + 8 \times 10^{-6}\,V$ and $V = 3.3$ V. Hence $Q_1 = C_1V = 4 \times 10^{-6} \times 3.3 = 1.3 \times 10^{-5}$ C and $Q_2 = C_2V = 8 \times 10^{-6} \times 3.3 = 2.6 \times 10^{-5}$ C. (b) $C = C_1 + C_2 = 4 + 8 = 12\mu$F.

**4.7** Initially $1/C = 1/2 + 1/4$ and total capacitance $C = 1.3\mu$F. Thus charge on this single capacitor or on each of the individual capacitors in the series arrangement is $Q = CV = 1.3 \times 10^{-6} \times 12 = 1.6 \times 10^{-5}$ C. When they are connected in parallel the total charge is the sum of the charges on each capacitor, i.e. $2 \times 1.6 \times 10^{-5} = 3.2 \times 10^{-5}$ C. In the parallel arrangement the total capacitance is $C = 2 + 4 = 6\mu$F and so the potential difference across this, or the individual capacitors in the parallel arrangement, is $V = Q/C = 3.2 \times 10^{-5}/6 \times 10^{-6} = 5.3$ V.

**4.8** (a) The charge on the initial capacitor is shared between the two capacitors, and as they are of equal capacitance this means that the charge on each capacitor is half the charge on the initial capacitor.
(b) The combined capacitance is twice the capacitance of an individual capacitor and thus $V = (\frac{1}{2}Q)/(2C) = \frac{1}{4}(Q/C)$ or one quarter of the initial potential difference.
(c) Energy $= \frac{1}{2} \times$ charge $\times$ p.d. and thus in this case the final energy $= \frac{1}{2}(\frac{1}{2}Q)(\frac{1}{4}V) = \frac{1}{2}(\frac{1}{8}QV)$. This compares with an initial energy of $\frac{1}{2}QV$ and so the energy stored has been reduced by a factor of one eighth. The energy lost in making this connection of the capacitors appears as heat when the charge flows through the connecting wires as it is redistributed between the capacitors.

**4.9** (a) Combined capacitance is given by $1/C_{tot} = 1/2 + 1/3$ and so $C_{tot} = 1.2\mu$F. Charge $Q = CV = 1.2 \times 10^{-6} \times 6 = 7.2 \times 10^{-6}$ C. This is the charge on each capacitor.

Thus energy $= \frac{1}{2}QV = \frac{1}{2}Q^2/C = \frac{1}{2} \times (7.2 \times 10^{-6})^2/(2 \times 10^{-6}) = 1.3 \times 10^{-5}$ J for one capacitor and $\frac{1}{2} \times (7.2 \times 10^{-6})^2/(3 \times 10^{-6}) = 8.6 \times 10^{-6}$ J for the other. The total energy is $2.2 \times 10^{-5}$ J.
(b) The total charge is $14.4 \times 10^{-6}$ J. This is shared between the two capacitors so that $Q = C_1V + C_2V$ and hence $14.4 \times 10^{-6} = 2 \times 10^{-6}\,V + 3 \times 10^{-6}\,V$ and $V = 2.88$ V. Thus energy $= \frac{1}{2}CV = \frac{1}{2} \times 2 \times 10^{-6} \times 2.88 = 2.88 \times 10^{-6}$ J for one capacitor and $\frac{1}{2} \times 3 \times 10^{-6} \times 2.88 = 4.32 \times 10^{-6}$ J for the other. The total energy $= 7.2 \times 10^{-6}$ J.
(c) The difference is a loss of $1.5 \times 10^{-5}$ J.
(d) Current in the connecting wires when the two capacitors are connected in parallel produces heat.

**4.10** (a) It redistributes itself between the two spheres and as they have the same capacitance this means that the charge on each sphere is half the charge on the initially charged sphere.
(b) They become like-charged and so repel each other.

# Lesson 5 Charging and discharging capacitors

## Aims

After completing this lesson you should be able to

- describe the way in which the current, charge and potential difference vary with time for a capacitor being (a) charged and (b) discharged
- derive the exponential equations for the charging and discharging processes.

## Prerequisites

Lessons 3 and 4 on capacitors are required, and also a knowledge of basic calculus and the exponential function.

## STUDY

**Read** about the way in which the current, charge and potential difference vary with time for a capacitor being charged and discharged.
References: *B* PP ch. 5, ch. 11, *B* ST4 ch. 2, *D* ch. 12, *N* ch. 29, *W et al.* ch. 38(26), *Wh&H* ch. 50.

**Self-assessment questions**

**5.1** Figure 5.1 shows how the current varies with time after the switch in the circuit shown is closed and the capacitor starts to charge.
(a) If initially the current in the circuit is $I$ and there is no potential difference across the capacitor, what is the potential difference across the resistor?
(b) When the capacitor is fully charged and the current has ceased, what must be the potential difference across the capacitor?
(c) Current is the rate of movement of charge ($I = \mathrm{d}Q/\mathrm{d}t$). Thus how could the charge–time graph be derived from the current–time graph?
(d) $V = Q/C$; thus how is the potential difference–time graph related to the charge–time graph?
(e) Sketch the basic forms of the charge–time and potential difference–time graphs for a capacitor when it is being charged.

****5.2** Consider the charging of the capacitor in the circuit shown in Figure 5.1. We can write, for some instant of time, an equation relating the potential differences in the circuit:

$$V = \text{p.d. across } R + \text{p.d. across } C$$

(a) If the current at that instant of time is $I$, what is the potential difference across the resistor?
(b) If at that instant of time the capacitor has a charge $Q$, what is the potential difference across the capacitor?
(c) Current $I = \mathrm{d}Q/\mathrm{d}t$. Hence obtain the differential equation

$$RC\frac{\mathrm{d}Q}{\mathrm{d}t} = CV - Q$$

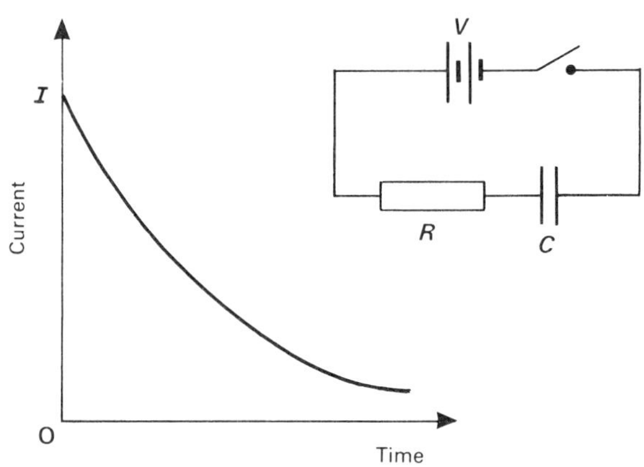


**Figure 5.1**

(d) The maximum charge $Q_{max}$ that is possible with the applied potential difference $V$ is $CV$. Thus arrive at the solution to the differential equation:

$$Q = Q_{max}(1 - e^{-(1/RC)t})$$

(e) Now arrive at equations for the current–time relationship and the potential difference–time relationship.

**5.3** If you were given the charge–time graph for a capacitor when being charged, how could you obtain the current–time graph from it?

**5.4** Figure 5.2 shows the current–time graph for a capacitor discharging through a resistor. The current is shown to be negative because it is in the opposite direction to that described in Figure 5.1.
(a) Current is the rate of movement of charge ($I = dQ/dt$). Thus how could the charge–time graph be derived from the current–time graph?
(b) $V = Q/C$; thus how is the potential difference–time graph related to the charge–time graph?
(c) Sketch the basic forms of the charge–time and potential difference–time graphs for a capacitor when it is being discharged.

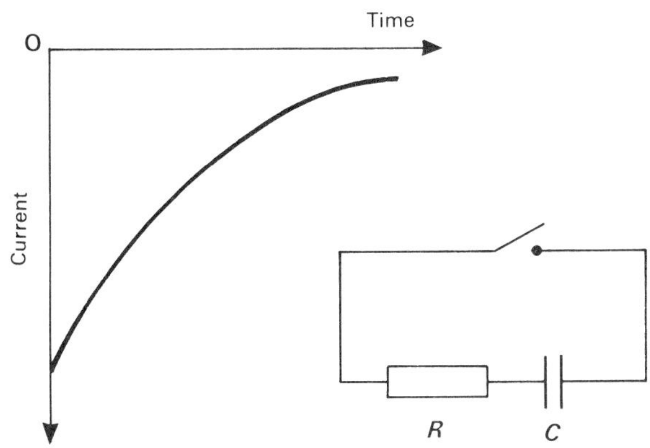


**Figure 5.2**

****5.5** For the discharging circuit shown in Figure 5.2, the discharge current $I$ at any instant equals the rate of loss of charge from the capacitor. Thus $I = -dQ/dt$.
(a) Why the minus sign in the above expression?
(b) The potential difference $V$ across the resistor when there is a current flowing is $IR$. How is the potential difference across the capacitor at the same instant related to this value?
(c) As $Q = CV$, arrive at the differential equation

$$\frac{dQ}{dt} = -\frac{1}{CR}Q$$

(d) Hence obtain the solution to the equation of

$$Q = Q_0 e^{-(t/CR)}$$

where $Q_0$ is the initial charge at time $t = 0$.
(e) Now arrive at the equations for the current–time relationship and the potential difference–time relationship.

**5.6** The product $CR$ is known as the *time constant* for either a charging or a discharging capacitor circuit.
(a) Show that the unit of $CR$ is the second.
(b) For a discharging capacitor $CR$ is the time taken for the charge to fall to 37% of its initial value. Using the relationship arrived at in 5.5(d) show that this statement is correct.
(c) What percentage of the initial charge will be on the discharging capacitor after a time of $2CR$?

**5.7** What is the effect of the circuit resistance on (a) the charging and (b) the discharging of a capacitor?

# SELF-ASSESSMENT

## Key points

When a capacitor is charged through a resistance $R$

$$I = \frac{V}{R}e^{-(1/RC)t}$$

$$Q = Q_{max}(1 - e^{-(1/RC)t})$$

$$V = V_{max}(1 - e^{-(1/RC)t})$$

When a capacitor is discharged through a resistance $R$

$$I = -\frac{1}{CR}Q_0 e^{-(1/RC)t}$$

$$Q = Q_0 e^{-(1/RC)t}$$

$$V = V_0 e^{-(1/RC)t}$$

The product $RC$ is known as the **time constant** and has the unit of the second.

## Answers

**5.1** (a) $IR$ (b) $V$
(c) For a particular time the charge will be the area between the zero time ordinate and the time ordinate concerned. By taking a number of measurements the charge–time graph can be derived.
(d) The potential difference–time graph must be of exactly the same form as the charge–time graph since the

potential difference is directly proportional to the charge.
(e) See Figure A5.1.

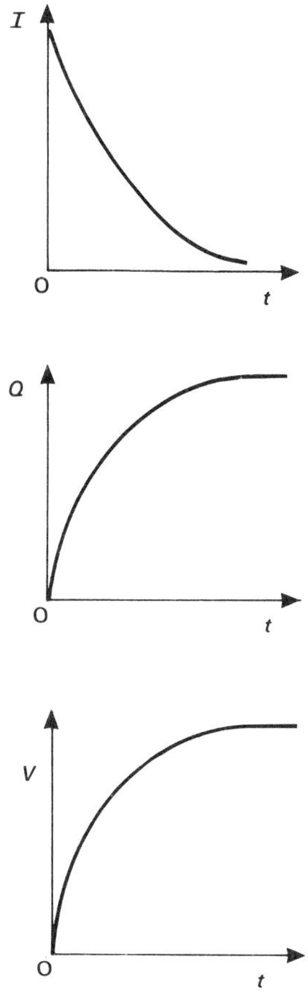


**Figure A5.1**

**5.2** (a) $IR$ (b) $Q/C$ (c) $V = IR + Q/C = R\,\mathrm{d}Q/\mathrm{d}t + Q/C$ and so $RC\,\mathrm{d}Q/\mathrm{d}t = CV - Q$
(d) $RC\,\mathrm{d}Q/\mathrm{d}t = Q_{\max} - Q$. Hence

$$\frac{1}{RC}\int_0^t \mathrm{d}t = \int_0^Q \frac{\mathrm{d}Q}{Q_{\max}-Q}$$

and $$\frac{t}{CR} = -\ln\left(\frac{Q_{\max}-Q}{Q_{\max}}\right)$$

Thus $Q = Q_{\max}(1 - \mathrm{e}^{-(1/RC)t})$.
(e) $I = \mathrm{d}Q/\mathrm{d}t = (Q_{\max}/RC)\,\mathrm{e}^{-(1/RC)t}$ and as $Q_{\max} = CV$ then $I = (V/R)\,\mathrm{e}^{-(1/RC)t}$. For the potential difference $V = Q/C = (Q_{\max}/C)(1 - \mathrm{e}^{-(1/RC)t})$. But as $Q_{\max} = CV_{\max}$, then $V = V_{\max}(1 - \mathrm{e}^{-(1/RC)t})$.

**5.3** As current is the rate of movement of charge then the current at any instant is the gradient of the charge–time graph at the time concerned. Thus by measuring the gradients at a number of times the current–time graph can be plotted.

**5.4** (a) By measuring the area under the graph between the zero time ordinate and the ordinate at a particular time, the charge on the capacitor at that time can be determined. Hence by taking a number of such area measurements the charge–time graph can be obtained.
(b) As $V$ is directly proportional to $Q$, the potential difference graph is of exactly the same form as the charge–time graph.
(c) See Figure A5.2.

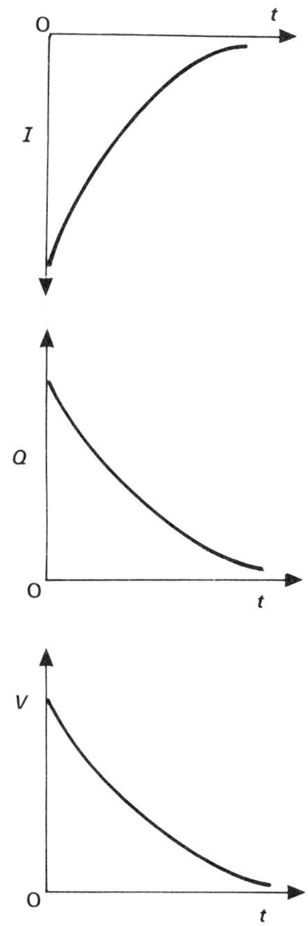


**Figure A5.2**

**5.5** (a) Because $Q$ is decreasing with increasing time.
(b) The same as they are in parallel.
(c) $V = IR = Q/C$ and so $I = Q/CR$. Thus $\mathrm{d}Q/\mathrm{d}t = -Q/CR$.

(d) $\int_0^t \frac{dt}{CR} = -\int_{Q_0}^{Q} \frac{dQ}{Q}$ and so $\frac{t}{CR} = \ln \frac{Q_0}{Q}$

and hence $Q = Q_0\,e^{-(1/CR)t}$.
(e) $I = -dQ/dt = (Q_0/CR)\,e^{-(1/CR)t}$. As $V = Q/C = (Q_0/C)\,e^{-(1/CR)t}$ and $V_0 = Q_0/C$ then $V = V_0\,e^{-(1/CR)t}$.

**5.6** (a) Farad = coulomb/volt, but coulomb = amp × second, thus farad = amp × second/volt. Ohm = volt/amp and so farad × ohm = (amp × second/volt) × (volt/amp) = second.
(b) When $t = CR$ then $Q = Q_0\,e^{-(1/CR)CR} = Q_0 e^{-1} = Q_0/e = Q_0/2.7 = 0.37Q_0$ or 37% of $Q_0$.
(c) 37% of 37% of $Q_0$, i.e. 13.7%.

**5.7** (a) The bigger the value of $CR$ the longer it takes to charge the capacitor.
(b) The bigger the value of $CR$ the longer it takes to discharge.

# Lesson 6 Magnetic fields

## Aims

After completing this lesson you should be able to

- describe the field patterns due to current in (a) a long straight conductor, (b) a flat circular coil and (c) a long solenoid
- define *magnetic flux density B* in terms of the force acting on a current element, i.e. $B = F/IL$
- state the relationship between the direction of the magnetic field, the direction of the current and the direction of the force
- explain how $B$ can be measured using a current balance
- calculate the torque on a current-carrying coil in a magnetic field
- describe the principles of the moving-coil meter, explaining the need for a radial field and the factors affecting the current and voltage sensitivities.

## Prerequisites

A basic knowledge of magnetism is assumed.

## STUDY

**I Read** about the magnetic field patterns due to current in a long straight conductor, a flat circular coil and a long solenoid.
References: *B* PP ch. 11, *B* ST4 ch. 3, *D* ch. 13, *N* ch. 33, *W et al.* ch. 30(27, 34), *Wh&H* ch. 52.

### Self-assessment questions

**6.1** What is represented by a magnetic line of force?

**6.2** The magnetic field lines due to a current in a long straight conductor are circles, concentric with the conductor. The *right-hand screw* or *corkscrew rule* can be used to predict the direction of the field for a given current direction: if a right-handed screw moves forward in the direction of the current, then the direction of rotation of the screw gives the directions of the field lines. Using this, or another rule, predict the directions of the field lines surrounding a current-carrying wire normal to the surface of the earth and in which the current is moving downwards towards the earth. Neglect the earth's magnetic field.

**6.3** Sketch the magnetic field pattern due to current in (a) a long straight conductor, (b) a flat circular coil and (c) a long solenoid.

**II Read** about the equation $B = F/IL$, Fleming's left-hand rule, the current balance and the definition of the ampere.
References: *B* PP ch. 11, *B* ST4 ch. 3, *D* ch. 13, *N* ch. 33, *W et al.* ch. 30(27), *Wh&H* ch. 52.

### Self-assessment questions

***6.4** Describe an experiment which can be used to show that the force acting on a wire lying at right angles to a magnetic field is directly proportional to the current in the wire and to the length of the wire in the field, and is also dependent on the strength of the magnetic field.

**6.5** The *magnetic flux density B* can be defined in terms of the force acting per unit length on a conductor which carries unit current and is at right angles to the direction of the field, i.e. $B = F/IL$. What are the units that can be used for $B$?

**6.6** The relative directions of current, magnetic field and the force acting on the current-carrying element can be described by *Fleming's left-hand rule*. This states that if the thumb and first two fingers of the left hand are held each at right angles to one another, the **f**irst finger pointing in the direction of the **f**ield and the se**c**ond finger in the direction of the **c**urrent, then the **thu**mb predicts the direction of the **thru**st or force. For the arrangement shown in Figure 6.1, what is the direction of the force?

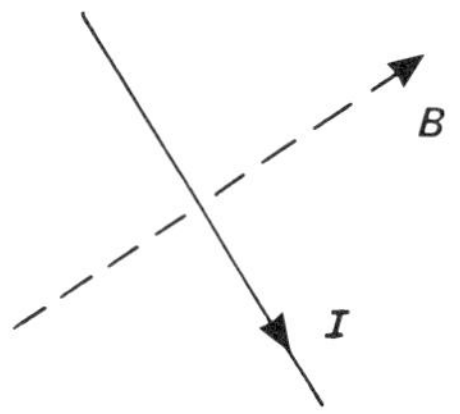


**Figure 6.1** Both *B* and *I* are in the horizontal plane

**6.7** A wire carries a current of 4 A, and 120 mm of its length is in a magnetic field of flux density $1.5 \times 10^{-2}$ T. What is the force on the wire if it is (a) parallel to the field direction, (b) perpendicular to the field direction, (c) at 30° to the field direction?

***6.8** Figure 6.2 shows a cylindrical metal bar resting on two horizontal metal rails which are connected to a battery.
(a) When a current is flowing through the metal bar and there is a magnetic field acting perpendicularly to the paper and into it, in which direction will the bar move?
(b) If the magnetic field has a flux density of 0.05 T and the current through the bar is 2.0 A, what is the size of the force acting on the bar?
(c) At what angle to the horizontal should the rails be tilted to keep the bar stationary? The metal bar has a mass of 4.0 g. ($g = 9.8\,\text{m s}^{-2}$)

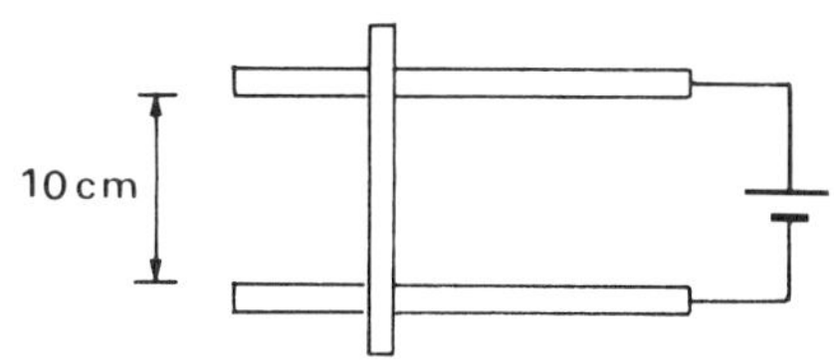


**Figure 6.2**

**III** **Read** about the derivation of the equation for the torque acting on a current-carrying coil in a magnetic field and the application of this effect in the moving-coil meter.
References: *B* ST4 ch. 3, *D* ch. 13, *N* ch. 33, *W et al.* ch. 30(34), *Wh&H* ch. 52, ch. 54.

**Self-assessment questions**

**6.9** Figure 6.3 shows a current-carrying coil in a magnetic field.
(a) What are the directions of the forces acting on the sides of the coil?
(b) What is the size of the force on a vertical side? The coil has *N* turns.
(c) The torque, or moment of the couple, for rotation about the vertical axis $YY^1$, is the product of the force acting on one of the vertical sides and the perpendicular distance between the lines of action of the forces acting on the two vertical sides. Hence obtain a relationship for the torque in terms of *B* when the coil is at the angle shown.

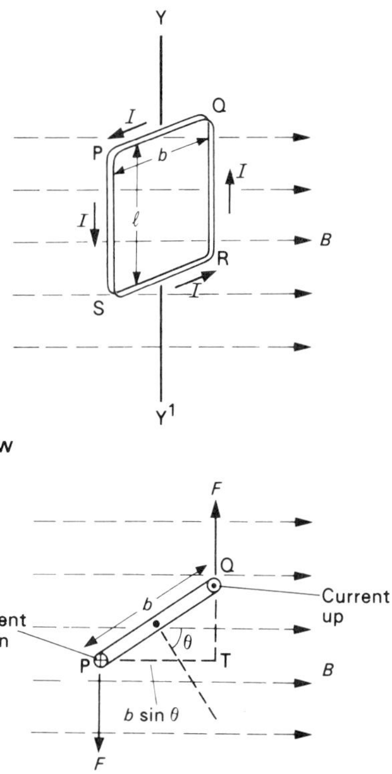


**Figure 6.3**

***6.10** A vertical rectangular coil consists of 100 turns of fine wire and has an area of $2.0\,\text{cm}^2$. What is the maximum torque acting on the coil when a current of 80 mA flows through it and it is in a horizontal magnetic field of flux density 0.4 T?

***6.11** Describe the structure and mode of action of a moving-coil galvanometer, explaining how a linear scale can be achieved.

***6.12** A moving-coil galvanometer has a rectangular coil of $N$ turns, each of area $A$, in a radial magnetic field of flux density $B$.
(a) What is the torque acting on the coil when a current $I$ flows through it?
(b) Rotation of the coil is opposed by springs. If they give an opposing torque of $c\alpha$, where $c$ is a constant for the springs and $\alpha$ the deflection of the coil, derive a relationship between the current $I$ and the deflection $\alpha$ of the coil.
(c) The current sensitivity of the meter is the deflection per unit current. How may the current sensitivity of a meter be made as large as possible?
(d) The voltage sensitivity of the meter is the deflection per unit potential difference. Derive an expression for the voltage sensitivity of a meter having a resistance $R$.

***6.13** The coil of a moving-coil meter has 20 turns and a resistance of 6 Ω. If the coil is removed and replaced by another coil having 30 turns and a resistance of 9 Ω, how will (a) the current sensitivity and (b) the voltage sensitivity be changed?

***6.14** A galvanometer coil has 50 turns, each turn having an area of 1.0 $cm^2$. If the coil is in a radial field of $1.0 \times 10^{-2}$ T and suspended by a suspension having a torsional constant of $2.0 \times 10^{-9}$ N m per degree, what current is needed to give a deflection of 30°?

# SELF-ASSESSMENT

## Key points

A **magnetic field** can be described by magnetic **lines of force**, or **field lines**, the tangent to such a line at any point representing the direction of the field at that point.
The **magnetic flux density** $B$ can be defined as the force acting on unit length of a conductor carrying unit current if it is at right angles to the direction of the field, i.e. $B = F/IL$. The unit of $B$ is N $A^{-1}$ $m^{-1}$ or Wb $m^{-2}$ (weber per square metre) or T (tesla) with 1 N $A^{-1}$ $m^{-1}$ = 1 Wb $m^{-2}$ = 1 T.
**Fleming's left-hand rule** can be used to describe the relative directions of the field, current and force. If the thumb and first two fingers of the left hand are held each at right angles to the other, when the **f**irst finger points in the direction of the **f**ield and the se**c**ond finger in the direction of the **c**urrent, then the **thu**mb predicts the direction of the **thr**ust or force.
The **torque** on a current-carrying coil in a magnetic field is $NBIA \sin\theta$, where $N$ is the number of turns, each of area $A$, $I$ the current and $B \sin\theta$ the flux density component at right angles to the plane of the coil, $\theta$ being the angle between the field direction and the normal to the plane of the coil. For a moving-coil meter with a *radial* field, i.e. $\theta = 0°$, torque $= NBIA = c\alpha$, where $\alpha$ is the deflection of the coil and $c$ the torsional constant of the suspension or springs. The current sensitivity $= \alpha/I = NBA/c$, the voltage sensitivity $= \alpha/V = NBA/cR$.

## Answers I

**6.1** See the key points.

**6.2** Clockwise, viewed downwards.

**6.3** See Figure A6.1.

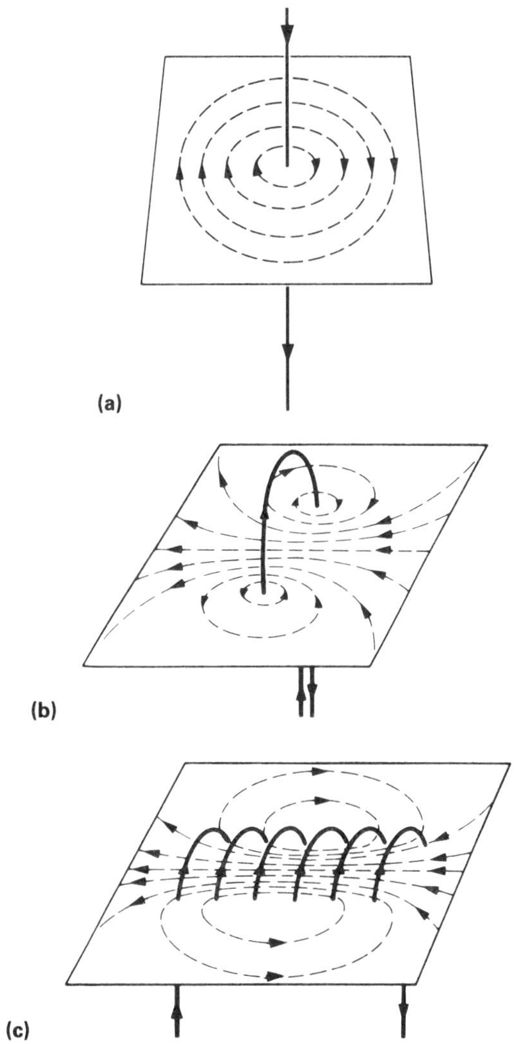


**Figure A6.1**

## Answers II

**6.4** See Figure A6.2. A current balance could be used, the force being given by the gravitational force *mg* on the rider which has to be added to maintain the balance horizontal. Different lengths of wire in the field can be obtained by using two identical sets of magnets, different strengths of magnetic field by using different magnets.

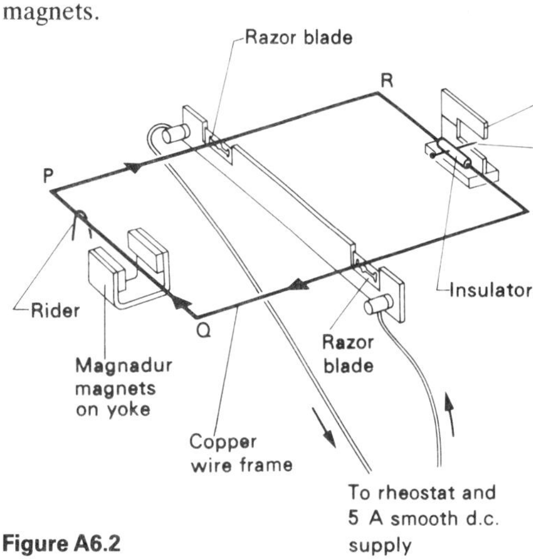


**Figure A6.2**

**6.5** See the key points.

**6.6** Upwards out of the paper.

**6.7** (a) Zero as there is no magnetic flux density component at right angles to the wire.
(b) $F = BIL = 1.5 \times 10^{-2} \times 4 \times 0.120 = 7.2 \times 10^{-3}$ N
(c) The flux density component at right angles to the wire is $B \sin 30°$ and so $F = BIL \sin 30° = 3.6 \times 10^{-3}$ N.

**6.8** (a) The current flows anticlockwise round the circuit. Thus the force acts horizontally to the right.
(b) $F = BIL = 0.05 \times 2.0 \times 0.10 = 0.01$ N
(c) The component of the weight acting down the incline, angle $\theta$ to the horizontal, is $mg \sin \theta$ (see Chapter 1, Lesson 1) and the component of the force due to the current that is up the incline is $F \cos \theta$ (see Figure A6.3). Thus for equilibrium $mg \sin \theta = F \cos \theta$ and so $\tan \theta = F/mg = 0.01/(4.0 \times 10^{-3} \times 9.8)$ and $\theta = 14.3°$.

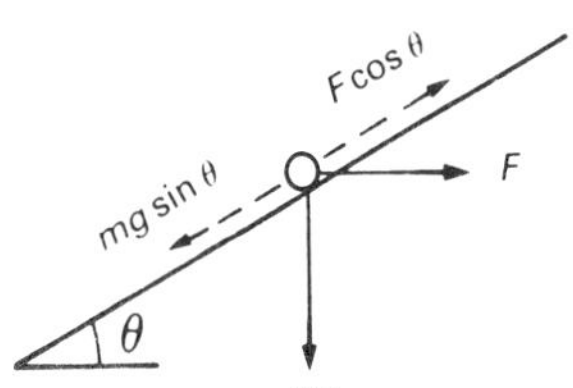


**Figure A6.3**

## Answers III

**6.9** (a) On PQ upwards in the Y direction, on RS downwards in the $Y^1$ direction. On the sides the forces are in the directions shown in Figure 6.3(b) and produce a torque.
(b) $F = NBIl$
(c) $T = NBIlb \sin \theta = NBIA \sin \theta$, where $A$ is the area of the coil.

**6.10** The maximum torque is when $\sin \theta = 1$. Then $T = NBIA$ (see 6.9 above) $= 100 \times 0.4 \times 80 \times 10^{-3} \times 2.0 \times 10^{-4} = 6.4 \times 10^{-4}$ N m.

**6.11** See Figure A6.4. The meter consists essentially of a coil of wire able to rotate in a radial magnetic field, this being produced in the air gap between shaped pole pieces of a permanent magnet and a fixed soft-iron cylinder. When a current passes through the coil it experiences a torque. Its rotation is opposed by hair springs and so the coil deflects until the torque due to the current is equal to, but in the opposite direction to, the

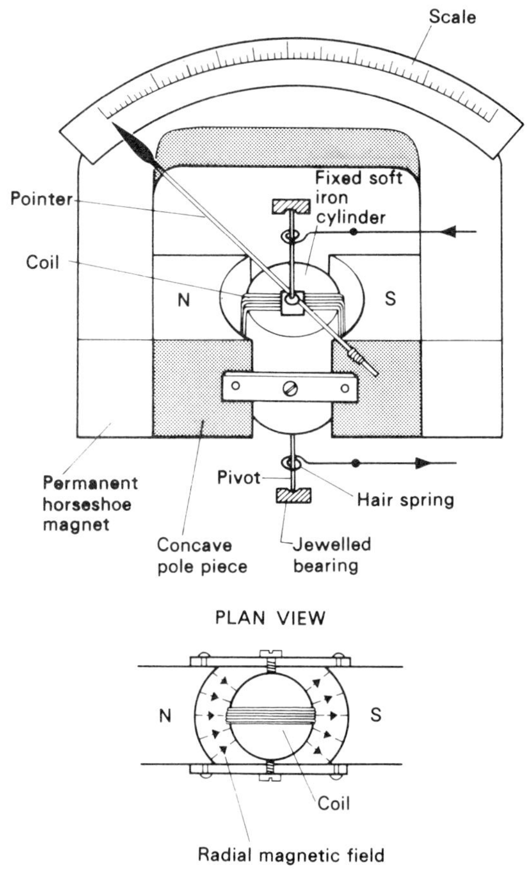


**Figure A6.4 Moving-coil galvanometer**

resisting torque. The deflection is proportional to the current. A linear scale is achieved by the use of the radial field; only then is the torque due to the current independent of the angle at which the coil is.

**6.12** (a) $NBIA$ (See the answer to SAQ 6.9.)
(b) $c\alpha = NBIA$ and so $I = (c/NBA)\alpha$.
(c) Current sensitivity $= \alpha/I = NBA/c$. Thus greater sensitivity can be obtained by a stronger magnetic field $B$, a larger number of turns $N$, a greater coil area $A$, but a lower value of $c$. In practice increasing $N$ and $A$ would increase the resistance of the coil, which is not desirable. Thus the sensitivity is mainly changed by changing the springs, or, where a suspension is used, the wire used for it.
(d) $V = IR = (cR/NAB)\alpha$. Thus voltage sensitivity $= \alpha/V = \mathrm{NAB}/cR$.

**6.13** See 6.12 above.
(a) Current sensitivity $= NBA/c$, thus increasing $N$ by a factor of 30/20 increases the current sensitivity by the same factor.
(b) Voltage sensitivity $= NAB/cR$ and thus increasing $N$ by a factor of 30/20 and increasing $R$ by a factor of 9/6 means a change in the voltage sensitivity of $(30/20) \div (9/6) = 1$, i.e. no change.

**6.14** See 6.12 above. $I = (c/NBA)\alpha = 2.0 \times 10^{-9} \times 30/(50 \times 1.0 \times 10^{-2} \times 1.0 \times 10^{-4}) = 1.2 \times 10^{-3}$ A.

# Lesson 7 Magnetic flux density

## Aims

After completing this lesson you should be able to

- state how the magnetic flux density (a) near a long straight conductor, (b) inside a circular coil and (c) inside a solenoid, is related to the current
- define *permeability*
- calculate the force between current-carrying conductors
- define the ampere
- apply the *Biot–Savart law* and *Ampere's law* to the determination of flux densities.

## Prerequisites

Lesson 6 on magnetic fields. Calculus is needed for some of the questions.

## STUDY

**I** **Read** about the factors determining the magnetic flux density near a long straight wire, inside a circular coil and inside a solenoid.
References: *B* PP ch. 11, *B* ST5 ch. 3, *D* ch. 13, *N* ch. 34, *W et al.* (ch. 37).

### Self-assessment questions

**7.1** Experimentally the magnetic flux density near a long straight wire is found to be proportional to the current $I$ and inversely proportional to the distance $a$ from the wire $B \propto I/a$.
(a) By what factor is the flux density near a long straight wire changed if the distance from the wire is doubled?
(b) The flux density $B = \mu_0 I/2\pi a$, where $\mu_0$ is the *permeability of free space* (value $4\pi \times 10^{-7}\,\mathrm{N\,A^{-2}}$). Hence calculate the flux density a distance of 20 mm from a wire carrying a current of 2 A.

**7.2** For a coil of $N$ turns with a radius $r$ the flux density at the centre of the coil is given by $B = \mu_0 NI/2r$. What flux density is produced at the centre of a coil of 50 turns, radius 10 cm and carrying a current of 1.0 A? ($\mu_0 = 4\pi \times 10^{-7}\,\mathrm{N\,A^{-2}}$)

**7.3** The flux density inside a solenoid is $\mu_0 nI$, where $n$ is the number of turns per unit length of solenoid and $I$ the current carried by the solenoid. What current will need to be passed through a solenoid of length 20 cm and with 1000 turns if a flux density inside the solenoid of $2.0 \times 10^{-3}$ T is to be produced? ($\mu_0 = 4\pi \times 10^{-7}\,\mathrm{N\,A^{-2}}$)

***7.4** A compass needle at the centre of a flat circular coil of wire points north when no current flows through the coil. When current flows through the coil the compass needle points north-east. If the coil has its plane along the north–south axis and has 100 turns of radius 10.0 cm, what is the current carried by the coil if the horizontal component of the earth's field is $2.0 \times 10^{-5}$ T? ($\mu_0 = 4\pi \times 10^{-7}\,\mathrm{N\,A^{-2}}$)

***7.5** (a) Solenoid X has twice the number of turns that solenoid Y has and is twice as long. For the same current how does the flux density in X compare with that in Y?
(b) A solenoid is in the form of a spring. If the solenoid is stretched to twice its length how is the flux density affected, the current remaining constant?

**II Read** about the definition of the ampere and calculations involving forces between current-carrying conductors.
References: *B* PP ch. 11, *B* ST5 ch. 3, *D* ch. 13, *N* ch. 34, *Wh&H* ch. 53.

**Self-assessment questions**

***7.6** A current of 2 A is passed in the same direction through each of two parallel long wires, a distance of 30 mm apart.
(a) What is the flux density produced at one wire by the current in the other?
(b) What is the force acting on a wire, per metre length, due to it being in the magnetic field of the other wire? ($\mu_0 = 4\pi \times 10^{-7}\,\mathrm{N\,A^{-2}}$)

***7.7** Derive an equation for the force acting on unit length of a long straight wire carrying a current $I_1$ when a distance $a$ from a parallel long straight wire carrying a current $I_2$.

***7.8** Define the ampere.

**III Read** about the use of the Biot–Savart law and Ampere's law for the determinations of flux densities.
References: Biot–Savart law – *D* ch. 13, *N* ch. 34, *W et al.* ch. 37, *Wh&H* ch. 53. *B* PP ch. 11 without stating the law effectively uses it for a solenoid. Ampere's law – *B* ST5 ch. 3, *N* ch. 34, *W et al.* (ch. 37), *Wh&H* ch. 53.

**Self-assessment questions**

****7.9** According to the *Biot–Savart law* a conductor is considered as consisting of a number of very short lengths of conductor with each length contributing to the total field at some point. For a very short length $\delta l$ of conductor carrying a current $I$, the flux density $\delta B$ produced at a point a distance $r$ from the element is

$$\delta B = \frac{\mu_0 I\,\delta l \sin\theta}{4\pi r^2}$$

where $\theta$ is the angle between the length $\delta l$ and the line joining it to the point concerned.
Consider a circular coil, just one turn, divided into segments of length $\delta l$ (Figure 7.1).

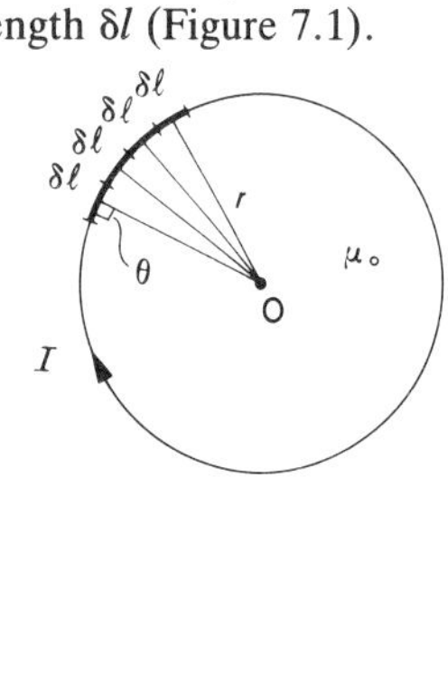


**Figure 7.1**

The magnetic flux density at the centre of the coil can be obtained by considering the flux density produced by each of these segments and then summing them. Taking into account that $\sin\theta$ has the value 1 and $\Sigma\,\delta l = 2\pi r$, derive the equation for the flux density at the centre of the single turn and then the flux density when there are $N$ turns.

****7.10** *Ampere's law* can be written as

$$\oint B\,\mathrm{d}l = \mu_0 I$$

The $\oint$ symbol means that the values of $B\,dl$ are summed all the way round a closed loop, that loop enclosing the current $I$.
(a) Consider a circular loop of radius $r$ drawn concentrically round a long straight wire carrying a current $I$ (Figure 7.2). The flux density $B$ is constant all the way round this circular loop, having a value $B$. Thus we can write

$$B\oint \mathrm{d}l = \mu_0 I$$

What is the total length of the loop? Hence derive the relationship for the flux density at a distance $r$ from the wire.

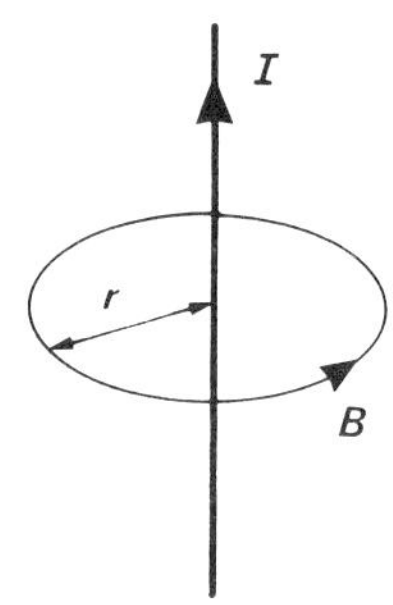


**Figure 7.2**

(b) Consider a solenoid of length $L$, with $N$ turns and carrying a current $I$. For the closed loop shown in Figure 7.3 derive a value for the integral and equate it to $\mu_0 NI$. Outside the solenoid the field can be considered to have zero value. Hence obtain a relationship for the flux density on the axis of the solenoid.

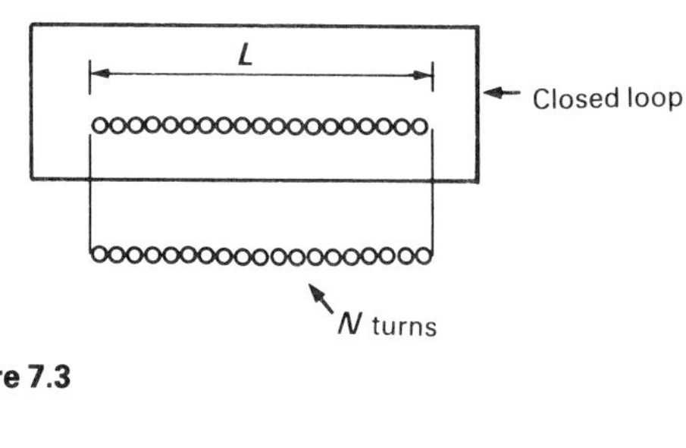


**Figure 7.3**

# SELF-ASSESSMENT

## Key points

For a long straight wire $B = \mu_0 I/2\pi a$ at a distance $a$ from the wire. For a flat coil of $N$ turns and radius $r$, at its centre $B = \mu_0 NI/2r$. For a solenoid of $n$ turns per unit length, inside the solenoid $B = \mu_0 nI$. $\mu_0$ is called the **permeability of free space**.

The force on a length $L$ of current-carrying wire by a long current-carrying wire parallel to it and a distance $a$ from it is $\mu_0 I_1 I_2 L/2\pi a$.

**Biot–Savart law**: for a very short length $\delta l$ of conductor carrying a current $I$, the flux density $\delta B$ produced at a point a distance $r$ from the element is

$$\delta B = \frac{\mu_0 I\,\delta l \sin\theta}{4\pi r^2}$$

where $\theta$ is the angle between the length $\delta l$ and the line joining it to the point concerned.

**Ampere's law** can be written as

$$\oint B\,\mathrm{d}l = \mu_0 I$$

## Answers I

**7.1** (a) halved (b) $B = 4\pi \times 10^{-7} \times 2/(2\pi \times 20 \times 10^{-3}) = 2.0 \times 10^{-5}$ T

**7.2** $B = 4\pi \times 10^{-7} \times 50 \times 1.0/(2 \times 10 \times 10^{-2}) = 3.1 \times 10^{-4}$ T

**7.3** $B = \mu_0 nI = 4\pi \times 10^{-7} \times 1000/(20 \times 10^{-2}) \times I = 2.0 \times 10^{-3}$, hence $I = 0.32$ A.

**7.4** Flux density due to the coil $= \mu_0 NI/2r = 4\pi \times 10^{-7} \times 100 \times I/(2 \times 10.0 \times 10^{-2}) = 6.28 \times 10^{-4} I$. The direction of the field is from west to east. The resultant field due to this combining with the earth's field is at an angle of 45° to either field (in north-east direction). Thus $B_e/B = \tan 45° = 1$ and so $6.28 \times 10^{-4} I = 2.0 \times 10^{-5}$ and $I = 0.032$ A.

**7.5** (a) As the number of turns per unit length is the same the flux densities are the same.
(b) The number of turns per unit length is halved, thus the flux density is halved.

## Answers II

**7.6** (a) $B = \mu_0 I/2\pi a = 4\pi \times 10^{-7} \times 2/(2\pi \times 30 \times 10^{-3}) = 1.3 \times 10^{-5}$ T (b) $F = BIL = 1.3 \times 10^{-5} \times 2 \times 1 = 2.6 \times 10^{-5}$ N

**7.7** $B = \mu_0 I_1/2\pi a$, so $F = BI_2 L = \mu_0 I_1 I_2 L/2\pi a$.

**7.8** *The ampere is that steady current which, flowing in two infinitely long straight parallel conductors of negligible circular cross-sectional area placed 1 m apart in a vacuum, causes each wire to exert a force of $2 \times 10^{-7}$ N on each metre of the other wire.*

## Answers III

**7.9** $B = \Sigma \dfrac{\mu_0 I\,\delta l \sin\theta}{4\pi r^2} = \dfrac{\mu_0 I \sin\theta}{4\pi r^2}\Sigma\,\delta l$

But $\Sigma\,\delta l = 2\pi r$ and $\sin\theta = 1$, hence

$$B = \frac{\mu_0 I\, 2\pi r}{4\pi r^2} = \frac{\mu_0 I}{2r}$$

For $N$ turns we thus have $B = \mu_0 NI/2r$.

**7.10** (a) $B \oint \mathrm{d}L = B \times 2\pi r = \mu_0 I$, hence $B = \mu_0 I/2\pi r$.
(b) $\oint B\,\mathrm{d}l = BL$ for that part within the solenoid and as the field is assumed to be zero outside it $B\,\mathrm{d}L$ is zero for all other parts of the loop. Thus $BL = \mu_0 NI$ and $B = \mu_0 NI/L = \mu_0 nI$ where $n = N/L$.

# Lesson 8 Charged particles in fields

## Aims

After completing this lesson you should be able to
- describe and explain Millikan's experiment
- describe and carry out calculations on the behaviour of beams of charged particles in electric and magnetic fields.

## Prerequisites

Lessons 1 and 2 on electric fields, Lesson 6 on magnetic fields. For the Millikan experiment a knowledge of viscosity and Stokes' law is required (Lesson 3 of Chapter 2). A basic knowledge is assumed of linear motion, velocity as a vector, energy, and motion in a circle (Chapter 1, Lessons 1, 2, 5 and 7 respectively).

# STUDY

**I Read** about Millikan's experiment.
References: *B* PP ch. 10, *B* ST5 ch. 4, *D* ch. 20, *N* ch. 38, *W et al.* ch. 43(31), *Wh&H* ch. 60.

**Self-assessment questions**

**8.1** Without giving any experimental or theoretical details, explain how the results of Millikan's experiment led to the idea that charge comes in 'packets', the size of the smallest packet being the charge carried by an electron.

**8.2** (a) One part of Millikan's experiment involved watching a charged oil drop falling freely under gravity in air. Three forces act on the drop, these being the weight, the viscous drag and the Archimedean upthrust. The viscous drag force depends on the velocity of the falling drop, increasing as the velocity increases. When these forces balance and there is no net force the drop falls with a constant velocity, called the *terminal* velocity. By measuring this velocity, and with other known quantities, Millikan was able to calculate the radius of the oil drop. Derive the relationship from which the radius can be calculated.
(b) To obtain the charge carried by the oil drop Millikan applied an electric field acting in such a direction that he was able to hold the drop stationary. When this happens the force acting on the drop due to the electric field must be opposite and equal to the (weight−upthrust) force. The electric field was determined by a measurement of the potential difference applied between two parallel plates a distance $d$ apart, the oil drop being between the plates. Derive the relationship between the charge on the oil drop and the potential difference.
(c) Hence explain how Millikan was able to measure the charge carried by the oil drop.

***8.3** In a form of Millikan's experiment an oil drop was observed to fall with a constant velocity of $2.5 \times 10^{-4}\,\text{m s}^{-1}$ in the absence of any electric field. When, however, a potential difference of 1000 V was applied between the horizontal plates, 10 mm apart, the drop remained stationary between them. If the oil has a density of $900\,\text{kg m}^{-3}$, air a density of $1.20\,\text{kg m}^{-3}$ and a viscosity of $1.8 \times 10^{-5}\,\text{N s m}^{-2}$, what is (a) the radius of the oil drop and (b) the number of electronic charges it carries? ($e = 1.6 \times 10^{-19}\,\text{C}$, $g = 9.81\,\text{m s}^{-2}$)

***8.4** In a Millikan oil-drop experiment the two horizontal parallel plates were 1.2 cm apart. An oil drop of mass $4.90 \times 10^{-15}$ kg was held stationary between the plates by a potential difference of 1.2 kV. What is the charge on the oil drop? Neglect the density of air in comparison with that of the oil. ($g = 9.8\,\text{m s}^{-2}$)

**II Read** about the movement of charged particles in electric and magnetic fields.
References: *B* PP ch. 13, *B* ST5 ch. 4, *D* ch. 20, *N* ch. 38, *W et al.* ch. 43(33), *Wh&H* ch. 60.

**Self-assessment questions**

**8.5** Electrons, each carrying a charge of $1.6 \times 10^{-19}$ C, are emitted from a filament and accelerated to an anode by a potential difference of 100 V maintained between the anode and the filament.
(a) What is the energy gained by a particle carrying a charge $q$ when accelerated through a potential difference of $V$?
(b) By equating the kinetic energy of the electron when it has reached the anode with the energy gained by the acceleration, calculate the velocity with which the electron reaches the anode. The electron has a mass of $9.1 \times 10^{-31}$ kg.

**8.6** Consider a charged particle with a charge $q$ moving with a velocity $v$ in a magnetic field where the flux density $B$ is at right angles to the path of the particle.
(a) What distance will the charged particle cover in a time $t$?
(b) Current is the rate of movement of charge, thus what is the current due to this charged particle in the time $t$?
(c) The force experienced by a current element in a magnetic field is $BIL$. If we consider the answer to (a) to represent the effective length of the current element, what is the force acting on the charged particle in the magnetic field?

**8.7** Consider a charged particle moving with a velocity $v$ in a magnetic field of flux density $B$, this being always at right angles to the path of the particle.
(a) In what direction will the force acting on the charged particle be?
(b) The force causes the particle to move in a circular path of radius $r$. What is the relationship between this radius and the flux density $B$?

***8.8** Electrons are accelerated from rest through a potential difference of 3000 V and then travel as a narrow beam through a magnetic field of $2.0 \times 10^{-3}$ T which is perpendicular to the beam. What is (a) the speed of the electrons entering the magnetic field and (b) the radius of their circular path in that field? ($e/m = 1.8 \times 10^{11}\,\text{C kg}^{-1}$)

**8.9** (a) An electric field $E$ is applied at right angles to the path of a beam of charged particles, each carrying a charge $q$. What is the force on a charged particle in this beam?
(b) A magnetic field of flux density $B$ is applied at right angles to the path of the beam of charged particles. What is the force on a charged particle in this beam due to the magnetic field?
(c) If the electric field and the magnetic field cause deflections of the beam in opposite directions, what is the relationship between $E$ and $B$ if the net result of simultaneously applying the fields is no deflection of the beam?
(d) What are the directions of the electric and magnetic fields for this situation to occur?

***8.10** What is the velocity of charged particles if a magnetic field of flux density $1.2 \times 10^{-2}$ T at right angles to an electric field of $2.0 \times 10^{4}$ V m$^{-1}$, both being at right angles to the velocity, results in no deflection of the charged particles from their straight-line path?

**8.11** Figure 8.1 shows the path of a beam of negatively charged particles, each carrying a charge $q$, on passing through an electric field.

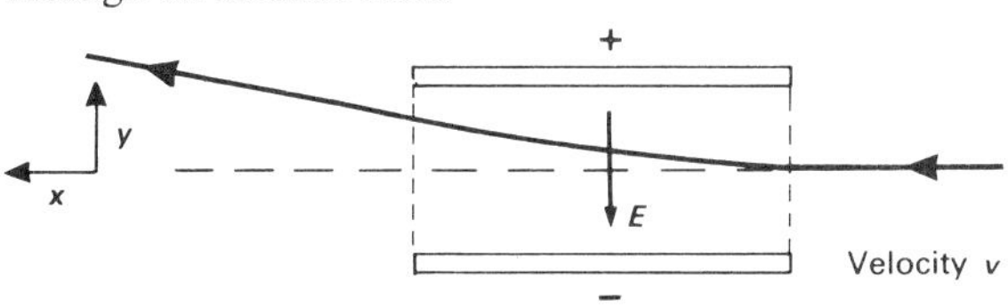


**Figure 8.1**

(a) Neglecting forces due to gravity, what is the force in the $y$-direction acting on the charged particles in the electric field?
(b) If each charged particle has a mass $m$, what is the vertical acceleration of each charged particle?
(c) What is the vertical displacement $y$ of a charged particle after a time $t$ with this acceleration?
(d) During this time $t$ the particle is also moving in the $x$-direction with a constant velocity. What is the distance covered in this direction in the time?
(e) Obtain a relationship between $y$ and $x$ for the beam while in the electric field and hence the equation of the parabolic path described by the beam.
(f) When the beam leaves the electric field what can be said about its path?

***8.12** Two horizontal parallel metal plates 40 mm long and 30 mm apart, with a potential difference of 250 V between them, have a beam of electrons with a velocity of $1.0 \times 10^{7}$ m s$^{-1}$ injected horizontally midway between the plates in a direction parallel to the 40 mm edge. What is the vertical deflection of the beam when it emerges from between the plates? ($e/m = 1.8 \times 10^{11}$ C kg$^{-1}$)

# SELF-ASSESSMENT

## Key points

Millikan's oil-drop experiment gave evidence that charge comes in 'packets' of a predetermined size and enabled the basic unit, i.e. the charge on the electron, to be determined.
For charged particles, each carrying a charge $q$, accelerated through a potential difference $V$ the gain in kinetic energy $= \frac{1}{2}mv^2 = qV$.
A charged particle moving with a velocity $v$ in a magnetic field, flux density $B$ at right angles to the path of the charge, experiences a force $F = Bqv$. The force is at right angles to both $B$ and $v$.
Charged particles passing through an electric field experience a force, in the direction of the field if they are positively charged, of $F = Eq$.

## Answers I

**8.1** The charge on a large number of oil drops was measured and found to be an integral multiple of a basic charge, that carried by the electron.

**8.2** (a) $\frac{4}{3}\pi r^3 \rho g = \frac{4}{3}\pi r^3 \sigma g + 6\pi\eta r v$ and so $r^2 = 9\eta v/2(\rho-\sigma)g$.
(b) $Eq = Vq/d = \frac{4}{3}\pi r^3 g(\rho-\sigma)$
(c) The first part (a) was used to determine the radius of an oil drop. This same oil drop was then used in part (b) to enable the charge $q$ to be determined. The density $\rho$ and viscosity $\eta$ of the oil needed to be determined, also the density $\sigma$ of the air. The only other quantities requiring measurement were the velocity $v$, the potential difference $V$, and $d$, the separation between the parallel plates.

**8.3** Using the expressions derived in 8.2 above, $r^2 = 9 \times 1.8 \times 10^{-5} \times 2.5 \times 10^{-4}/(2 \times 898.8 \times 9.81)$ and so $r = 1.52 \times 10^{-6}$ m. Therefore $q = F/E = 4\pi \times (1.52 \times 10^{-6})^3 \times 9.81 \times 10 \times 10^{-3} \times 898.8/(3 \times 1000) = 1.30 \times 10^{-18}$ C. Hence the number of charge carriers is $1.30 \times 10^{-18}/(1.6 \times 10^{-19}) = 8$.

**8.4** $mg = qE = qV/d$ and hence $q = 4.90 \times 10^{-15} \times 9.8 \times 1.2 \times 10^{-2}/(1.2 \times 10^{3}) = 4.8 \times 10^{-19}$ C.

## Answers II

**8.5** (a) energy $= qV$ (b) $\frac{1}{2}mv^2 = qV$ and so $v^2 = 2 \times 1.6 \times 10^{-19} \times 100/(9.1 \times 10^{-31})$ and $v = 5.9 \times 10^6\,\mathrm{m\,s^{-1}}$.

**8.6** (a) $vt$ (b) $q/t$ (c) $F = B(q/t)vt = Bqv$

**8.7** (a) At right angles to both the field and the velocity. Fleming's left-hand rule can be used, but you need to remember that the current direction is the direction of movement of *positive* charges and thus electrons move in the opposite direction to the direction indicated for the current.
(b) The force is always at right angles to the velocity and provides a centripetal force, hence $mv^2/r = Bqv$ and thus $r = mv/Bq$.

**8.8** (a) $\frac{1}{2}mv^2 = eV$ and so $v^2 = 2 \times 3000 \times 1.8 \times 10^{11}$ and $v = 3.3 \times 10^7\,\mathrm{m\,s^{-1}}$. (b) $mv^2/r = Bev$, hence $r = v/(Be/m) = 3.3 \times 10^7/(2.0 \times 10^{-3} \times 1.8 \times 10^{11}) = 0.092\,\mathrm{m}$.

**8.9** (a) $F = Eq$ (b) $F = Bqv$ (c) $Eq = Bqv$ and $v = E/B$. (d) If the beam is horizontal, in the plane of this paper, the electric field is at right angles in the same plane and the magnetic field is vertical, i.e. at right angles to the plane of the paper.

**8.10** See 8.9 above; $v = E/B = 2.0 \times 10^4/(1.2 \times 10^{-2}) = 1.7 \times 10^6\,\mathrm{m\,s^{-1}}$.

**8.11** (a) $F = Eq$ (b) $F = ma$ and so $a = Eq/m$. (c) Using $s = ut + \frac{1}{2}at^2$ with $u$ in the vertical direction zero, then $y = \frac{1}{2}at^2 = \frac{1}{2}Eqt^2/m$. (d) $x = vt$
(e) Eliminating $t$ between the equations gives $y = (Eq/2mv^2)x^2$. (f) It is a straight line in the absence of any further force.

**8.12** Using the equation developed in 8.11 above, $y = (Eq/2mv^2)x^2$ with $E = V/d = 250/(30 \times 10^{-3}) = 8.3 \times 10^3\,\mathrm{V\,m^{-1}}$. Thus $y = 8.3 \times 10^3 \times 1.8 \times 10^{11} \times (40 \times 10^{-3})^2/(2 \times (1.0 \times 10^7)^2) = 0.012\,\mathrm{m}$.

# Lesson 9 Current as charge movement

## Aims

After completing this lesson you should be able to
- derive the relationship between current and the velocity of charge carriers, and use the relationship in problems
- explain the Hall effect, deriving the equation relating the Hall voltage, the magnetic field and the charge carried by the charge carriers.

## Prerequisites

Lessons 1 and 2 on electric fields, Lesson 6 on magnetic fields and Lesson 8 on charged particles in fields. Lesson 4 of Chapter 2, on the structure of matter, is also needed.

## STUDY

**I Read** about the derivation of the equation $I = nvAq$ for the relationship between current and the velocity of charge carriers.
References: *B* PP ch. 10, *B* ST5 ch. 4, *D* ch. 3, *N* ch. 30, *W et al.* ch. 44(32), *Wh&H* ch. 48.

### Self-assessment questions

**9.1** Consider a model for a conductor in which the current is produced by a stream of charge carriers passing along the conductor (Figure 9.1).

(a) If the charge carriers have a velocity $v$ what will be the distance covered by one in a time $t$?
(b) All the charge carriers in a certain volume of the conductor will be able to pass through a particular cross-section XY of the conductor in a time $t$. What is this volume?
(c) If the number of charge carriers per unit volume is $n$ and each carries a charge $q$, what is the amount of charge passing through XY in time $t$?
(d) As current is the rate of movement of charge, derive a relationship for the current in terms of the velocity of the charge carriers.

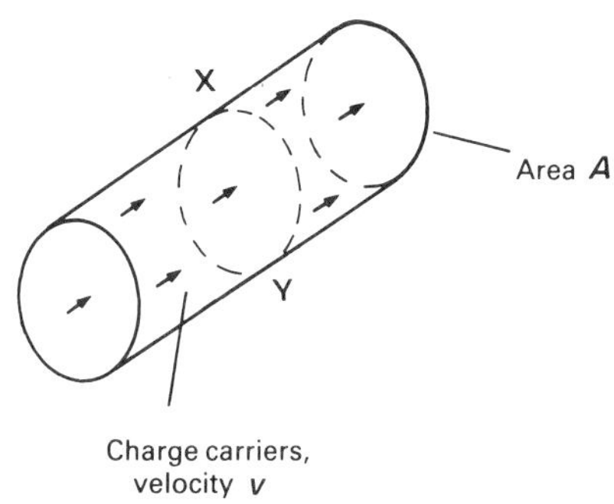


**Figure 9.1**

***9.2** On the basis of the model of a conductor given in SAQ 9.1, what is the velocity of the charge carriers when a current of 1 A passes through a copper wire with a cross-sectional area of 1 $mm^2$? Assume that each copper atom donates one electron for conduction purposes and the atomic mass for copper is 63.6 and its density 8930 kg $m^{-3}$. (Avogadro's number = $6.0 \times 10^{23}$ and $e$ = $1.6 \times 10^{-19}$ C.)

***9.3** A metal wire has a cross-sectional area of 1.0 $mm^2$ and contains $2.0 \times 10^{28}$ free electrons per cubic metre. If the electrons move with a velocity of 1.2 mm $s^{-1}$ along the wire when a potential difference is applied between its ends, what is the current? ($e = 1.6 \times 10^{-19}$ C)

**II Read** about the Hall effect and the derivation of the equation for the Hall voltage.
References: *B* PP ch. 11, *B* ST5 ch. 4, *D* ch. 13, *N* ch. 33, *W et al.* ch. 44(32), *Wh&H* ch. 52.

**Self-assessment questions**

**9.4** What is the Hall effect?

**9.5** Figure 9.2 shows a slice of conducting material with a current $I$ passing through it and a magnetic field of flux density $B$ at right angles to the plane of the slice. Note that the direction of the current, according to normal convention, is opposite to that followed by the negative charge carriers, charge $q$.

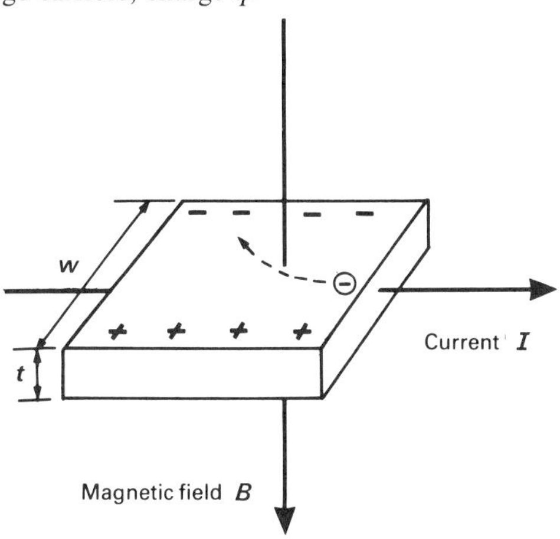


**Figure 9.2**

(a) The charge carriers passing through the slice are acted on by a force due to the presence of the magnetic field. What is this force?
(b) The force causes the charged particles to become deflected from straight-line paths and one edge of the slice becomes negatively charged and the other positively charged. This charge separation produces an electric field $E$. What is the force acting on a charge carrier due to this field?
(c) When equilibrium is reached the force resulting from the magnetic field is balanced by the force resulting from the electric field. Hence derive a relationship between the velocity of the charge carriers and $E$ and $B$.
(d) Because there is an electric field between the two edges of the slice there must be a potential difference. How is the potential difference $V$ related to the electric field?
(e) The current $I$ through the slice is given by $I = nvAq$ (see SAQ 9.1). Hence derive a relationship for this Hall voltage in terms of directly measurable quantities.

***9.6** Calculate the number of charge carriers per unit volume, and hence the number of charge carriers per atom, for aluminium given that in a Hall-effect experiment a strip of aluminium foil of thickness 0.060 mm carrying a current of 5.0 A gave a Hall voltage of 7.1 μV when the magnetic flux density was 2.0 T. (Avogadro constant = $6.0 \times 10^{23}$, $e = 1.6 \times 10^{-19}$ C, atomic mass = 27, density = 2700 kg $m^{-3}$.)

***9.7** Explain how the Hall effect can be used to determine whether the predominant charge carriers in a material are positively or negatively charged.

# SELF-ASSESSMENT

## Key points

The **current** $I$ for charge carriers moving with a velocity $v$ is $I = nvAq$, where $A$ is the cross-sectional area of the conductor, $n$ the number of charge carriers per unit volume and $q$ their charge.
The **Hall effect** is that a transverse potential difference is produced in a conductor when a current passes through it in the presence of a magnetic field at right angles to the current. The **Hall voltage** $V = BI/ntq$, where $B$ is the flux density at right angles to the conductor, $n$ the number of charge carriers per unit volume, $q$ their charge and $t$ the thickness of the conductor.

## Answers I

**9.1** (a) $vt$ (b) $vtA$ (c) $nvtAq$ (d) $I = nvAq$

**9.2** $6.0 \times 10^{23}$ copper atoms have a mass of 63.6 g. This is a volume of $63.6 \times 10^{-3}/8930 = 7.1 \times 10^{-6}$ $m^3$. Hence the number of atoms per $m^3 = 6.0 \times 10^{23}/(7.1 \times 10^{-6}) = 8.5 \times 10^{28}$. Thus as $I = nvAq$ we have $v = 1/(8.5 \times 10^{28} \times 1 \times 10^{-6} \times 1.6 \times 10^{-19}) = 7.4 \times 10^{-5}$ m $s^{-1}$.

**9.3** $I = nvAq = 2.0 \times 10^{28} \times 1.2 \times 10^{-3} \times 1.0 \times 10^{-6} \times 1.6 \times 10^{-19} = 3.8$ A

## Answers II

**9.4** See the key points.

**9.5** (a) $F = Bqv$ (b) $F = Eq$ (c) $Eq = Bqv$, hence $E = Bv$ (d) $E = V/w$ (e) $V/w = Bv = BI/(nAq)$. But $A = wt$, hence $V = BI/ntq$.

**9.6** $6.0 \times 10^{23}$ aluminium atoms have a mass of 27 g. This is a volume of $27 \times 10^{-3}/2700 = 1 \times 10^{-5}\,m^3$. Hence the number of atoms per $m^3 = 6.0 \times 10^{23}/(1 \times 10^{-5}) = 6.0 \times 10^{28}$. Hall voltage $= BI/ntq$ (see 9.5 above), thus $n = 2.0 \times 5.0/(7.1 \times 10^{-6} \times 0.060 \times 10^{-3} \times 1.6 \times 10^{-19}) = 1.47 \times 10^{29}$. Thus number of charge carriers per atom $= 1.47 \times 10^{29}/(6.0 \times 10^{28}) = 2.45$. The number of free electrons per atom is likely therefore to be 2 or 3 to give this average value of 2.45.

**9.7** The Hall voltage is in the opposite sense for positive charge carriers to that with negative charge carriers; see Figure A9.1.

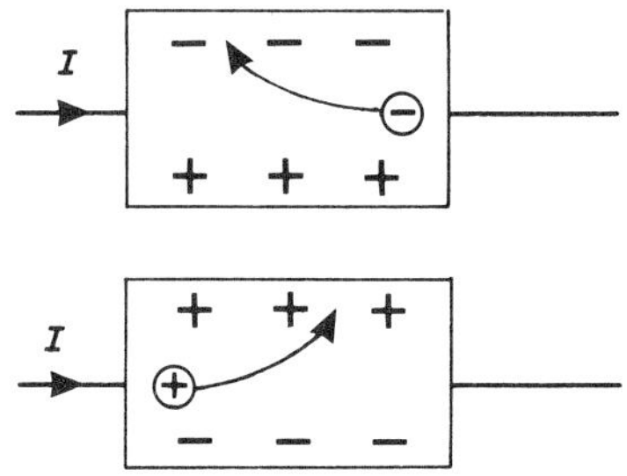


**Figure A9.1** The magnetic field is vertical and into the paper

# Lesson 10 Electromagnetic induction

## Aims

After completing this lesson you should be able to

- explain electromagnetic induction in terms of flux linkage
- state, and use in problems, *Faraday's law* and *Lenz's law*
- derive and use the equation

$$\text{induced e.m.f.} = BLv$$

## Prerequisites

Lesson 6 on magnetic fields, and aspects of Lesson 8 and Chapter 1, Lesson 8. Basic calculus is required.

# STUDY

**I** **Read** about the electromagnetic induction phenomena and the concept of flux and flux linkage. Also read about Faraday's and Lenz's laws.
References: *B* PP ch. 11, *B* ST4 ch. 3, *D* ch. 14, *N* ch. 35, *W et al.* ch. 31 (35), *Wh&H* ch. 55.

### Self-assessment questions

**10.1** Define *flux* and state its unit.

**10.2** Give examples of two ways by which electromagnetic induction can be demonstrated, one involving a permanent magnet and the other a current in a conductor. In both cases explain the production of an induced e.m.f. in terms of flux linkage.

**10.3** State *Faraday's law* and *Lenz's law*.

**10.4** A coil has $N$ turns, each with an area $A$. What is the flux linked by such a coil when the flux density is at an angle $\theta$ to the normal to the plane of the coil?
If the coil now rotates with a steady angular velocity $\omega$, derive an equation for the rate of change of flux linked by the coil and hence the e.m.f. induced in the coil.

***10.5** What is the maximum e.m.f. induced in a coil of 500 turns, each with an area $4.0\,cm^2$, which makes 50 revolutions per second in a uniform magnetic field of flux density 0.040 T?

***10.6** A flat circular coil with 50 turns, each of area $8.0 \times 10^{-3}\,m^2$, lies on a horizontal surface. If the earth's field has a vertical component of $4.0 \times 10^{-5}$ T, what will be the size of the pulse of e.m.f. induced in the coil when it is flipped over in a time of 0.10 s?

**10.7** One method of determining flux density is to use a search coil with a galvanometer.
(a) Such a search coil has $N$ turns, each of cross-sectional area $A$. What is the e.m.f. induced in the coil when it is moved from a place where the flux density is $B$, at right angles to the plane of the coil, to where it is zero, in a time $t$?
(b) The induced e.m.f. causes a current $I$ to flow through the galvanometer. If the circuit has a resistance $R$, how is the current related to the flux density $B$?
(c) The current will only flow for a short interval of time. As $I = dQ/dt$, what is the amount of charge that flows through the galvanometer in the time $t$?

(d) The galvanometer used is generally a *ballistic* galvanometer. Such a galvanometer gives an initial deflection which is proportional to the charge passed through it, when current pulses of short duration are involved. Hence, how is the deflection of the galvanometer related to the flux density being measured?
(e) A search coil with 20 turns, each of area $2.0 \times 10^{-4}\,m^2$, is connected to a ballistic galvanometer, the total circuit resistance being $100\,\Omega$. What charge will flow through the galvanometer when the coil is moved from a region where the flux density at right angles to the plane of the coil is 0.10 T to a region where it is negligible?

**II** **Read** about the derivation and use of the equation

$$\text{induced e.m.f.} = BLv$$

References: *B* PP ch. 11, *D* ch. 14, *N* ch. 35, *W et al.* (ch. 35), *Wh&H* ch. 55.

**Self-assessment questions**

**10.8** Consider a conducting rod of length $L$ moving with a steady velocity $v$ through, and at right angles to, a uniform magnetic field of flux density $B$ (Figure 10.1). The induced e.m.f. $\mathscr{E}$ causes a current $I$ to flow in the circuit of which the rod is a part.

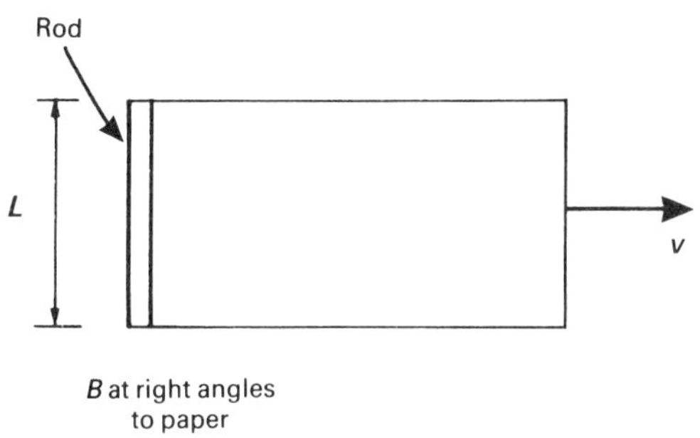


**Figure 10.1**

(a) What is the force acting on the rod due to the current flowing through it?
(b) Because the rod is in equilibrium, it having a constant velocity, what must be the work done per second, i.e. the power?
(c) The electrical power is $\mathscr{E}I$, hence as energy is conserved derive a relationship between the induced e.m.f. $\mathscr{E}$ and the velocity $v$.

**10.9** This is an alternative to the derivation given in SAQ 10.8. Consider a conducting rod of length $L$ moving with a steady velocity $v$ through, and at right angles to, a uniform magnetic field of flux density $B$.
(a) The rod contains free electrons, each carrying a charge $e$. What is the magnetic force acting on these electrons?
(b) The magnetic force will cause the electrons to move along the rod and so produce a charge separation, one end of the rod becoming negatively charged and the other end positively charged. This will produce an electric field. The electric field will be the potential gradient. What is the potential gradient if there is a potential difference of $V$ produced between the ends of the rod?
(c) The electrons will only continue to move along the rod until an equilibrium is produced when the force on them due to the magnetic field is balanced by the force on them due to the electric field. Hence derive a relationship between the potential difference $V$ and the velocity $v$. As the rod is on open circuit this potential difference equals the induced e.m.f. (see p. 138). Hence, what is the relationship between the induced e.m.f. and the velocity $v$?

***10.10** An aircraft is flying horizontally at $200\,m\,s^{-1}$ through a region where the vertical component of the earth's magnetic field is $4.0 \times 10^{-5}$ T. If the aircraft has a wing span of 40 m, what will be the potential difference produced between the wing tips?

**10.11** This question concerns a rotating metal disc with an area of it near its circumference rotating through a magnetic field (see Figure 10.2). The arrangement is sometimes known as a *homopolar* or *disc generator*.
We can consider the disc to be effectively a large number of spokes, each spoke being a conducting rod moving through a magnetic field.

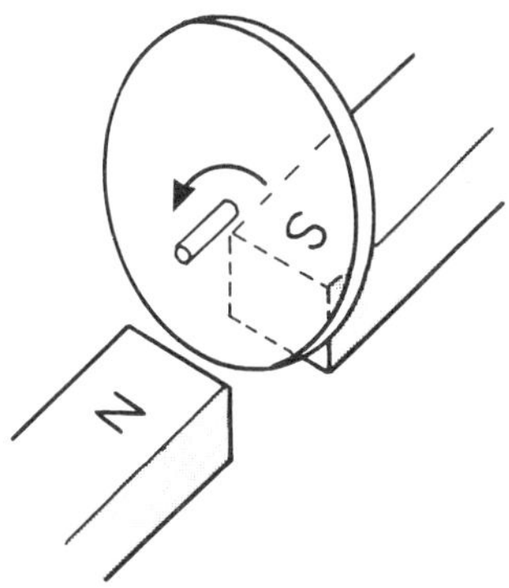


**Figure 10.2**

(a) If the disc has an angular velocity $\omega$, what is the linear velocity of a point a distance $r$ from the centre?
(b) The centre of the disc, and hence one end of the rod, has zero velocity. What therefore is the average velocity of the rod?

(c) Induced e.m.f. = $BLv$. What is the induced e.m.f. between the centre and the end of the rod a distance $r$ out from the centre?
(d) The angular velocity $\omega = 2\pi f$, where $f$ is the frequency of the rotation. Hence derive a relationship for the induced e.m.f. between the centre of the disc and the edge, radius $r$, in terms of the frequency.

***10.12** A circular metal disc with a radius of 10 cm rotates at 10 revolutions per second. If the disc is in a uniform magnetic field of 0.020 T at right angles to the plane of the disc, what will be the e.m.f. induced between the centre and the rim of the disc?

***10.13** A cycle wheel has metal spokes of length 40 cm. If the wheel is held with its plane horizontal and rotated at 2.0 revolutions per second, what will be the e.m.f. induced between the rim and the axle if the earth's magnetic field has a vertical component of $4.0 \times 10^{-5}$ T?

# SELF-ASSESSMENT

## Key points

**Magnetic flux** $\Phi = BA$ where $B$ is the flux density perpendicular to the area $A$. The unit of flux is the weber (Wb). The magnetic flux passing through a loop or coil is said to be *linking* the loop. **Flux linkage** = $N\Phi$, where $N$ is the number of turns of the coil.
**Faraday's law**: the induced e.m.f. is directly proportional to the rate of change of flux linkage or rate of flux cutting.
**Lenz's law**: the direction of the induced e.m.f. is such that it tends to oppose the flux change causing it, and does oppose it if an induced current flows.

$$\therefore\ \mathscr{E} = -N\frac{d\Phi}{dt}$$

For a conducting rod of length $L$ moving with a velocity $v$ at right angles to flux density $B$, induced e.m.f. = $BLv$.

## Answers I

**10.1** See the key points.

**10.2** Example 1: a permanent magnet moving in the vicinity of a coil of wire, the coil being connected to a galvanometer. The flux linked by the coil is changing due to the movement of the magnet.
Example 2: A current passing through a coil which is in the vicinity of, or perhaps wound on top of, another coil which is connected to an oscilloscope. When the current in first coil changes the flux produced by it changes and hence the flux linked by the other coil changes and an e.m.f. is induced.

**10.3** See the key points.

**10.4** $\Phi = NAB\cos\theta$. $\mathscr{E} = -d(NAB\cos\theta)/dt = -NAB\,d(\cos\omega t)/dt = NAB\,\omega\sin\omega t$. Maximum e.m.f. = $NAB\,\omega$ and hence we can write $\mathscr{E} = \mathscr{E}_{max}\sin\omega t$.

**10.5** Maximum e.m.f. = $NAB\omega$ (see 10.4 above) = $500 \times 4.0 \times 10^{-4} \times 0.040 \times 2\pi \times 50 = 2.5$ V.

**10.6** Change in flux linked in 0.10 s is $2 \times 50 \times 8.0 \times 10^{-3} \times 4.0 \times 10^{-5}$ Wb and so rate of change of flux linked = induced e.m.f. = $3.2 \times 10^{-4}$ V.

**10.7** (a) Average induced e.m.f. = $NAB/t$ (b) $IR = NAB/t$ (c) $I = Q/t = NAB/Rt$ and so $Q = NAB/R$ (d) Deflection is proportional to $B$. (e) $Q = NAB/R = 20 \times 2.0 \times 10^{-4} \times 0.10/100 = 4 \times 10^{-6}$ C

## Answers II

**10.8** (a) $F = BIL$ (b) power = $Fv$ (i.e. $F \times$ distance covered per second) (c) $\mathscr{E}I = Fv = BILv$ and so $\mathscr{E} = BLv$.

**10.9** (a) $F = Bev$ (b) $E = V/L$ (c) $E = Bev$ and so $(V/L)e = Bev$ and $V = BLv$. Hence e.m.f. = $BLv$.

**10.10** p.d. = $BLv = 4.0 \times 10^{-5} \times 40 \times 200 = 0.32$ V

**10.11** (a) $v = r\omega$ (b) $\frac{1}{2}r\omega$ (c) e.m.f. = $Br(\frac{1}{2}r\omega) = \frac{1}{2}Br^2\omega$ (d) e.m.f. = $B\pi r^2 f$

**10.12** See 10.11 above; e.m.f. = $B\pi r^2 f = 0.020 \times \pi \times 0.10^2 \times 10 = 6.3 \times 10^{-3}$ V.

**10.13** See 10.11 above; e.m.f. = $B\pi r^2 f = 4.0 \times 10^{-5} \times \pi \times 0.40^2 \times 2.0 = 4.0 \times 10^{-5}$ V.

# Lesson 11 Applications of electromagnetism

## Aims

After completing this lesson you should be able to

- explain the principle of and solve problems involving the transformer
- explain eddy currents and their involvement in induction heating and galvanometer damping
- explain the principle of the d.c. motor, both series- and shunt-wound.

## Prerequisites

Lesson 6 on magnetic fields and Lesson 10 on electromagnetic induction.

# STUDY

**I Read** about the transformer, and the *ideal* transformer.
References: *B* PP ch. 11, *B* ST4 ch. 3, *D* ch. 14, *N* ch. 35, *W et al.* ch. 32(38), *Wh&H* ch. 56.

### Self-assessment questions

**11.1** Essentially a transformer consists of two coils, called the *primary* and the *secondary*, wound on an iron core.
(a) When an alternating potential difference is applied to the primary coil an alternating e.m.f. is produced in the secondary coil. Explain how this is produced.
(b) What would be produced in the secondary coil if the primary coil had a direct potential difference applied to it?
(c) Why are the coils wound on an iron core?

**11.2** An *ideal* transformer is one in which the primary coil has negligible resistance and all the flux in the iron core links both the primary and secondary windings, i.e. there is no flux leakage. Consider such a transformer, having $N_p$ turns in the primary and $N_s$ turns in the secondary.
(a) If $\Phi$ is the flux that is produced at some instant of time by the current in the primary coil, what is the flux linked by one turn of the secondary coil? What is the total flux linked by the secondary coil?
(b) What is the e.m.f. induced in the secondary coil if the rate of change of flux produced by the primary coil is $d\Phi/dt$?
(c) But the changing flux links both the primary and the secondary coils. What therefore is the e.m.f. induced in the *primary* coil by this? This is referred to as the *back* e.m.f.
(d) For the primary we must have $\mathscr{E}_p$ + back e.m.f. = $IR$, where $\mathscr{E}_p$ is the applied e.m.f., $I$ the current in the primary and $R$ its resistance. But if we take the transformer to be ideal then $R = 0$. Hence derive an equation relating $\mathscr{E}_p$ and $\mathscr{E}_s$.

**11.3** Consider the situation where the secondary coil is not on open circuit and a current $I_s$ flows in it.
(a) If for this condition there is a primary current $I_p$, what is the power input to the transformer?
(b) What is the power output from the transformer?
(c) If there is no power loss, what is the relationship between the ratio $I_p/I_s$ and the turns ratio?

***11.4** What is the turns ratio required for a transformer required to step down an alternating voltage from 240 V to 4 V?

***11.5** If the secondary circuit of a transformer is open the current in the primary is very small. Why is this, and also why does the current in the primary increase when the secondary circuit is closed and a secondary current flows?

**II Read** about eddy currents and their involvement in induction heating and galvanometer damping.
References: *D* ch. 14, *N* ch. 35, *W et al.* (ch. 38), *Wh&H* ch. 56.

### Self-assessment questions

**11.6** Explain what *eddy currents* are and how they are produced.

**11.7** With a transformer, eddy currents in the core lead to a loss in electrical energy due to their heating effect. How can such losses be reduced?

**11.8** In a moving-coil galvanometer the coil is wound on a metal frame in order to damp the motion of the coil and make it come rapidly to its final deflection without undue oscillation about that position. Explain how this is achieved.

**11.9** In induction heating a coil carrying a high-frequency alternating current surrounds the material to be heated, the material being either of conducting material or having conducting parts. Explain how the material is heated.

**III** **Read** about the principles of the d.c. motor and the forms of construction termed *series*- and *shunt-wound*. References: *B* PP ch. 11, *D* ch. 14, *N* ch. 35, *W et al.* ch. 30(35), *Wh&H* ch. 56.

**Self-assessment questions**

**11.10** Essentially a d.c. motor consists of a coil on an axle in a magnetic field. When the coil is supplied with a current it experiences a couple and rotates. Explain why a split-ring commutator has to be used.

**11.11** When the armature coil in a motor rotates, it cuts the magnetic flux of the field magnet and an e.m.f. is induced in it, this e.m.f. being called the *back* e.m.f.
(a) What is the relationship between the applied potential difference $V$, the induced e.m.f. $\mathscr{E}$ and the current $I$ through the armature coil, resistance $r$?
(b) Hence obtain an expression for the mechanical power output of the motor. Hint: this is the difference between the input power and that lost as heat in the windings of the coil.

***11.12** This follows from the previous question. Consider a motor with an input from the mains supply, 240 V a.c., and having an armature coil of resistance 1 Ω. What is the current in the armature when (a) the motor is just switched on, (b) the motor has attained full speed and the back e.m.f. has risen to 230 V? Hence explain why a starting resistor is used in series with the armature coil.

****11.13** With the series-wound motor the field and armature coils are in series. With the shunt-wound motor the coils are in parallel. With the series-wound motor the field coil is generally only a relatively few turns of wire and such a motor develops a high starting torque, which drops off as the motor speed picks up. With the shunt-wound motor the field coils have many turns of fine wire and such a motor develops a fairly constant torque which changes little with motor speed. Explain the reasons for these differences in motor performance.

# SELF-ASSESSMENT

## Key points

For a **transformer**:
(e.m.f. induced in the secondary)/(p.d. applied to the primary) = (number of secondary turns)/(number of primary turns).

If there is no energy loss:
(secondary current)/(primary current) = (primary p.d.)/(secondary e.m.f.).

$$\frac{I_s}{I_p} = \frac{\mathscr{E}_p}{\mathscr{E}_s} = \frac{N_p}{N_s}$$

**Eddy currents** are induced in pieces of conducting material when the material either moves in a magnetic field or is in a changing magnetic field. The currents arise from electromagnetic induction of an e.m.f. within the material and are in such a direction as to oppose the change producing them.
The **d.c. motor** is essentially a coil, the *armature coil*, on an axle in a magnetic field. When the coil is supplied with a current it experiences a couple and rotates, the rotation continuing provided the current sense is reversed twice in every rotation by a split-ring commutator. The magnetic field is generally provided by a current in a coil, referred to as the *field coil*. This coil is in series with the armature coil in the series-wound motor and in parallel in the shunt-wound motor. The series-wound motor develops a high starting torque which decreases as the motor speed increases; the shunt-wound motor gives a lower but fairly constant torque which does not change significantly with motor speed.

## Answers I

**11.1** (a) The alternating primary potential difference produces an alternating current in the primary and hence an alternating flux in the core. This flux links the secondary and an e.m.f. is induced in it.
(b) No e.m.f. is induced if there is no alternating flux, i.e. no alternating primary potential difference.
(c) The iron core ensures that virtually all the flux produced by the primary links the secondary.

**11.2** (a) $\Phi$ = flux per turn, hence total flux = $N_s\Phi$ (b) $\mathscr{E}_s = -N_s\mathrm{d}\Phi/\mathrm{d}t$ (c) back e.m.f. = $-N_p\mathrm{d}\Phi/\mathrm{d}t$ (d) $\mathscr{E}_p$ + back e.m.f. = 0 and so $\mathscr{E}_p = N_p\mathrm{d}\Phi/\mathrm{d}t$ and $\mathscr{E}_p/\mathscr{E}_s = -N_p/N_s$. The minus sign is because they are out of phase.

**11.3** (a) $I_p\mathscr{E}_p$ (b) $I_s\mathscr{E}_s$ (c) $I_p\mathscr{E}_p = I_s\mathscr{E}_s$ and so $I_p/I_s = \mathscr{E}_s/\mathscr{E}_p = N_s/N_p$

**11.4** $N_p/N_s = 240/4 = 60$

**11.5** With the open circuit for the secondary, the back e.m.f. in the primary is almost the same as the applied potential difference, thus the net e.m.f. causing the current through the primary resistance is very low. When a secondary current flows it is in such a direction as to oppose the change producing it, i.e. the flux change in the core. The reduced flux in the core however means that the back e.m.f. in the primary coil is reduced and thus the net e.m.f. is greater, hence the primary current is increased.

## Answers II

**11.6** *Eddy currents* are currents that are produced within a conducting material by either its movement in a magnetic field or its location in a changing magnetic field. The currents are the result of electromagnetic induction of an e.m.f. within the material.

**11.7** By laminating the core to increase its resistance to the eddy currents or in some other way breaking up the core, perhaps by the use of a core packed with iron dust.

**11.8** Eddy currents are produced in the frame by its movement in a magnetic field. These are in such a direction as to oppose the motion producing them and so damp the motion.

**11.9** The high-frequency alternating current produces a high-frequency alternating flux and so induces high-frequency eddy currents in the material. These currents dissipate energy through the resistance of the material.

## Answers III

**11.10** See Figure A11.1. If the current through the coil is not reversed every half cycle it would not keep rotating. When the plane of the coil is perpendicular to the field the split-ring commutator reverses the current in the coil, thus ensuring that the couple acting on the coil continues to act in the same direction.

**11.11** (a) $V - \mathscr{E} = Ir$ (b) $VI = \mathscr{E}I + I^2r$. Power supplied = power output + power dissipated as heat, hence power output $= \mathscr{E}I$.

**11.12** (a) When the motor is just switched on there is no, or little, rotation, and so the back e.m.f. is virtually zero. Hence the current is $V/r$ = 240/1 = 240 A.
(b) When the back e.m.f. has risen to 230 V the current becomes $(V - \mathscr{E})/r = (240 - 230)/1 = 10$ A. Thus to limit the current through the coil a series resistance is used when the motor is switched on, the resistance being reduced until it is zero when the motor is fully running.

**11.13** With the series-wound motor, the armature current flows through the field coil and because initially the back e.m.f. is very low this current is high. This means both a high field and a high armature current. Thus the torque is high. However as the rotation increases the back e.m.f. increases and so the current through both coils decreases and the torque decreases. With the shunt-wound motor, the current through the field coil is less than that through the armature coil because it has a larger resistance. If the load on the motor increases and reduces its speed then the back e.m.f. falls. This results in an increase in current and the motor develops more torque and so picks up speed.

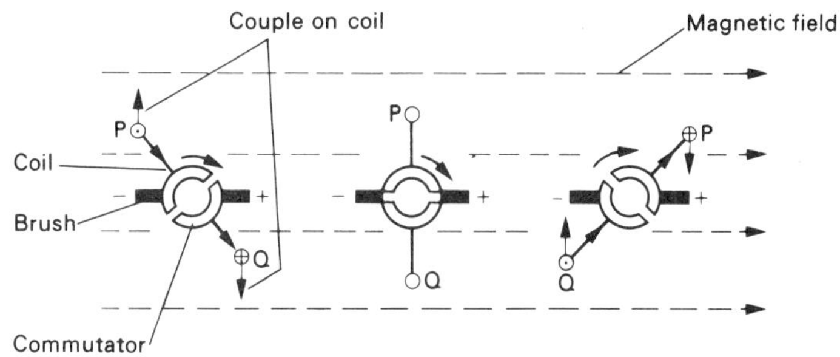


**Figure A11.1**

# Lesson 12 Inductance

## Aims

After completing this lesson you should be able to

- explain what is meant by *self-inductance*, define it and state its unit
- derive the relationship for the inductance of a solenoid
- explain the effect of inductance in a d.c. circuit
- explain what is meant by *mutual* inductance, define it and state its unit
- derive the relationship for the mutual inductance of two coils on a common axis.

## Prerequisites

Lesson 7 on magnetic flux density and Lesson 10 on electromagnetic induction. Some basic calculus is also required.

# STUDY

**I Read** about self-inductance, the derivation of the relationship for the inductance of a solenoid and the effect of inductance in a d.c. circuit.
References: *B* PP ch. 11, *B* ST4 ch. 3, *D* ch. 14, *N* ch. 35, *W et al.* ch. 31 (37), ch. 40, *Wh&H* ch. 57.

### Self-assessment questions

**12.1** Explain what is meant by *self-inductance.*

***12.2** The circuit shown in Figure 12.1 shows lamps monitoring the current through a resistor and an inductor. When the switch is closed the lamp in series with the resistor lights up a few seconds before the lamp in series with the inductor lights up. Explain why this occurs.

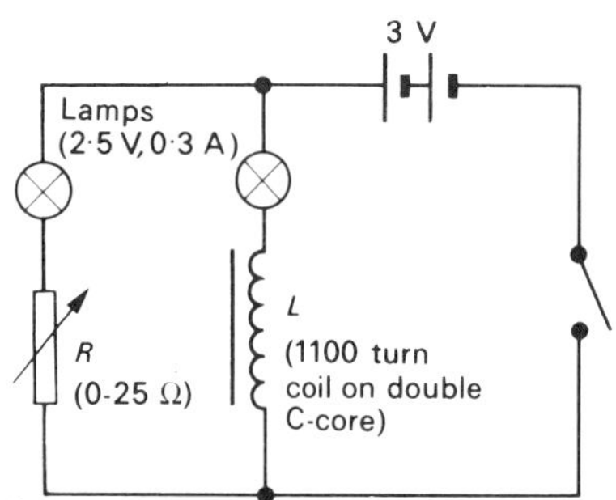


**Figure 12.1**

***12.3** Explain how the direct current changes in a circuit containing inductance when the current is (a) switched on and (b) switched off.

**12.4** Define *self-inductance* and state its unit.

***12.5** A coil having a self-inductance of 2 H and a resistance of 6Ω is connected to a battery of e.m.f. 6 V and negligible internal resistance. When the current is switched on what is (a) the greatest rate of change of current and (b) the greatest current?

**12.6** Consider an air-cored solenoid of length $l$, cross-sectional area $A$, having $N$ turns and carrying a current $I$.
(a) What is the flux density inside the solenoid?
(b) What is the flux linked by each turn of the solenoid?
(c) What is the total flux linked by the $N$ turns of the solenoid?
(d) By Faraday's law the induced e.m.f. in the solenoid when the current changes at the rate $\mathrm{d}I/\mathrm{d}t$ is $\mathscr{E} = -\mathrm{d}(N\Phi)/\mathrm{d}t$. Hence calculate this induced e.m.f. and the self-inductance of the solenoid.

**12.7** What is the inductance of a solenoid having 500 turns of wire in a length of 20 cm and a cross-sectional area of 60 cm$^2$? ($\mu_0 = 4\pi \times 10^{-7}$ H m$^{-1}$)

**II Read** about mutual inductance and the derivation of the relationship for the mutual inductance of a pair of coils on a common axis.
References: *B* PP ch. 11, *B* ST4 ch. 3, *D* ch. 14, *N* ch. 35, *W et al.* ch. 31, *Wh&H* ch. 57.

### Self-assessment questions

**12.8** Define *mutual inductance* and state its unit.

**12.9** (a) Consider an air-cored solenoid of length $l_p$, with $N_p$ turns and cross-sectional area $A$. What is the flux density inside this solenoid when it carries a current $I_p$?
(b) A short coil containing $N_s$ turns is wound over the middle section of the solenoid. What is the flux linked by this short coil due to the current in the solenoid?
(c) If the current in the solenoid changes at the rate $\mathrm{d}I_p/\mathrm{d}t$, what will be the e.m.f. induced in the short coil? Hence determine a relationship for the mutual inductance of the coils.

***12.10** An air-cored solenoid has a cross-sectional area of 10 cm$^2$ and 1600 turns of wire in a length of 40 cm. A short coil with 500 turns is wound over the middle region of the solenoid. What is (a) the self-inductance of the solenoid, (b) the mutual inductance of the pair of coils, (c) the e.m.f. induced in the short coil when the current in the solenoid is changing at the rate of 14 A s$^{-1}$? ($\mu_0 = 4\pi \times 10^{-7}$ H m$^{-1}$)

# SELF-ASSESSMENT

## Key points

When the current in a coil changes the resulting changing flux causes an e.m.f. to be induced in the coil, the e.m.f. opposing the change producing it. This effect is known as **self-inductance**. Self-inductance $L$ is defined as $L = -\mathscr{E}/(\mathrm{d}I/\mathrm{d}t)$, where $\mathscr{E}$ is the e.m.f. induced in the coil when the rate of change of the current is $\mathrm{d}I/\mathrm{d}t$. The unit of self-inductance is the henry (H). For an air-cored solenoid $L = \mu_0 A N^2/l$, where $A$ is the cross-sectional area, $N$ the number of turns and $l$ the length.

When the current in one coil changes and induces an e.m.f. in another coil the coils are said to be showing **mutual inductance**. Mutual inductance is defined by $M = -\mathcal{E}/(\mathrm{d}I/\mathrm{d}t)$, where $\mathcal{E}$ is the e.m.f. induced in one coil by a rate of change of current in the other coil of $\mathrm{d}I/\mathrm{d}t$. The unit of mutual inductance is the henry (H). For a short coil wound on the middle section of a long air-cored solenoid $M = \mu_0 A N_p N_s / l_p$, where $N_p$ is the number of turns on the primary coil (the solenoid) of length $l_p$, $A$ the cross-sectional area and $N_s$ the number of turns on the secondary (short) coil.

## Answers I

**12.1** See the key points.

**12.2** Because in the inductor branch of the circuit a back e.m.f. is produced which opposes the current and so slows down the current in its growth to its maximum value.

**12.3** (a) With no inductance the current would have risen very rapidly to its maximum value, the inductance however opposes the current growth and so the current grows more slowly to its maximum value.
(b) With no inductance the current would have fallen very rapidly to zero, the inductance however opposes the current collapse and so the current falls more slowly to its zero value.
In both the growth and decay of current there is an e.m.f. induced in the inductance which opposes the change producing it.

**12.4** See the key points.

**12.5** (a) When the current is first switched on the current changes at its greatest rate, the rate declining to zero when the steady current is achieved. The inductor can be thought of as an inductance $L$ in series with a resistance $R$. Then the applied e.m.f. $\mathcal{E}$ minus the back e.m.f. $L\,\mathrm{d}I/\mathrm{d}t$ must equal the p.d. across $R$, i.e. $\mathcal{E} - L\,\mathrm{d}I/\mathrm{d}t = IR$. When $I$ is zero $\mathcal{E} = L\,\mathrm{d}I/\mathrm{d}t$ and so $\mathrm{d}I/\mathrm{d}t = 6/2 = 3\ \mathrm{A\,s^{-1}}$.
(b) The greatest current is when $\mathrm{d}I/\mathrm{d}t$ is zero. Thus $\mathcal{E} = IR$ and so $I = 6/6 = 1$ A.

**12.6** (a) $B = \mu_0 NI/l$ (b) $BA = \mu_0 NAI/l$
(c) $\mu_0 N^2 AI/l$ (d) $\mathcal{E} = -(\mu_0 N^2 A/l)\mathrm{d}I/\mathrm{d}t = -L\,\mathrm{d}I/\mathrm{d}t$ and so $L = \mu_0 N^2 A/l$.

**12.7** Using the expression derived in 12.6 above, $L = 4\pi \times 10^{-7} \times 500^2 \times 60 \times 10^{-4}/0.20 = 9.4 \times 10^{-3}$ H.

## Answers II

**12.8** See the key points.

**12.9** (a) $B = \mu_0 N_p I_p / l_p$ (b) $N_s A \mu_0 N_p I_p / l_p$ (c) $\mathcal{E}_s = -(\mu_0 A N_p N_s / l_p)\,\mathrm{d}I_p/\mathrm{d}t = -M\mathrm{d}I_p/\mathrm{d}t$ and so $M = \mu_0 A N_p N_s / l_p$.

**12.10** (a) $L = \mu_0 N^2 A/l = 4\pi \times 10^{-7} \times 1600^2 \times 10 \times 10^{-4}/0.40 = 8.0 \times 10^{-3}$ H (b) $M = \mu_0 A N_p N_s / l_p = 4\pi \times 10^{-7} \times 10 \times 10^{-4} \times 1600 \times 500/0.40 = 2.5 \times 10^{-3}$ H
(c) $\mathcal{E} = -M\mathrm{d}I/\mathrm{d}t = -2.5 \times 10^{-3} \times 14 = -0.035$ V

# CHAPTER 7

# Current electricity

## Lesson 1 D.c. resistive circuits

### Aims

After completing this lesson you should be able to

- define *resistance* and *resistivity*
- state *Ohm's law* and describe situations in which the law is obeyed and also those in which it is not obeyed, and in addition the effects of temperature
- solve problems involving networks of resistors
- describe energy conversion in a resistance
- define *e.m.f.* and solve circuit problems involving cells with internal resistance.

### Prerequisites

An elementary knowledge of d.c. circuits is assumed, also the basic concepts of energy and power (as in Lesson 5 of Chapter 1).

## STUDY

**I Read** about the definition of resistance and resistivity, and Ohm's law.
References: *B* PP ch. 10, *B* ST4 ch. 1, *D* ch. 3, *N* ch. 30, *W et al.* ch. 25(23), *Wh&H* ch. 48.

**Self-assessment questions**

**1.1** Define *resistance* and state its unit.

**1.2** Define *resistivity* and state its unit.

**1.3** What is the resistance of a length of wire if a potential difference of 0.10 V across it occurs when a current of 150 mA passes through it?

**1.4** Constantan has a resistivity of $5.0 \times 10^{-7}\ \Omega$ m; what therefore is the resistance of a 20 cm length of constantan wire with a diameter of 1.0 mm?

**1.5** State *Ohm's law* and give examples of components that obey the law, and those that do not. In both types of example, sketch the current–potential difference graphs.

**1.6** Describe an experiment by which you could determine whether a component obeys Ohm's law.

**1.7** Describe the effects of temperature changes on the resistance of a metal wire.

**1.8** Define the temperature coefficient of resistance.

**1.9** A length of copper wire has a resistance of 1.00 Ω at 0 °C; what will be its resistance at 30 °C if the temperature coefficient of resistance of copper is $4.0 \times 10^{-3}\ K^{-1}$?

**II Read** about the solution of problems involving networks of resistors, i.e. combinations of resistors in series and parallel.
References: *B* PP ch. 10, *B* ST4 ch. 1, *D* ch. 3, *N* ch. 30, *W et al.* ch. 25(25), *Wh&H* ch. 48.

**Self-assessment questions**

**1.10** Three resistors $R_1$, $R_2$ and $R_3$ are in series.
(a) If $I$ is the current through $R_1$, what is the current through the other resistors?
(b) How is the potential difference $V$ across all three resistors related to the potential differences across each resistor, i.e. $V_1$, $V_2$ and $V_3$?
(c) The combined resistance of the three resistors is $V/I$, where $V$ is the potential difference across all three resistors and $I$ the current through them. Hence derive a relationship for $R$ in terms of $R_1$, $R_2$ and $R_3$.

**1.11** Resistor $R_1$ is in parallel with resistor $R_2$, which in turn is in parallel with resistor $R_3$.
(a) How is the total current $I$ entering the arrangement related to the currents through each resistor?
(b) How is the potential difference across the arrangement related to the potential differences across each resistor?
(c) The combined resistance $R$ of the arrangement is $V/I$, where $V$ is the potential difference across the arrangement and $I$ is the total current entering it. Hence derive a relationship between $R$ and $R_1$, $R_2$ and $R_3$.

**1.12** Calculate the combined resistance for the following networks of resistors:
(a) 100 Ω, 50 Ω and 25 Ω in series,
(b) three 10 Ω resistors in parallel,
(c) a 20 Ω and a 40 Ω resistor in series, with a 10 Ω resistor in parallel with the series arrangement.

***1.13** For the circuit shown in Figure 1.1, what is (a) the current through the 6 Ω resistor, (b) the potential difference across the 12 Ω resistor?

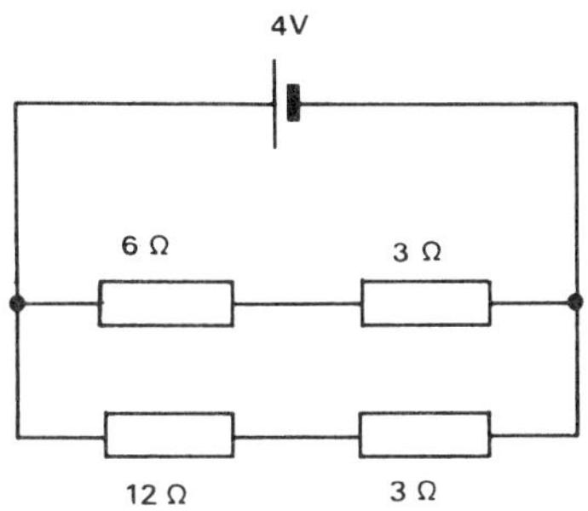


**Figure 1.1**

***1.14** For the circuit shown in Figure 1.2, what is the potential difference across the 100 Ω resistor? What would be the potential difference across that resistor if an additional 200 Ω resistor were connected in parallel with it?

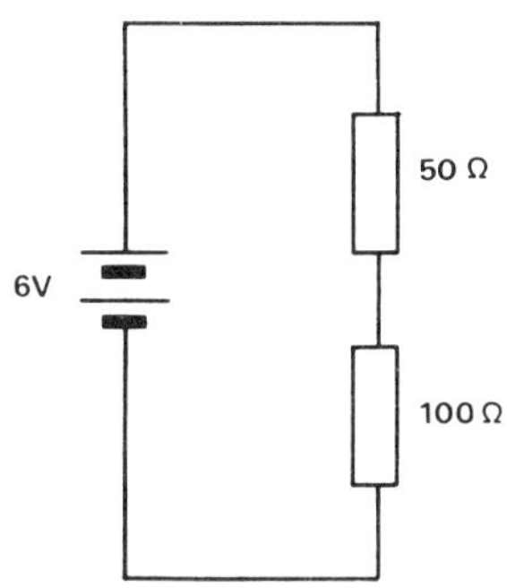


**Figure 1.2**

***1.15** A d.c. supply of 4.0 V, having negligible internal resistance, is connected in series with a resistance of 200 Ω and an unknown resistance $R$. When a voltmeter is connected across the 200 Ω resistance the reading given is 3.2 V. If the voltmeter has an infinite resistance what would be its reading when connected across $R$ and what is the value of this resistance?

**III Read** about energy conversion in a resistance and the use of the equation

$$\text{power} = IV$$

References: *B* PP ch. 10, *B* ST4 ch. 1, *D* ch. 3, *N* ch. 30, *W et al.* ch. 26 (24), *Wh&H* ch. 49.

**Self-assessment questions**

**1.16** What is the rate at which electrical energy is dissipated as heat when a current $I$ flows through a resistance $R$?

**1.17** What is the maximum current that can be used with a 100 Ω resistor if the maximum power that can be dissipated is $\frac{1}{4}$ W?

***1.18** A heating coil dissipating 20 W is required and is to be made from nichrome wire, resistivity $1.0 \times 10^{-6}$ Ω m, of cross-sectional area 0.4 mm$^2$. What length of wire would be required if the power is to be produced when there is a potential difference of 24 V across the coil?

***1.19** An electric heater has two coils which can be used alone, in series or in parallel. If each coil when used alone on a 240 V supply has a power of 100 W, what is the power when the coils are used (a) in series and (b) in parallel from the same voltage supply?

***1.20** An electrical circuit consists of a heating element of resistance 50 Ω connected to the supply by copper wires having a total resistance of 0.1 Ω. What percentage of the power supplied is dissipated by the heating element?

**IV Read** about the e.m.f. of a cell, its internal resistance and the solution of problems involving these terms.
References: *B* PP ch. 10, *B* ST4 ch. 1, *D* ch. 3, *N* ch. 30, *W et al.* ch. 26 (24), *Wh&H* ch. 50.

**Self-assessment questions**

**1.21** Define *e.m.f.*

***1.22** Explain why, when a current is taken from a battery, the reading given by a high-resistance voltmeter connected across the battery terminals decreases as the current taken increases.

***1.23** Two batteries, each of e.m.f. 2.0 V and internal resistance 0.5 Ω, are (a) connected in series and (b) connected in parallel. In each case state the e.m.f. and the internal resistance of the arrangement. Assume that when the cells are connected in series the positive terminal of one cell is connected to the negative of the other, but when they are in parallel the positive terminals are connected together, as are the negative terminals.

***1.24** A cell has an e.m.f. of 1.5 V and an internal resistance of 0.8 Ω. What will be the potential difference between its terminals when a current of 1.2 A is taken by a load?

***1.25** Two cells, each of e.m.f. 1.50 V, are connected in series with a resistance of 6.0 Ω. When a high-resistance voltmeter is connected across one cell a reading of 0.70 V is obtained and when connected across the other cell it reads 0.72 V. What are the internal resistances of the two cells?

# SELF-ASSESSMENT

## Key points

The **resistance** of a component is the potential difference across the component divided by the current through it, i.e. $R = V/I$, and the unit used is the ohm (Ω).

The **resistivity** $\rho$ of a material is defined by the equation $\rho = RA/L$, where $A$ is the cross-sectional area of the piece of material, $L$ its length and $R$ its resistance. The unit is Ω m.

For many materials, provided the temperature does not change, $V/I$ is a constant. This statement is known as **Ohm's law**.

For resistors in series the combined resistance $R = R_1 + R_2 + R_3 +$ etc. For resistors in parallel $1/R = 1/R_1 + 1/R_2 + 1/R_3 +$ etc.

The rate at which electrical energy is dissipated when passing through a resistor is $I^2R = IV$.

The **electromotive force (e.m.f.)** of a source is the energy converted into electrical energy when unit charge passes through it. It is the value of the potential difference between the terminals of the source when no current is drawn from it.

## Answers I

**1.1** See key points.

**1.2** See key points.

**1.3** $R = V/I = 0.10/(150 \times 10^{-3}) = 0.67\,\Omega$

**1.4** $R = \rho L/A = 5.0 \times 10^{-7} \times 0.20/(\frac{1}{4}\pi\ 0.0010^2) = 0.13\,\Omega$

**1.5** See the key points and Figure A1.1.

**1.6** Probably the simplest is the ammeter–voltmeter method; an ammeter is used to measure the current through the component and a voltmeter to measure the potential difference across it. A graph of $V$ against $I$ is a straight line through the origin if Ohm's law is obeyed.

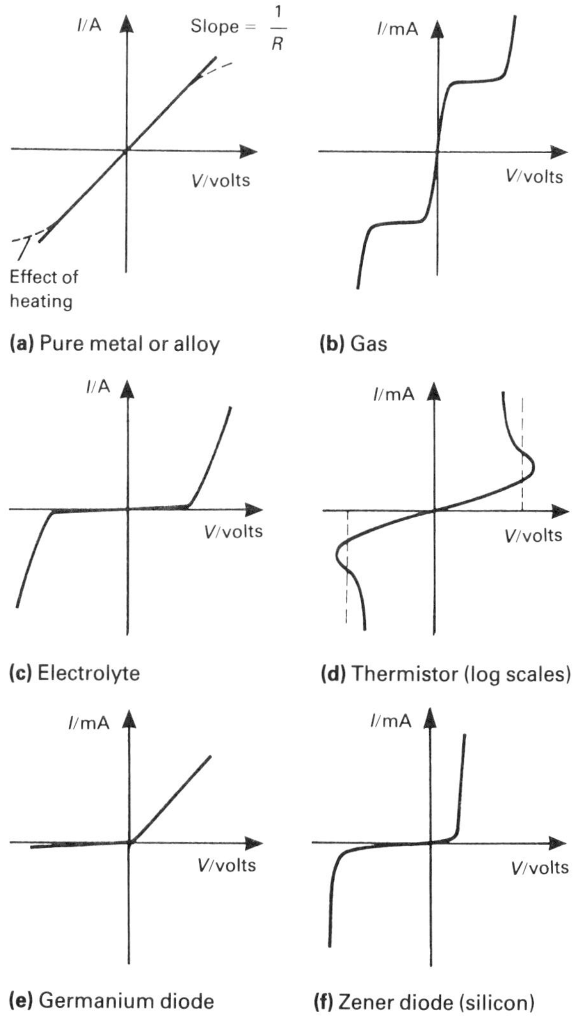


**Figure A1.1 Characteristics of conductors**

**1.7** The resistance increases with an increase in temperature, the relationship being virtually linear.

**1.8** The temperature coefficient of resistance is the fractional increase in the resistance at 0 °C per unit rise in temperature, i.e. $\alpha = (R_\theta - R_0)/R_0\theta$.

**1.9** $R_\theta = R_0(1 + \alpha\theta) = 1.00\,(1 + 4.0 \times 10^{-3} \times 30) = 1.12\,\Omega$

## Answers II

**1.10** (a) The same, i.e. $I$. (b) $V = V_1 + V_2 + V_3$ (c) $(V/I) = (V_1/I) + (V_2/I) + (V_3/I)$ and so $R = R_1 + R_2 + R_3$.

**1.11** (a) $I = I_1 + I_2 + I_3$ (b) The same for all. (c) $(I/V) = (I_1/V) + (I_2/V) + (I_3/V)$ and so $1/R = 1/R_1 + 1/R_2 + 1/R_3$.

**1.12** (a) $100 + 50 + 25 = 175\,\Omega$ (b) $1/R = 3/10$ and so $R = 3\frac{1}{3}\,\Omega$. (c) $1/R = 1/60 + 1/10$ and so $R = 8.6\,\Omega$.

**1.13** (a) For the arm with the $6\,\Omega$ resistor, total resistance is $9\,\Omega$ and hence $I = V/R = 4/9 = 0.44$ A. (b) For the other arm, total resistance is $15\,\Omega$ and so current is $4/15 = 0.27$ A. Hence p.d. across the $12\,\Omega$ resistor is $0.27 \times 12 = 3.2$ V.

**1.14** Total resistance $= 150\,\Omega$ and so current $= 6/150 = 0.04$ A. P.d. across resistor $= 100 \times 0.04 = 4.0$ V. If a $200\,\Omega$ resistor is connected in parallel with the $100\,\Omega$ resistor the combined resistance is given by $1/R = 1/100 + 1/200$ and so $R = 66.7\,\Omega$. Thus total resistance $= 116.7\,\Omega$ and current $= 6/116.7 = 0.051$ A. Hence p.d. $= 0.051 \times 66.7 = 3.4$ V.

**1.15** Current $= 4.0/(200 + R)$ and so, as p.d. across $R$ is $4.0 - 3.2 = 0.8$ V, then $0.8 = R \times 4.0/(200 + R)$ and $R = 50\,\Omega$.

## Answers III

**1.16** Power $= I^2R$

**1.17** $P = I^2R$ and so $I^2 = \frac{1}{4}/100$ and $I = 0.05$ A.

**1.18** $P = IV = V^2/R$ and so $R = 24^2/20 = 28.8\,\Omega$. As $\rho = RA/L$ then $L = 28.8 \times 0.4 \times 10^{-6}/1.0 \times 10^{-6} = 11.52$ m.

**1.19** (a) With one coil $P = I^2R = 100$ W. When combined in series, resistance $= 2R$ and current $= I/2$, so combined power $= (I/2)^2 \times 2R = I^2R/2 = 50$ W. (b) When combined in parallel, resistance $= \frac{1}{2}R$ and current $= 2I$, so combined power $= (2I)^2(\frac{1}{2}R) = 2I^2R = 200$ W. This assumes the internal resistance of the supply is zero.

**1.20** Percentage power $= [(50I^2 - 0.1I^2)/50.1I^2] \times 100 = 99.6\%$

## Answers IV

**1.21** See the key points.

**1.22** The cell has an internal resistance $r$. Thus $\mathscr{E} = I(r + R)$, where $R$ is external circuit load. The potential difference across the battery is the same as the potential difference $V$ across the load, thus $\mathscr{E} = Ir + V$. Thus $V = \mathscr{E} - Ir$. The bigger $I$ the greater the p.d. across the internal resistance and so the lower $V$.

**1.23** (a) E.m.f. in series $= \mathscr{E}_1 + \mathscr{E}_2 = 4.0$ V. Total internal resistance $= r_1 + r_2 = 1.0\,\Omega$. (b) Total e.m.f. when in parallel is the same as when the cell is on its own, i.e. 2.0 V. Total internal resistance is given by $1/r = 1/r_1 + 1/r_2$ and so $r = 0.25\,\Omega$.

**1.24** Using the equation developed in 1.22 above, $V = \mathscr{E} - Ir = 1.5 - 0.8 \times 1.2 = 0.54$ V.

**1.25** Total e.m.f. $= 3.00$ V. Thus $3.00 = I(6.0 + r_1 + r_2)$. Hence $3.00 = 6.0I + 0.70 + 0.72$ and so $I = 0.26$ A. Thus $r_1 = 0.70/0.26 = 2.66\,\Omega$ and $r_2 = 0.72/0.26 = 2.77\,\Omega$.

# Lesson 2 Use of meters

## Aims

After completing this lesson you should be able to

- calculate shunt and multiplier values for use with a meter to give different current and voltage ranges
- solve circuit problems in which meter resistance is involved.

## Prerequisites

Lesson 1 on d.c. resistive circuits.

## STUDY

**I** **Read** about the calculation of shunt and multiplier values for use with a meter to give different current and voltage ranges.

References: *B* PP ch. 10, *B* ST4 ch. 1, *D* ch. 3, *N* ch. 30, *W et al.* (ch. 25), *Wh&H* ch. 51.

**Self-assessment questions**

**2.1** Consider a basic moving-coil meter with a resistance $R_g$ and capable of giving a full-scale deflection with a current $I_g$. A shunt resistor $R_s$ is connected in parallel with the meter.
(a) How is the potential difference across the meter related to that across the shunt?
(b) How is the total current $I$ entering the arrangement related to the current through the meter and the current through the shunt $I_s$?
(c) Hence, by using your answers to (a) and (b), arrive at a relationship involving the shunt resistance, the galvanometer resistance, the full-scale current through the meter, and the current which enters the arrangement to give this current through the meter.

***2.2** A meter has a resistance of 80 Ω and gives a full-scale reading with 100 μA. What is the resistance of the shunt that should be used with the meter in order that a full-scale reading corresponds to 1 A?

**2.3** A basic moving-coil meter has a resistance of $R_g$ and gives a full-scale reading with a current of $I_g$.
(a) What potential difference across the meter corresponds to this full-scale reading?
(b) A resistor $R_m$ is connected in series with the meter. What potential difference across the combination will give rise to a full-scale reading on the meter?

***2.4** A meter has a resistance of 100 Ω and gives a full-scale reading with 1 mA. What resistance must be placed in series with the instrument so that the meter gives a full-scale reading when 5 V is connected across the combination?

***2.5** A moving-coil meter has a resistance of 30 Ω and gives a full-scale reading with a current of 2.0 mA. How can this meter be converted into (a) an ammeter giving a full-scale reading with 10 A, (b) a voltmeter giving a full-scale reading with 2 V?

**II** **Read** about the effects of meter resistance on the measurements made with a meter.
References: *B* ST4 ch. 1, *D* ch. 3, *N* ch. 30.

**Self-assessment questions**

***2.6** A voltmeter having a resistance of 5000 Ω is connected across the terminals of a battery, e.m.f. 3.0 V and internal resistance 1 Ω. What will be the reading of the instrument? What effect has the voltmeter resistance on the reading obtained?

***2.7** A resistor of 1000 Ω and one of 3000 Ω are in series with a 20 V supply. What will be the reading given by a voltmeter of resistance 1000 Ω when placed across (a) the 1000 Ω and (b) the 3000 Ω resistor?

***2.8** A resistor of 400 Ω is connected across the terminals of a d.c. supply of 20 V which has an internal resistance of 10 Ω. What will be the reading obtained on a voltmeter of resistance 1000 Ω connected across the 400 Ω resistor? What was the potential difference across the resistor before the voltmeter was connected into the circuit?

***2.9** A cell has an e.m.f. of 1.52 V and an internal resistance of 5 Ω. What will be the readings given by voltmeters, connected each in turn across the cell terminals, having resistances of (a) 500 Ω, (b) 5000 Ω, (c) 1 MΩ?

# SELF-ASSESSMENT

## Key points

Resistors in parallel, termed **shunts**, are used to change the current ranges of meters. Resistors in parallel, termed **multipliers**, are used to change the voltage range of meters. Shunts have very low resistances, multipliers have very high resistances. When a voltmeter is used to measure a potential difference, unless the resistance of the voltmeter is very high when compared with that of the resistance across which the measurement is made then a significant current is taken by the meter and this affects the voltage reading obtained.

## Answers I

**2.1** (a) The same. (b) $I = I_g + I_s$ (c) $I_s R_s = I_g R_g$ and so $I = I_g + I_g R_g / R_s = I_g(1 + R_g/R_s)$.

**2.2** Using the expression derived in 2.1(c), $1 = 100 \times 10^{-6}(1 + 80/R_s)$ and so $R_s = 8.00 \times 10^{-3}\,\Omega$.

**2.3** (a) $I_g R_g$ (b) $I_g(R_g + R_m)$

**2.4** Using the expression given in 2.3(b), $V = 5 = 1 \times 10^{-3}(R_m + 100)$ and so $R_m = 4900\,\Omega$.

**2.5** (a) Using the expression given in 2.1(c) above, $10 = 2.0 \times 10^{-3}(1 + 30/R_s)$ and so $R_s = 6.00 \times 10^{-3}\,\Omega$. (b) Using the expression given in 2.3(b) above, $V = 2 = 2.0 \times 10^{-3}(R_m + 30)$ and so $R_m = 970\,\Omega$.

## Answers II

**2.6** This follows from SAQ 1.22. $\mathscr{E} = Ir + IR$, where $r$ is the internal resistance. Thus $3.0 = I(1 + 5000)$ and so

$I = 3.0/5001$. Hence $V = Ir = (3.0/5001) \times 5000 = 2.999$ V. The instrument will effectively give the e.m.f. The smaller the voltmeter resistance the more the voltmeter reading will depart from the e.m.f.

**2.7** (a) 1000 Ω in parallel with 1000 Ω gives a combined resistance of 500 Ω. Thus the circuit current $I = 20/3500$ and $V = IR = (20/3500) \times 500 = 2.86$ V. (b) 1000 Ω in parallel with 3000 Ω gives a combined resistance of 750 Ω. Thus the circuit current $I = 20/1750$ and $V = (20/1750) \times 750 = 8.57$ V.

**2.8** 1000 Ω in parallel with 400 Ω gives a combined resistance of 285.7 Ω. Thus $\mathscr{E} = Ir + IR = 20 = I(285.7 + 10)$ and $I = 20/295.7$. Hence $V = IR = (20/295.7) \times 285.7 = 19.3$ V. With no voltmeter present $\mathscr{E} = Ir + IR = 20 = I(400 + 10)$ and so $I = 20/410$. Hence $V = IR = (20/410) \times 400 = 19.5$ V.

**2.9** $\mathscr{E} = Ir + IR$, thus $I = \mathscr{E}/(r + R)$ and hence $V = \mathscr{E}R/(r + R)$. (a) $V = 1.52 \times 500/505 = 1.505$ V (b) $V = 1.52 \times 5000/5005 = 1.518$ V (c) $V = 1.52 \times 1 \times 10^6/(1 \times 10^6 + 5) = 1.520$ V.

# Lesson 3 Bridges and potentiometers

## Aims

After completing this lesson you should be able to

- explain the principle of the Wheatstone bridge and solve problems involving it
- explain the principle of the potentiometer and how it can be used for the measurement of e.m.f., potential difference, thermoelectric e.m.f., resistance and current
- solve problems involving potentiometer circuits.

## Prerequisites

Lesson 1 on d.c. resistive circuits.

## STUDY

**I Read** about the Wheatstone bridge and its metre-bridge version.
References: *B* PP ch. 10, *B* ST4 ch. 1, *D* ch. 3, *N* ch. 3, *W et al.* ch. 27(25), *Wh&H* ch. 51.

### Self-assessment questions

***3.1** For the circuit shown in Figure 3.1, what is (a) the potential difference across the 100 Ω resistor, (b) the potential difference across the 50 Ω resistor, (c) the potential difference between points A and B?

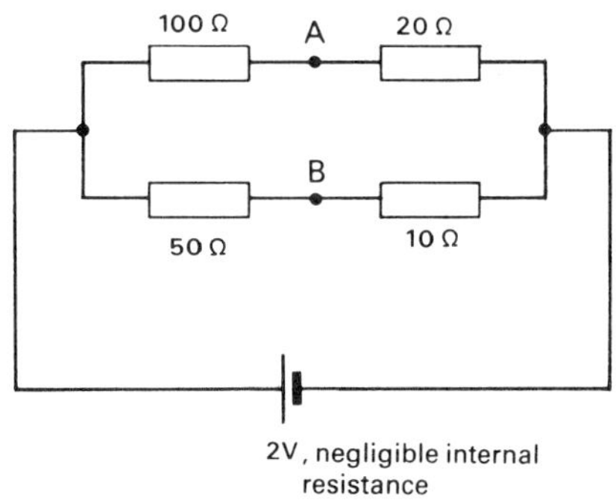


**Figure 3.1**

**3.2** Figure 3.2 shows the basic form of the Wheatstone bridge.
(a) When there is no current through the galvanometer, what is the potential difference between B and D?
(b) What is the potential difference across *P*, i.e. between A and B?
(c) What is the potential difference across *R*, i.e. between A and D?
(d) What must be the relationship between the potential differences across *P* and across *R*? Hence derive a relationship between the resistances *P* and *R*.
(e) Derive a similar relationship for the resistances *Q* and *S*.
(f) Hence obtain a relationship between the resistances *P*, *Q*, *R* and *S* when there is no current detected by the galvanometer.

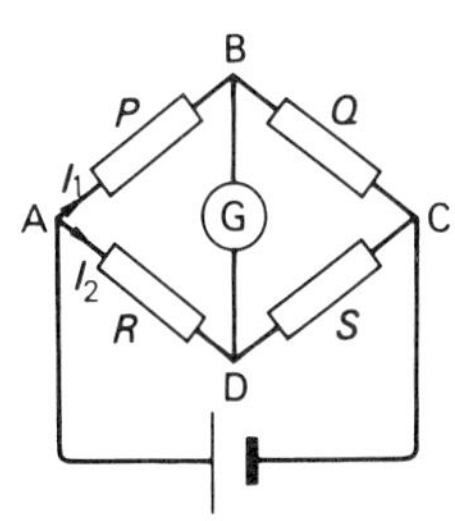


**Figure 3.2**

**3.3** Figure 3.3 shows one form of a Wheatstone bridge, known as the metre bridge. AC is a uniform length of resistance wire. The contact, point D, is moved along the wire until no current is detected by the galvanometer. What is then the relationship between the resistances $P$ and $Q$, and the length AD and the total length of the wire AC?

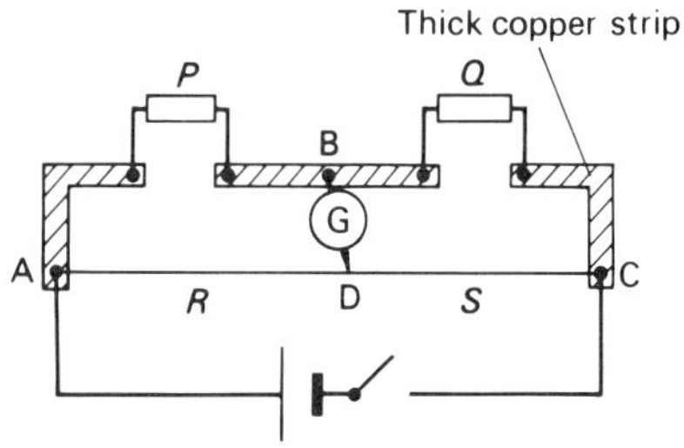


**Figure 3.3**

**3.4** With a metre bridge employing a uniform resistance wire of length 100 cm, what will be the position of the contact on the wire when no current flows through the galvanometer, if the left-hand gap in the bridge is occupied by a resistor of 20 Ω and the right-hand gap by 5 Ω?

***3.5** With a metre bridge employing a uniform resistance wire of length 100 cm, a coil of copper wire in the left-hand gap and a 10 Ω resistor in the right-hand gap, what will be the position of the contact on the wire at 20 °C if the copper has a resistance of 4.00 Ω at that temperature? By how much will the position of the contact have to be changed if the temperature of the copper coil is increased to 40 °C? Temperature coefficient of resistance of copper = $0.0043\ K^{-1}$.

**II Read** about the potentiometer circuit and the various uses to which it can be put.
References: *B* PP ch. 10, *B* ST4 ch. 1, *D* ch. 3, *N* ch. 31, *W et al.* ch. 27(25), *Wh&H* ch. 51.

**Self-assessment questions**

***3.6** Explain the principle of the potentiometer method for the measurement of potential differences.

***3.7** A student sets up a potentiometer and finds that no matter where along the potentiometer wire contact is made the galvanometer deflection is always in the same direction and it is impossible to obtain a zero current reading. Give two reasons why this might occur.

***3.8** When used to determine the e.m.f. of a cell, a potentiometer gives a balance length of 62.5 cm. When the unknown cell is replaced by a standard cell giving an e.m.f. of 1.02 V the balance length is found to be 45.0 cm. What is the unknown e.m.f.?

***3.9** A cell gives a balance length of 48.5 cm of wire with a potentiometer. When a 6.0 Ω resistor is connected in parallel with the cell the balance length becomes 46.0 cm. What is the internal resistance of the cell?

***3.10** Explain how a potentiometer can be used to calibrate a voltmeter.

***3.11** Figure 3.4 shows a potentiometer circuit that can be used to measure the e.m.f. of a thermocouple.
(a) Why is the resistance $R$ in series with the potentiometer wire?
(b) If the driver cell has an e.m.f. of 1.5 V, negligible internal resistance, and the potentiometer wire has a length of 100 cm and a total resistance of 2.0 Ω, what is the e.m.f. of the thermocouple if the balance length is 26.0 cm when $R$ is 1000 Ω?

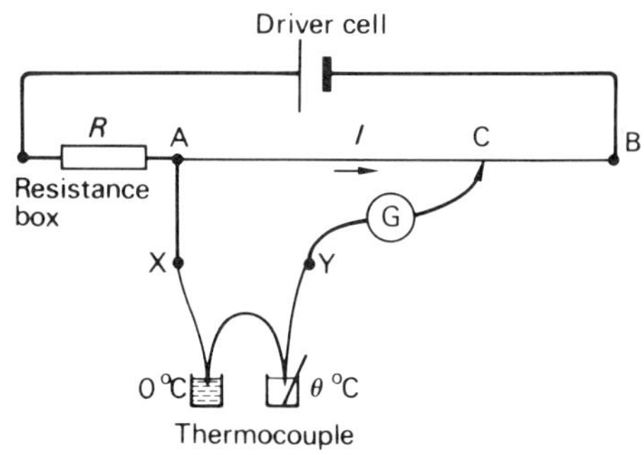


**Figure 3.4**

***3.12** A potentiometer wire is 100 cm long and has a resistance of 3.0 Ω. Explain how you could obtain a potential difference of 0.02 mV per cm of the wire if a driver cell of e.m.f. 1.5 V and negligible internal resistance is used, and a resistance box is available.

***3.13** Explain how a potentiometer can be used to compare the resistances of two resistors.

***3.14** Figure 3.5 shows a potentiometer circuit that can be used to compare two resistances.
(a) Which of the resistances are being compared? Identify the function of the other resistors in the circuit.
(b) With the connections as indicated in the figure, the balance length was found to be 46.5 cm. When point B was connected to F instead of E, and the galvanometer connected to G instead of F, the balance point was found to be 63.4 cm. What is the ratio of the resistances? State which resistor has the higher resistance.

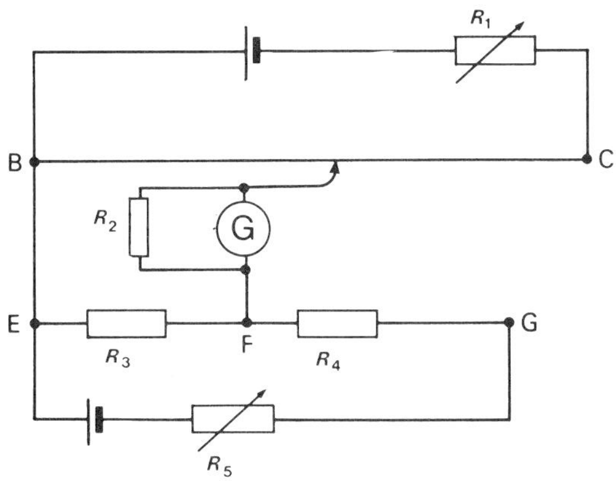


**Figure 3.5**

***3.15** Explain how a potentiometer can be used to calibrate an ammeter.

# SELF-ASSESSMENT

## Key points

Both the **Wheatstone bridge** and the **potentiometer** are *null* methods in that they involve determining when zero current flows through a galvanometer. The condition for such a balance with the Wheatstone bridge is that $P/Q = R/S$, where these are the resistances in the four arms of the bridge. A potentiometer is a circuit used for the measurement, with accuracy, of a potential difference. The potential difference for one item is compared with the potential difference for another, the ratio of the potential differences being the ratio of the balance lengths, i.e. $V_1/V_2 = L_1/L_2$.

## Answers I

**3.1** (a) For the upper branch of the circuit $V = 2 = IR = I \times 120$ and so $I = 2/120$ A. Therefore the p.d. across the 100 Ω is $(2/120) \times 100 = 1.67$ V.
(b) For the lower branch $V = 2 = IR = I \times 60$ and so $I = 2/60$ A. Thus the p.d. across the 50 Ω is $(2/60) \times 50 = 1.67$ V.
(c) A and B must be both at the same potential since the p.d. across the 100 Ω is the same as that across the 50 Ω, hence the p.d. between A and B is zero.

**3.2** (a) Zero. This is the only condition which satisfies $V = IR$ when $R$ has a finite value and $I = 0$.
(b) $I_1P$ (c) $I_2R$ (d) They must be the same, thus $I_1P = I_2R$. (e) $I_1Q = I_2S$ (f) $P/Q = R/S$

**3.3** $P/Q = L/(L_t - L)$, if the total length is $L_t$ and the balance length $L$.

**3.4** $20/5 = L/(100 - L)$ and so $L = 80$ cm.

**3.5** $4.00/10 = L/(100 - L)$ and so $L = 28.6$ cm. New resistance $R_\theta = R_0(1 + \alpha\theta)$. Thus $R_{20} = R_0(1 + 20\alpha)$ and $R_{40} = R_0(1 + 40\alpha)$. Thus $R_{40}/R_{20} = (1 + 40\alpha)/(1 + 20\alpha)$ and so $R_{40} = 4.32$ Ω. Hence new balance length is given by $4.32/10 = L/(100 - L)$, and $L = 30.2$ cm. Thus the balance contact will have to be moved 1.6 cm.

## Answers II

**3.6** A driver cell is used to maintain a constant potential difference across a length of wire. As a result of this there is a potential difference between any two points in the wire which is proportional to their distance apart, provided the wire has a constant resistance per unit length. The unknown potential difference is balanced against this potential difference until there is no net current flow, i.e. the unknown potential difference has the same value as the potential difference between the two points on the wire.
The wire is generally calibrated by making an initial balance with a standard cell. Then the balance is found with the unknown potential difference. The ratio of the potential differences is the ratio of their balance lengths.

**3.7** The polarity of the driver cell is not opposing the polarity of the unknown potential difference, i.e. positive terminal is not connected to positive terminal. Another possibility is that the unknown potential difference has a greater value than that of the driver cell.

**3.8** $\mathscr{E}/1.02 = 62.5/45.0$ and so $\mathscr{E} = 1.42$ V.

**3.9** $\mathscr{E}/V = 48.5/46.0$, where $V$ is the p.d. between the terminals of the cell when it drives a current through the 6.0 Ω resistor. $V = IR = R\mathscr{E}/(r + R)$, where $I$ is the current in the cell–6.0 Ω resistor circuit. Hence $48.5/46.0 = (r + R)/R = (r + 6.0)/6.0$ and $r = 0.33$ Ω.

**3.10** See Figure A3.1. The potential divider circuit is used to give a number of potential differences which are measured by the potentiometer and compared with the reading given by the voltmeter. A standard cell would be used to give the initial calibration of the potentiometer wire.

**3.11** (a) To increase the effective length of the wire so that the p.d. per unit length of the wire is small enough.
(b) Current through potentiometer wire = 1.5/1002 A. Resistance of 26.0 cm is $26.0 \times 2.0/100$ Ω. Hence p.d. across 26.0 cm $= IR = 1.5 \times 26.0 \times 2.0/(100 \times 1002) = 7.78 \times 10^{-4}$ V.

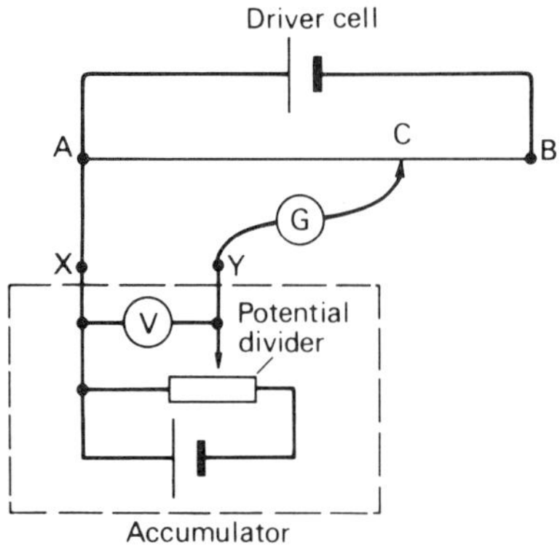


**Figure A3.1**

**3.12** A resistance $R$ could be put in series with the wire. Then $\xi = 1.5 = I(R + 3.0)$ and thus $I = 1.5/(R + 3.0)$. Resistance per cm of wire = $3.0/100\,\Omega$ and hence p.d. per cm = $(3.0/100) \times 1.5/(R + 3.0) = 0.02 \times 10^{-3}$. Hence $R = 2247\,\Omega$.

**3.13** See Figure 3.5 on page 143. The potential difference across one resistor is compared with the potential difference across the other when the same current is flowing through the two resistors.

**3.14** (a) $R_3$ and $R_4$ are being compared. $R_1$ is to vary the potential difference applied to the potentiometer wire and hence the potential difference per unit length of the wire. $R_2$ is a shunt to change the range of the galvanometer. $R_5$ is to vary the current through the two resistors that are being compared.
(b) Ratio = $46.5/63.4 = 0.733 = R_3/R_4$. $R_4$ has the higher resistance.

**3.15** See Figure A3.2. The potentiometer is used to measure the potential difference across a standard resistor, hence the current through the resistor can be calculated. The resistor is in series with the ammeter. A series of readings can be taken for different currents through the resistor.

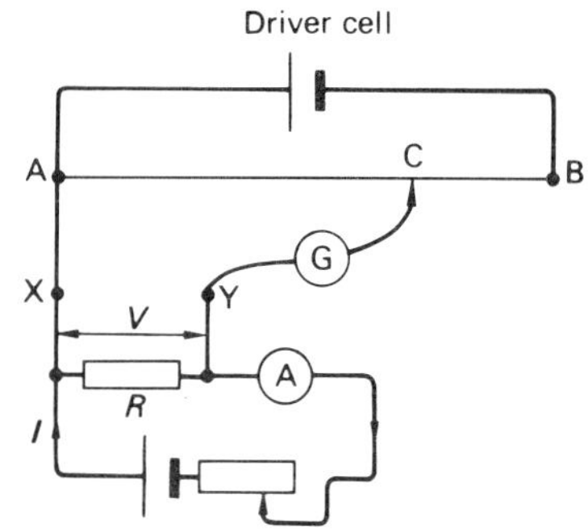


**Figure A3.2**

# Lesson 4 Alternating current

## Aims

After completing this lesson you should be able to

- explain what is meant by *root-mean-square* values of a.c. and know how to calculate them
- explain the terms *phase*, *phase difference*, *phasors*, *lagging* and *leading* when applied to a.c.
- explain the effects of pure resistance, pure inductance and pure capacitance in a.c. circuits
- define *reactance*
- calculate voltages, currents, and impedances for a.c. circuits containing pure resistance, pure inductance or pure capacitance.

## Prerequisites

Lesson 1 on d.c. resistive circuits. Lessons 3 and 12 in Chapter 6 on capacitance and inductance, and a knowledge of waves, e.g. Lesson 11 in Chapter 1 on simple harmonic motion or Lesson 1 in Chapter 5 on wave motion. A basic knowledge of calculus is necessary for many of the questions.

# STUDY

**I** **Read** about root-mean-square values for a.c.
References: *B* PP ch. 10, *B* ST4 ch. 1 or ST8 ch. 2, *D* ch. 15, *N* ch. 37, *W et al.* ch. 32(36), *Wh&H* ch. 58.

### Self-assessment questions

**4.1** What is the average value of a sinusoidally varying current over one complete cycle?

**4.2** Define *root-mean-square* value.

**4.3** Figure 4.1 shows how a sinusoidally varying current varies with time and how the square of such a current varies with time. Hence explain why the root-mean-square current is $I_0/\sqrt{2}$, where $I_0$ is the maximum value of the current.

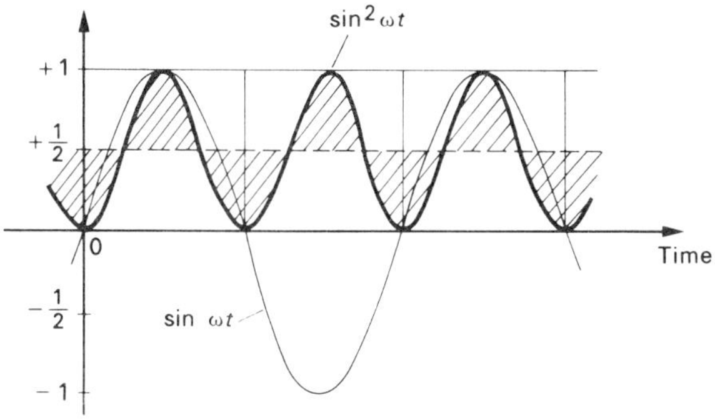


**Figure 4.1**

**4.4** For the British mains supply, which is sinusoidal, the root-mean-square voltage is 240 V. What is the maximum voltage?

**4.5** What is the root-mean-square current for the square waveform shown in Figure 4.2?

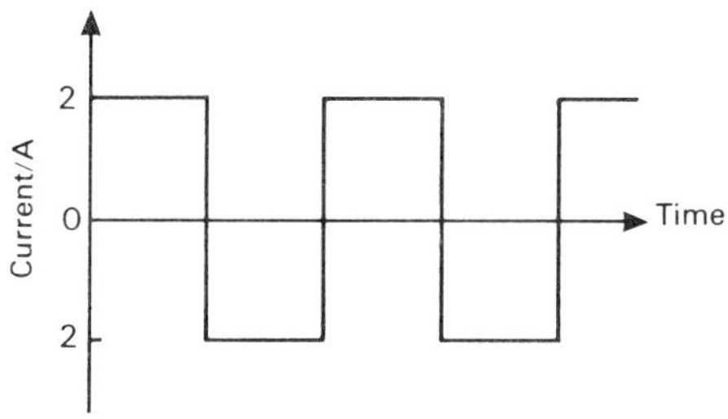


**Figure 4.2**

**II Read** about the effects of pure resistance, pure inductance and pure capacitance in a.c. circuits, including the concepts of phase, phase difference, lagging and leading, phasors and reactance.
References: *B* PP ch. 10, *B* ST8 ch. 2, *D* ch. 15, *N* ch. 37, *W et al.* ch. 32(36), *Wh&H* ch. 58.

**Self-assessment questions**

**4.6** Consider a circuit involving a source of a.c. and just pure resistance.
(a) If the applied potential difference to the resistor is given by $V = V_0 \sin \omega t$, what is the current variation with time in view of the relationship $V = IR$?
(b) Are the potential difference and the current in phase?
(c) Have the current and potential difference the same frequency?
(d) The instantaneous power is the product of the instantaneous current and the potential difference. What therefore is the mean value of this power? Note that the mean value of $\sin^2 \omega t$ is $\frac{1}{2}$, as found in SAQ 4.3.

**4.7** Consider a circuit involving a source of a.c. and just pure capacitance.
(a) For a capacitor $V = Q/C$. Thus, as $I = dQ/dt$, how is the current related to the potential difference across the capacitor?
(b) If the potential difference applied to the capacitor is given by $V = V_0 \sin \omega t$, derive a relationship showing how the current varies with time.
(c) What is the phase difference between the current and the potential difference?
(d) Have the current and the potential difference the same frequency?
(e) The *reactance* of a device is defined as being $(V_0/I_0$ (or $V_{rms}/I_{rms})$; hence derive the relationship for the capacitive reactance.
(f) The instantaneous power is the product of the instantaneous current and potential difference. What therefore is the mean value of this power? Note that $2 \sin \omega t \cos \omega t = \sin 2\omega t$.

**4.8** Consider a circuit involving a source of a.c. and just pure inductance.
(a) When there is a changing current in an inductor a back e.m.f. is produced of $L dI/dt$, where $L$ is the inductance. Since there is no resistance in the circuit this back e.m.f. and the applied potential difference of $V = V_0 \sin \omega t$ are the only potential differences in the circuit. Hence obtain a relationship between $dI/dt$ and time.
(b) By integration obtain the relationship between current and time.
(c) What is the phase difference between the current and the potential difference?
(d) Have the current and the potential difference the same frequency?
(e) The reactance of a device is defined as being $V_0/I_0$ (or $V_{rms}/I_{rms}$); hence derive the relationship for the inductive reactance.
(f) The instantaneous power is the product of the instantaneous current and the potential difference. What therefore is the mean value of this power? Note that $2 \sin \omega t \cos \omega t = \sin 2\omega t$.

**4.9** Calculate the reactance of each of the following:
(a) a 2 μF capacitor at 50 Hz,
(b) a 2 μF capacitor at 50 kHz,
(c) an inductor of 2 mH at 50 Hz,
(d) an inductor of 2 mH at 50 kHz.

**4.10** Sketch graphs showing how the reactance of a

capacitor and the reactance of an inductor vary with frequency.

**4.11** A sinusoidal quantity can be represented by the projection of a rotating vector onto a fixed line. Thus if there is a vector of magnitude $V_0$ rotating at a constant angular speed $\omega$, the $y$-component of the vector is $V = V_0 \sin \omega t$. The vector in this case is often referred to as a *phasor*.
(a) Draw a pair of phasors to represent the current and the potential difference for a pure resistor. The phasors must give current and potential difference variations with time that are always in phase, i.e. both reach their maxima at the same time.
(b) Draw a pair of phasors to represent the current and the potential difference for a pure capacitor. The current must always lead the potential difference by 90° (or $\frac{1}{2}\pi$).
(c) Draw a pair of phasors to represent the current and the potential difference for a pure inductor. The current must always lag behind the potential difference by 90° (or $\frac{1}{2}\pi$).

# SELF-ASSESSMENT

## Key points

The **root-mean-square value** of an alternating quantity is the steady direct value of that quantity which converts electrical energy to other forms of energy at the same rate as the alternating quantity. For a sinusoidal current $I_{rms} = I_0/\sqrt{2}$, for a potential difference $V_{rms} = V_0/\sqrt{2}$, where $I_0$ and $V_0$ are the maximum values of those quantities.
When an alternating potential difference is applied to a pure resistor the alternating current is in phase with the p.d. and of the same frequency. $V_0/I_0 = V_{rms}/I_{rms} = R$, the resistance. The mean power developed $= I_{rms}V_{rms} = I_{rms}^2 R$.
For an alternating potential difference applied to a pure capacitor, the current leads the potential difference by 90° and is of the same frequency. $V_0/I_0 = 1/\omega C = X_C$, the **capacitive reactance**. The average power over a complete cycle is zero.
For an alternating potential difference applied to a pure inductor, the current lags behind the potential difference by 90° and is of the same frequency. $V_0/I_0 = \omega L = X_L$, the **inductive reactance**. The average power over a complete cycle is zero.

## Answers I

**4.1** Zero current. This is because there are equal amounts of positive and negative current.

**4.2** See the key points.

**4.3** The mean value of the $\sin^2 \omega t$ graph is the line which gives equal areas above and below the line, the shaded areas. This occurs at half the maximum value, i.e. $\frac{1}{2}I_0^2$. Hence $I_{rms}^2 = \frac{1}{2}I_0^2$ and $I_{rms} = I_0/\sqrt{2}$.

**4.4** $V_{rms} = V_0/\sqrt{2}$ and so $V_0 = V_{rms} \times \sqrt{2} = 240 \times \sqrt{2} = 339$ V.

**4.5** When this wave is squared a straight line is produced with the value of 4 $A^2$. Hence $I_{rms}^2 = 4$ and $I_{rms} = 2$ A.

## Answers II

**4.6** (a) $I = (V_0/R) \sin \omega t$. The current is a maximum when $\sin \omega t = 1$, thus $I_0 = V_0/R$ and so $I = I_0 \sin \omega t$.
(b) Yes, both have the same $\sin \omega t$ terms.
(c) Yes (same $\omega = 2\pi f$).
(d) Power $= IV = I_0 \sin \omega t \times V_0 \sin \omega t = I_0 V_0 \sin^2 \omega t$. Hence the mean value $= \frac{1}{2}I_0V_0 = (I_0/\sqrt{2})(V_0/\sqrt{2}) = I_{rms}V_{rms}$.

**4.7** (a) $I = C\, dV/dt$
(b) $I = C\, d(V_0 \sin \omega t)/dt = \omega C V_0 \cos \omega t = I_0 \cos \omega t$, where the maximum value of the current $= I_0 = \omega C V_0$.
(c) The current leads the potential difference by 90°, the cosine function reaching its maximum value 90° ahead of the sine function.
(d) Yes, both involve $\omega t$.
(e) As in (b) we have $I_0 = \omega C V_0$, so the capacitive reactance $X_C = V_0/I_0 = 1/\omega C$.
(f) Power $= IV = I_0 \cos \omega t \times V_0 \sin \omega t = \frac{1}{2}I_0V_0 \sin 2\omega t$. The mean value of a sine term is zero over a complete cycle. Thus mean power = 0. During one half-cycle the capacitor is charged up, during the next half-cycle it discharges. Thus the energy it gains in one half-cycle it gives up in the next half.

**4.8** (a) $V - L\, dI/dt = 0$ and so $V = L\, dI/dt = V_0 \sin \omega t$.
(b) $dI = (V_0/L) \sin \omega t\, dt$. Hence $I = -(V_0/\omega L) \cos \omega t = -I_0 \cos \omega t$, where $I_0 = V_0/\omega L$.
(c) The current lags behind the potential difference by 90°.
(d) The same frequency.
(e) $X_L = V_0/I_0 = \omega L$
(f) Power $= IV = -I_0 \cos \omega t \times V_0 \sin \omega t = -\frac{1}{2}I_0V_0 \sin 2\omega t$. The mean value of the sine term is zero and so the mean power is zero. During one half-cycle energy is stored in the magnetic field of the inductor, during the next half-cycle this energy is returned by the field to the circuit.

**4.9** (a) $X_C = 1/\omega C = 1/2\pi fC = 1/(2\pi \times 50 \times 2 \times 10^{-6}) = 1.59 \times 10^3\,\Omega$
(b) Increasing the frequency by a factor of 1000 for a capacitor decreases the reactance by a factor of 1000 to give $X_C = 1.59\,\Omega$.
(c) $X_L = \omega L = 2\pi fL = 2\pi \times 50 \times 2 \times 10^{-3} = 0.628\,\Omega$.
(d) Increasing the frequency by a factor of 1000 for an inductor increases the reactance by a factor of 1000 to $628\,\Omega$.

**4.10** See Figure A4.1.

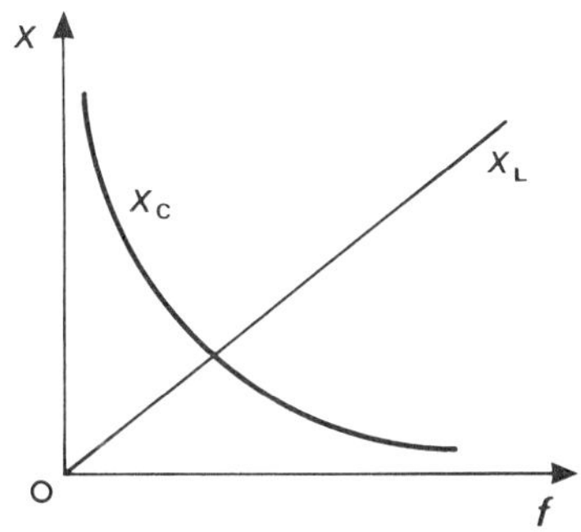


**Figure A4.1**

**4.11** See Figure A4.2.

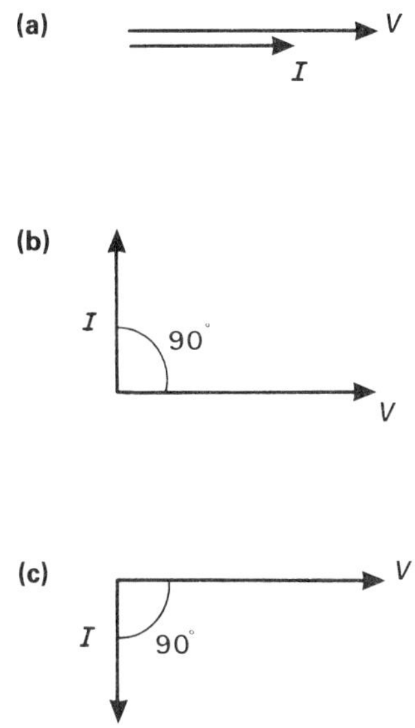


**Figure A4.2**

# Lesson 5 A.c. circuits

## Aims

After completing this lesson you should be able to
- solve problems involving (a) resistance and capacitance, (b) resistance and inductance, (c) resistance, capacitance *and* inductance in series in an a.c. circuit
- define *impedance*
- explain *series resonance*
- determine the power in an a.c. circuit involving resistance, capacitance and inductance, explaining the term *power factor*.

## Prerequisites

Lesson 4 on alternating current, also vectors in Lesson 1 of Chapter 1.

## STUDY

**I** **Read** about series circuits involving (a) resistance and capacitance, (b) resistance and inductance, (c) resistance, capacitance *and* inductance, and the definition of impedance.

References: *B* ST8 ch. 2, *D* ch. 15, *N* ch. 37, *W et al.* ch. 32(36), *Wh&H* ch. 58.

**Self-assessment questions**

**5.1** Consider a circuit supplied with an alternating potential difference and having a resistor in series with a capacitor.
(a) The current will be the same at every instant of time for both components. Therefore we take the direction of the current phasor as being the reference direction. What are the directions of the phasors for the potential differences across (i) the resistor and (ii) the capacitor, with reference to the current phasor?
(b) Draw the phasors for the potential differences across the resistor and across the capacitor and find their resultant (in the same way as the resultant of two vectors—such as forces—would be determined).
(c) The resultant $V = \sqrt{(V_R^2 + V_C^2)}$. Since $V_R = IR$ and $V_C = IX_C$, rewrite the equation in terms of $R$ and $X_C$.
(d) The *impedance* $Z$ of a circuit is defined by the equation $Z = V_0/I_0 = V_{rms}/I_{rms}$. Hence obtain a relationship for the impedance of the resistor in series with the capacitor.

(e) Remembering that the current is in phase with the potential difference across the resistor, what is the phase difference between the potential difference across the series combination, i.e. $V$, and the current?

***5.2** A resistance of 3000 Ω is in series with a 2 μF capacitor when a potential difference of 20 V and frequency 100 Hz is applied across the arrangement. What is (a) the impedance of the circuit, (b) the potential difference across the resistor, (c) the potential difference across the capacitor, (d) the current?

**5.3** Consider a circuit supplied with an alternating potential difference and having a resistor in series with an inductor.
(a) The current will be the same at every instant of time for both components. Therefore we take the direction of the current phasor as being the reference direction. What are the directions of the phasors for the potential differences across (i) the resistor and (ii) the inductor?
(b) Draw the phasors for the potential differences across the resistor and the inductor and find their resultant.
(c) Rewrite the resulting equation for the resultant in terms of resistance and inductive reactance.
(d) Hence obtain a relationship for the impedance of the resistor in series with the inductor.
(e) Remembering that the current is in phase with the potential difference across the resistor, what is the phase difference between the potential difference across the series combination and the current?

***5.4** A 100 Ω resistor is in series with a 0.5 H inductor. If an alternating potential difference of 12 V, frequency 50 Hz, is applied across the arrangement, what is (a) the impedance of the circuit, (b) the potential difference across the resistor, (c) the potential difference across the inductor, (d) the current?

**5.5** Consider a circuit consisting of a resistor, a capacitor and an inductor in series, and the arrangement being supplied with an alternating potential difference.
(a) The current will be the same at every instant of time for all the components, since they are in series. Therefore take the direction of the current phasor as being the reference direction. Draw the phasors for the potential differences across each of the three components and hence find the resultant.
(b) Obtain a relationship for the circuit impedance in terms of the resistance, the capacitive impedance and the inductive impedance.
(c) What is the phase angle between the resultant potential difference and the current?

***5.6** A resistor of 300 Ω, a capacitor of 4 μF and an inductor of 200 mH are in series. When a potential difference of 12 V, frequency 50 Hz, is connected across the arrangement, what is (a) the impedance, (b) the current, (c) the potential difference across each component?

**II** **Read** about series resonance.
References: *B* PP ch. 10, *B* ST8 ch. 2, *D* ch. 15, *N* ch. 37, *W et al.* ch. 32(36), *Wh&H* ch. 58.

**Self-assessment questions**

**5.7** For a series circuit containing a resistor, a capacitor and an inductor, what is the impedance when the capacitive reactance equals the inductive reactance? This is the minimum impedance. What is the phase angle between the potential difference across the arrangement and the current for this condition?

***5.8** A circuit consists of a 200 Ω resistor, a 2 μF capacitor and a 0.5 H inductor in series.
(a) At what frequency will the circuit impedance be a minimum?
(b) If the potential difference applied to the arrangement is 20 V r.m.s. at this *resonant* frequency, what will the current be?

***5.9** A capacitor, an inductor, a resistor and a lamp are connected in series with an alternating voltage. Describe what happens as the frequency of the voltage is varied from a low to a high value, the voltage being kept constant.

**III** **Read** about power in a series circuit having resistance, capacitance and inductance, including the term *power factor*.
References: *B* ST8 ch. 2, *D* ch. 15, *N* ch. 37, *W et al.* ch. 32, *Wh&H* ch. 58.

**Self-assessment questions**

**5.10** When a circuit is composed of resistance, capacitance and inductance, power is only absorbed by the resistive part of the circuit. Explain this statement.

**5.11** For an a.c. circuit supplied with a potential difference of $V = V_0 \sin \omega t$, a current will be produced having a phase angle $\phi$ relative to the potential difference, i.e. $I = I_0 \sin (\omega t + \phi)$.
(a) Write an equation for the instantaneous power.
(b) Since $\sin (\omega t + \phi) = \sin \omega t \cos \phi + \cos \omega t \sin \phi$, and the average value of a $\sin^2 \omega t$ (or $\cos^2 \omega t$) term is $\frac{1}{2}$ and that of $\sin \omega t \cos \omega t$ is zero, derive an equation for the average power.

(c) Average power = $\frac{1}{2}V_0I_0 \cos \phi$, the term $\cos \phi$ being known as the *power factor*. Rewrite this equation in terms of root-mean-square values.
(d) Consider the phasor diagram in SAQ 5.5 and hence state the relationship between $R$, $Z$ and $\phi$.
(e) Since $V_{rms}/I_{rms} = Z$, and using the relationship arrived at in (d), write the equation for the average power in terms of $I_{rms}$ and $R$.

***5.12** A circuit consists of a 200 Ω resistor, an 8 μF capacitor and a 400 mH inductor in series with a voltage supply of root-mean-square value 10 V and frequency 200 Hz. What is the average power developed?

# SELF-ASSESSMENT

## Key points

The **impedance** $Z$ of a circuit or device is defined as $Z = V_0/I_0 = V_{rms}/I_{rms}$.
For a resistor in series with a capacitor $Z = \sqrt{(R^2 + X_C^2)}$. The potential difference in such a circuit *lags* the current by $\phi$, where $\tan \phi = V_C/V_R = X_C/R$.
For a resistor in series with an inductor $Z = \sqrt{(R^2 + X_L^2)}$. The potential difference in such a circuit *leads* the current by $\phi$, where $\tan \phi = V_L/V_R = X_L/R$.
For a series circuit having a resistor, a capacitor and an inductor, $Z = \sqrt{[R^2 + (X_L - X_C)^2]}$. The potential difference in such a circuit leads the current by $\phi$, where $\tan \phi = (V_L - V_C)/V_R = (X_L - X_C)/R$. The minimum value of the impedance occurs when $X_L = X_C$ and so $Z = R$. At this condition, known as **resonance**, $\phi = 0$.
The mean power in a series circuit involving resistance, capacitance and inductance is $\frac{1}{2}V_0I_0 \cos \phi = V_{rms}I_{rms} \times \cos \phi = I_{rms}^2 R$, the power being determined purely by the resistance.

## Answers I

**5.1** (a) The phasor for the p.d. across the resistor $V_R$ is in phase with the current, the phasor for the capacitor $V_C$ lags the current by 90°. (b) See Figure A5.1; $V^2 = V_R^2 + V_C^2$. (c) $V = I\sqrt{(R^2 + X_C^2)}$ (d) $Z = V/I = \sqrt{(R^2 + X_C^2)}$ (e) $\tan \phi = V_C/V_R = IX_C/IR = X_C/R$.

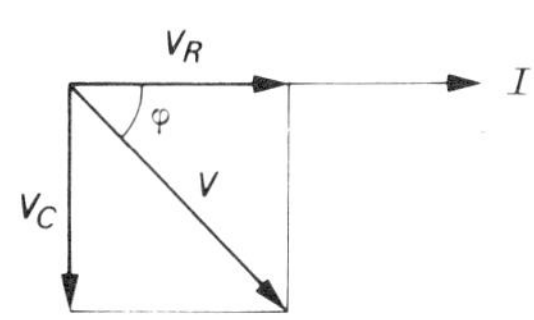


**Figure A5.1**

**5.2** (a) $Z = \sqrt{(R^2 + X_C^2)}$, where $X_C = 1/\omega C = 1/2\pi fC = 1/(2\pi \times 100 \times 2 \times 10^{-6}) = 7.96 \times 10^2 \Omega$. Hence $Z = \sqrt{(3000^2 + 795^2)} = 3104\ \Omega$.
(b) $V = IZ$ and so $I = 20/3104 = 6.44 \times 10^{-3}$ A. $V_R = IR = 6.44 \times 10^{-3} \times 3000 = 18.4$ V.
(c) $V_C = IX_C = 6.44 \times 10^{-3} \times 7.96 \times 10^2 = 5.1$ V. Notice that the simple sum of the potential differences across the resistor and the capacitor does not give the source potential difference.
(d) As earlier $I = 6.44 \times 10^{-3}$ A.

**5.3** (a) The potential difference across the resistor is in phase with the current, the potential difference across the inductance leads the current by 90°. (b) See Figure A5.2; $V = \sqrt{(V_R^2 + V_L^2)}$. (c) $V = I\sqrt{(R^2 + X_L^2)}$ (d) $Z = V/I = \sqrt{(R^2 + X_L^2)}$ (e) $\tan \phi = V_L/V_R = IX_L/IR = X_L/R$

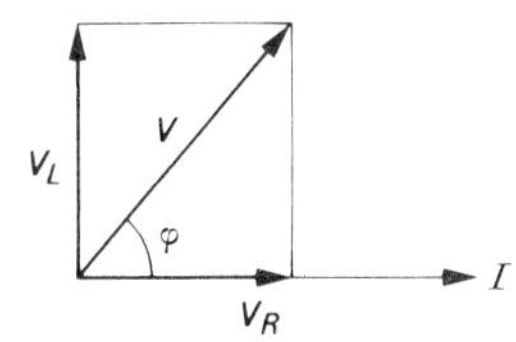


**Figure A5.2**

**5.4** (a) $Z = \sqrt{(R^2 + X_L^2)}$, where $X_L = 2\pi fL = 2\pi \times 50 \times 0.5 = 157\ \Omega$. Hence $Z = \sqrt{(100^2 + 157^2)} = 186\ \Omega$.
(b) $V = IZ$ and so $I = 12/186 = 0.0645$ A. Hence $V_R = IR = 0.0645 \times 100 = 6.45$ V. (c) $V_L = IX_L = 0.0645 \times 157 = 10.1$ V (d) See earlier; $I = 0.0645$ A.

**5.5** (a) See Figure A5.3; $V^2 = (V_L - V_C)^2 + V_R^2$. (b) $V = I\sqrt{[R^2 + (X_L - X_C)^2]}$, hence $Z = V/I = \sqrt{[R^2 + (X_L - X_C)^2]}$ (c) $\tan \phi = (V_L - V_C)/V_R = (X_L - X_C)/R$

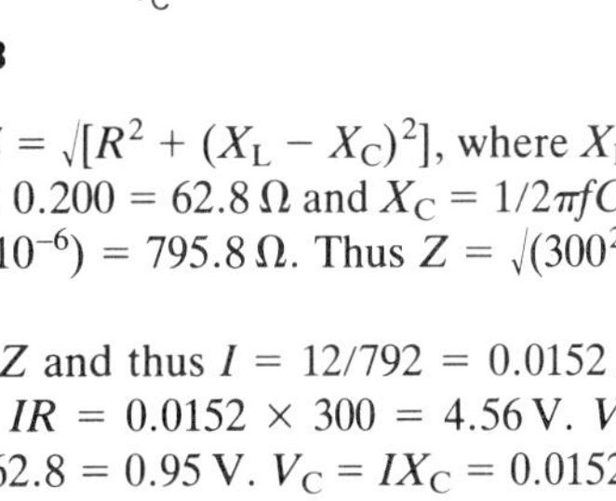


**Figure A5.3**

**5.6** (a) $Z = \sqrt{[R^2 + (X_L - X_C)^2]}$, where $X_L = 2\pi fL = 2\pi \times 50 \times 0.200 = 62.8\ \Omega$ and $X_C = 1/2\pi fC = 1/(2\pi \times 50 \times 4 \times 10^{-6}) = 795.8\ \Omega$. Thus $Z = \sqrt{(300^2 + 733^2)} = 792\ \Omega$.
(b) $V = IZ$ and thus $I = 12/792 = 0.0152$ A.
(c) $V_R = IR = 0.0152 \times 300 = 4.56$ V. $V_L = IX_L = 0.0152 \times 62.8 = 0.95$ V. $V_C = IX_C = 0.0152 \times 795.8 = 12.10$ V.

## Answers II

**5.7** At resonance $X_L = X_C$ and $Z$ has its minimum value of $R$; $\tan\phi = (X_L - X_C)/R = 0$, and so $\phi = 0$: the potential difference and the current are in phase.

**5.8** (a) $X_L = X_C$, hence $2\pi fL = 1/2\pi fC$ and so $f^2 = 1/(4\pi^2 LC) = 1/(4\pi^2 \times 0.5 \times 2 \times 10^{-6})$ and $f = 159$ Hz. (b) $V = IZ$, but $Z = R$, thus $I = 20/200 = 0.10$ A.

**5.9** Initially the current will be very low and the lamp will not glow. As the frequency is increased so the impedance decreases and the current grows. Near to the resonant frequency the lamp will start to glow as the impedance decreases even more. At the resonant frequency the impedance is a minimum and the lamp will be its brightest. Increasing the frequency yet more will increase the impedance and the lamp will become less bright.

## Answers III

**5.10** As has been indicated earlier, in Lesson 4, the average power for a capacitor or an inductor with a.c. is zero over complete cycles since during half the cycle the component takes power from the source and then in the next half-cycle it returns it. In the case of the capacitor the power absorbed is stored in the electric field, in the case of the inductor it is stored in the magnetic field, and in both cases it is then returned to the circuit. It is only the resistor which absorbs power during the entire cycle.

**5.11** (a) $P = V_0 \sin\omega t \times I_0 \sin(\omega t + \phi)$
(b) $P = V_0 I_0 \sin\omega t (\sin\omega t \cos\phi + \cos\omega t \sin\phi) = V_0 I_0 \cos\phi \sin^2\omega t + V_0 I_0 \sin\phi \sin\omega t \cos\omega t$. Average power $= \frac{1}{2} V_0 I_0 \cos\phi + 0$.
(c) Average power $= V_{rms} I_{rms} \cos\phi$
(d) $IR = IZ \cos\phi$ and so $R = Z\cos\phi$
(e) Average power $= I_{rms}{}^2 Z \cos\phi = I_{rms}{}^2 R$. The average power depends only on the resistance.

**5.12** $Z = \sqrt{[R^2 + (X_L - X_C)^2]}$, where $X_L = 2\pi fL = 2\pi \times 200 \times 0.400 = 503\,\Omega$ and $X_C = 1/2\pi fC = 1/(2\pi \times 200 \times 8 \times 10^{-6}) = 99\,\Omega$. Thus $Z = \sqrt{(200^2 + 404^2)} = 451\,\Omega$. Hence $I_{rms} = V_{rms}/Z = 10/451 = 0.022$ A. Average power $= I_{rms}{}^2 R = 0.022^2 \times 200 = 0.098$ W.

# Lesson 6 Rectification

## Aims

After completing this lesson you should be able to

- describe the characteristics of rectifiers
- explain (a) *half-wave* rectification and (b) *full-wave* rectification
- explain the functions of the components in smoothing circuits.

## Prerequisites

Lesson 4 on alternating current and Lessons 3, 10 and 11 of Chapter 6.

## STUDY

**I Read** about rectifiers, their characteristics, and their use in both half- and full-wave rectification.
References: *B* ST8 ch. 3, *D* ch. 15, *N* ch. 38, *W et al.* (ch. 36), *Wh&H* ch. 58.

### Self-assessment questions

**6.1** Rectifiers are components that have a low resistance to current flow in one direction and a high resistance to current flow in the opposite direction. State two types of component that have this property.

**6.2** When connection is made so that a rectifier conducts, it is said to be *forward-biased*. When the connection is such that it does not conduct because it is in the high-resistance mode, then it is said to be *reverse-biased*. The arrow-head on the symbol for a rectifier indicates the forward direction of conventional current flow. Hence explain for the circuit shown in Figure 6.1 which of the input terminals has to be positive and which negative for current flow.

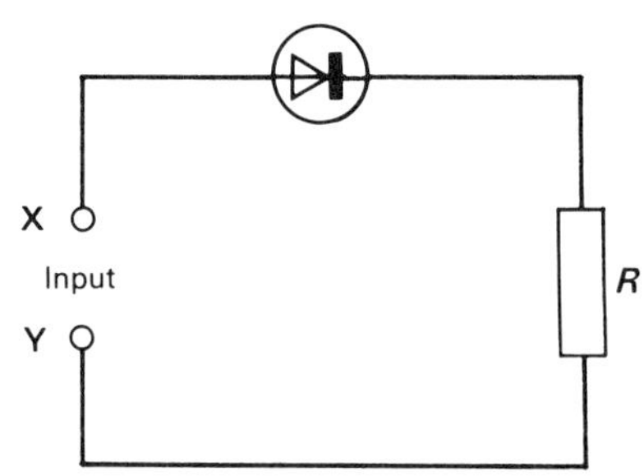


**Figure 6.1**

**6.3** For the circuit shown in Figure 6.1, describe how the current in the circuit and the potential difference across $R$ will vary with time when an alternating input is applied between X and Y.

**6.4** The arrangement described in SAQ 6.3 is said to be *half-wave rectification*. Why 'half-wave'?

**6.5** Figure 6.2 shows the circuit of a centre-tap full-wave rectifier. Explain how it works.

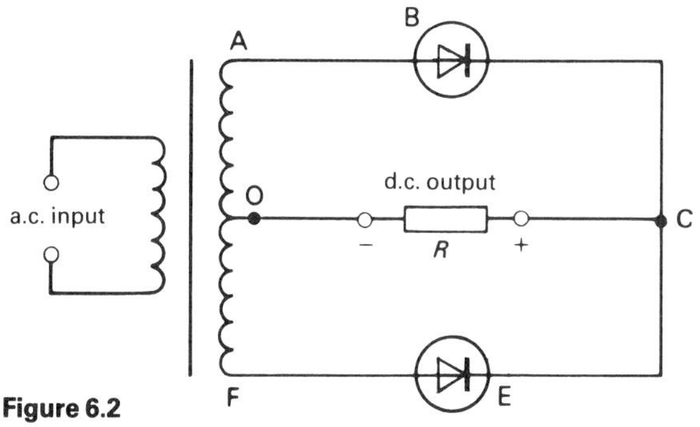


**Figure 6.2**

**6.6** Figure 6.3 shows the circuit of a bridge full-wave rectifier. Explain how it works.

**II** **Read** about smoothing circuits for rectifiers. References: *B* ST8 ch. 3, *D* ch. 15, *Wh&H* ch. 58.

**Self-assessment questions**

**6.7** What is the function of a smoothing circuit?

**6.8** Figure 6.4 shows a simple smoothing circuit consisting of a capacitor connected across the load. Explain how this smooths the rectifier output, drawing a graph showing the effect of the capacitor on the rectifier output.

**6.9** Figure 6.5 shows the circuit already shown in Figure 6.4, modified by the addition of what is called a *filter circuit*. Explain how this filter circuit operates. Hint: you might find it easier to think of the output from the Figure 6.4 circuit as consisting of two signals superimposed on each other, one being a d.c. signal and the other a ripple signal.

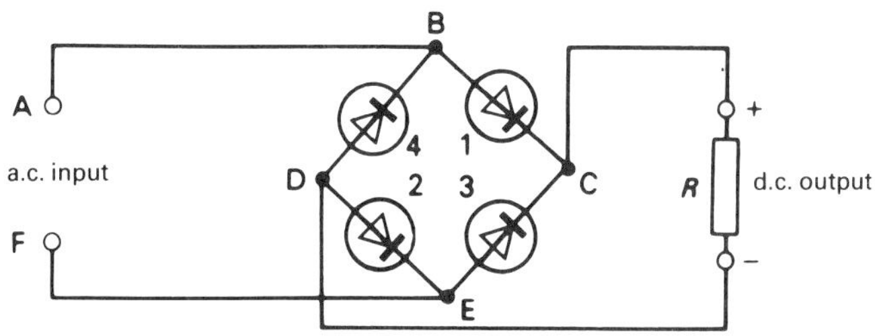


**Figure 6.3**

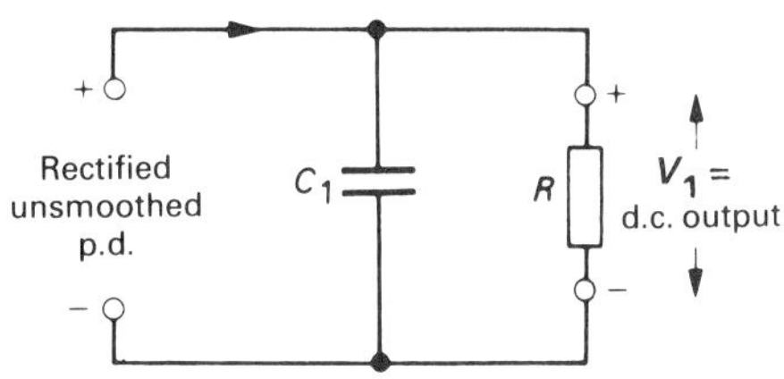


**Figure 6.4**

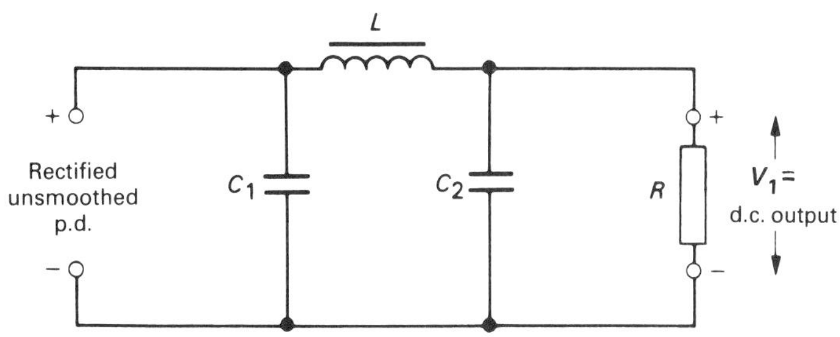


**Figure 6.5**

# SELF-ASSESSMENT

## Key points

**Rectifiers** are components that have a low resistance to current flow in one direction and a high resistance to current flow in the reverse direction. A **half-wave rectifier** circuit is one that gives a unidirectional output using just half of each cycle of the alternating input. A **full-wave rectifier** gives a unidirectional output using both halves of every cycle of input. **Smoothing circuits** are used to make the unidirectional output a more steady value.

## Answers I

**6.1** The thermionic diode valve, the semiconductor diode and the selenium or copper rectifier.

**6.2** X has to be positive and Y negative for a current to flow, the rectifier having a low resistance for current flow in this direction.

**6.3** See Figure A6.1.

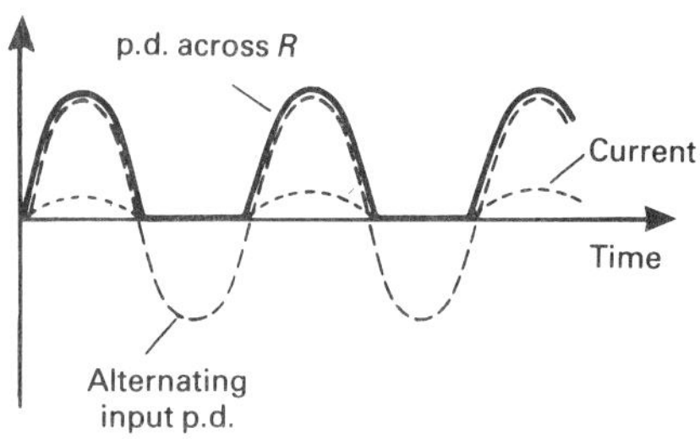


**Figure A6.1**

**6.4** Because only half of every cycle is used for the output.

**6.5** The centre-tap point O of the transformer has a potential half-way between that at A and that at F. It is thus convenient to think of O being always at zero potential and the other points A and F oscillating about it. Thus when A is positive F is negative. This means that B conducts and E does not. When however F becomes positive then A is negative and E conducts but B does not. Hence as the input alternates so the output switches from being conducted by one rectifier to being conducted by the other. The result is an output across the resistor $R$ which is unidirectional and makes use of each half cycle.

**6.6** When A is positive and F negative then rectifiers 1 and 2 conduct but 3 and 4 do not. When A is negative and F positive then 3 and 4 conduct but 1 and 2 do not. The result is that each half cycle gives a unidirectional output across $R$.

## Answers II

**6.7** To make a unidirectional but fluctuating signal smoother, i.e. a more steady value.

**6.8** See Figure A6.2. The rectifier input to the circuit causes current to flow through $R$ and the capacitor to become charged, to give the p.d. across the capacitor (and the load) indicated by A on the graph. At A the input p.d. from the rectifier begins to fall and the capacitor starts to discharge. It can only discharge through $R$, not through the rectifier since the polarity is wrong. The p.d. across $R$ only falls to B during the time taken for the rectifier output to start charging the capacitor up again.

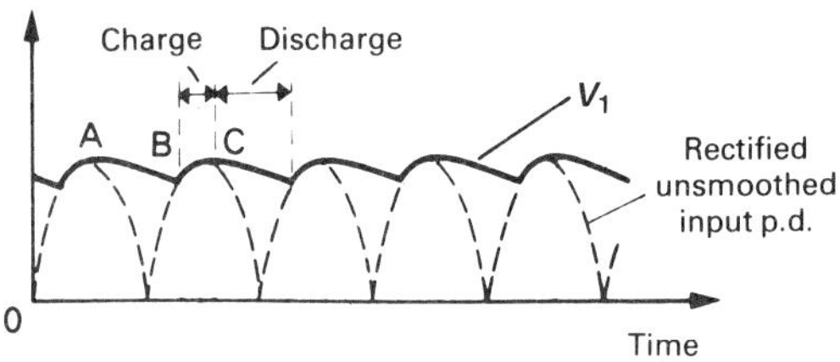


**Figure A6.2**

**6.9** Figure A6.3 shows an alternative way of considering the circuit. $C_1$ behaves the same as the capacitor in SAQ 6.8 and thus the potential difference across it is as shown in Figure A6.2. In thinking of this potential difference as being two signals, one a steady d.c. and the other a ripple, we can consider each of the signals in turn being applied to the series arrangement of $L$ and $C_2$. With a d.c. signal $C_2$ will have a very high resistance while $L$ will have a very small resistance. Thus virtually all the potential difference will be across $C_2$. With the ripple signal, $L$ will offer a greater reactance than $C_2$ and thus the greater part of the alternating potential difference will be across $L$. Hence the output across $C_2$ will be predominantly steady d.c.

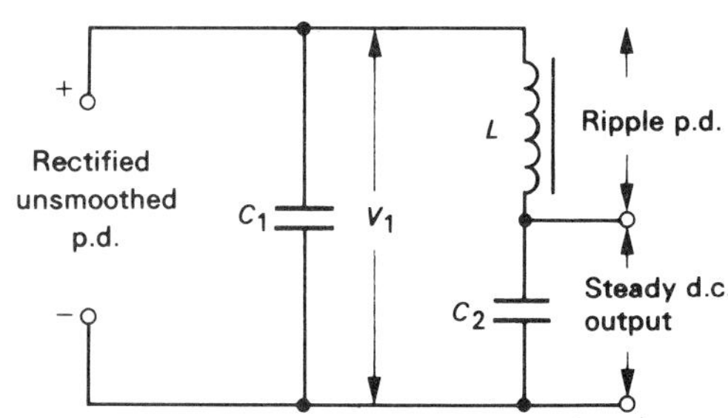


**Figure A6.3**

# Lesson 7 Solid-state electronics

## Aims

After completing this lesson you should be able to

- distinguish between p- and n-type semiconductors in terms of their main charge carriers
- explain the action of the p–n junction diode in terms of *hole* and electron movements and the *depletion layer*
- describe and explain the action of the junction transistor
- explain the operation of a transistor amplifier
- describe the output–input characteristic of a single-stage amplifier and its operation as a switch.

## Prerequisites

Lesson 1 on d.c. resistive circuits and Lesson 9 of Chapter 6 on current and charge movement. For some questions a basic knowledge of the action of capacitors is required, see Lesson 3 of Chapter 2.

# STUDY

**I Read** about p- and n-type semiconductors and the action of the junction diode, in terms of a simple model of conduction as the movement of holes and electrons.
References: *B* PP ch. 10, *B* ST8, ch. 4, *D* ch. 21, *N* ch. 39, *Wh&H* ch. 48, ch. 61.

**Self-assessment questions**

**7.1** At about room temperature how do metals, semiconductors and insulators compare with regard to (a) resistivity, (b) effect of temperature on resistivity and (c) number of charge carriers available for conduction?

**7.2** (a) What is the difference between *intrinsic* and *extrinsic* semiconductors?
(b) With intrinsic semiconductors what charge carriers are responsible for an electric current?
(c) What is meant by *doping* and what is the purpose of it?
(d) In (i) n-type and (ii) p-type semiconductors what charge carriers are responsible for an electric current?

**7.3** The junction diode is a crystal of germanium or silicon in which doping has been used to produce a p–n junction.
(a) Explain what the *depletion layer* is and how it is produced.
(b) The term *barrier*, or *contact*, or *junction* potential is used. What is this potential?
(c) What happens when (i) a forward bias and (b) a reverse bias is applied to the junction diode?
(d) Sketch the current–potential difference graph for a junction diode, for both forward and reverse bias.

**II Read** about the junction transistor and its operation as an amplifier and as a switch.
References: *B* PP ch. 10, *B* ST8 ch. 4, ch. 5, *D* ch. 21, *N* ch. 39, *W et al.* (ch. 41), *Wh&H* ch. 61.

**Self-assessment questions**

**7.4** A junction transistor can be considered to be two p–n junctions back to back, the region between the two junctions being very narrow. Figure 7.1 shows diagrammatically such a transistor in a circuit, the arrangement being known as a *common-emitter* connection.

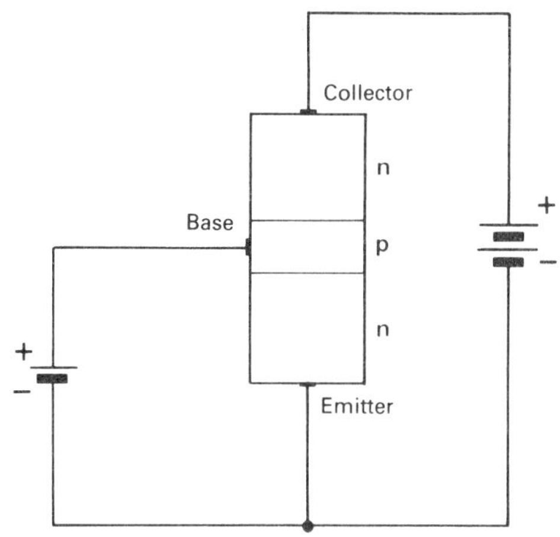


**Figure 7.1**

(a) For the collector–base junction and the emitter–base junction, how are they biased when transistor action is required?
(b) The term *transistor* means 'transfer-resistance', a current in a low-resistance circuit giving rise to a current in a high-resistance circuit. Explain this statement.
(c) Figure 7.2 shows how the collector current is related to the base current for a particular value of potential difference between collector and emitter. What is the d.c. current gain ($h_{FE}$, the forward current transfer ratio)?

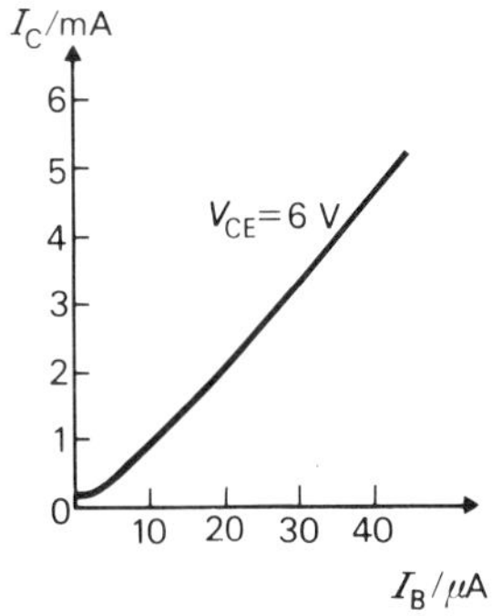


**Figure 7.2**

***7.5** Figure 7.3 shows a single-stage practical amplifying circuit. Explain the functions of the capacitors $C_1$ and $C_2$, and the resistors $R_B$ and $R_C$.

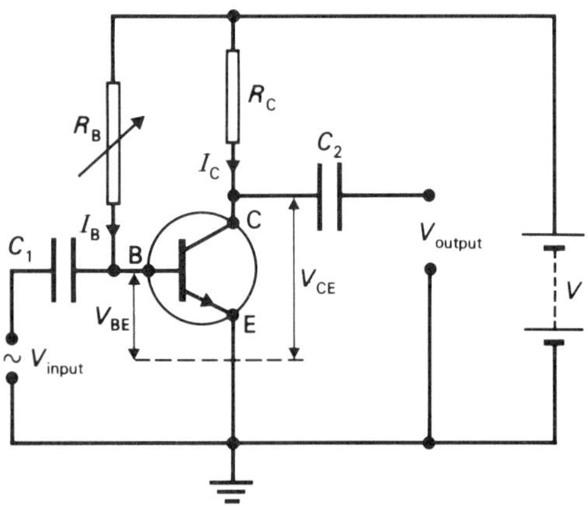


**Figure 7.3**

***7.6** Figure 7.4 shows a transistor circuit and Figure 7.5 the relationship between the input potential difference $V_{input}$ and the output potential difference $V_{output}$.

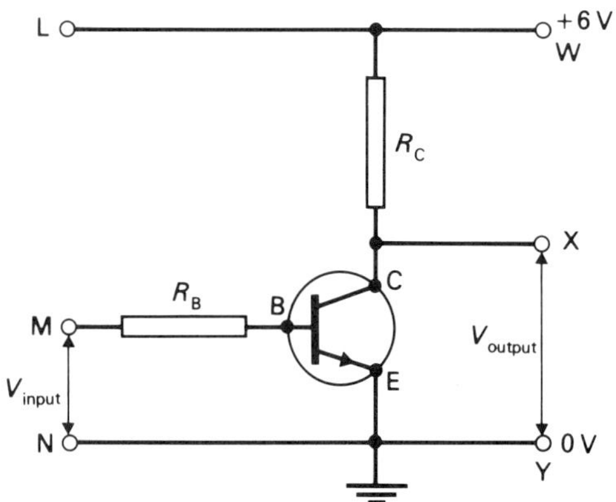


**Figure 7.4**

(a) What will be the outputs when (i) L is connected to M, (ii) M is connected to N?
(b) Explain how the circuit can be used as a switching circuit.
(c) Over what range of input voltage could this circuit be used as an amplifier?

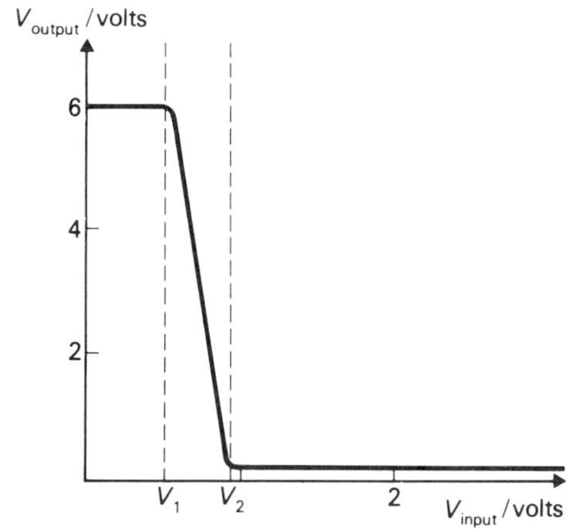


**Figure 7.5**

***7.7** Draw graphs showing the input and output characteristics for a transistor in the common-emitter mode of connection. On the output characteristic indicate the region of values for which the transistor would be operated as an amplifier.

# SELF-ASSESSMENT

## Key points

**Semiconductors** have resistivities of the order of a few ohm m, **metals** about $10^{-8}$ ohm m and **insulators** about $10^{12}$ ohm m. Whereas in a metal there is about one free charge carrier per atom, in a semiconductor there is about one per million atoms and in an insulator hardly any at all.

The introduction of small amounts of certain impurities into a semiconductor, such as germanium or silicon, is known as **doping** and significantly increases the number of charge carriers available for conduction. In **n-type** material extra electrons have been introduced and conduction is mainly by electrons, in **p-type** extra 'holes' have been introduced and conduction is mainly by movement of holes.

In the **junction diode**, a **p–n junction**, electrons and holes move across the junction and produce a **depletion layer** on either side of the junction. This results in a **barrier potential**.

A **junction transistor** can be considered to be two p–n junctions back to back, with the region between the junctions being very narrow. With the transistor in the **common-emitter** form of connection, a current in the base circuit, a low-resistance circuit, gives rise to a current in the collector circuit, a much higher-resistance circuit.

## Answers I

**7.1** (a) Resistivities: metals, $10^{-8}$ ohm m; semiconductors, a few ohm m; insulators, $10^{12}$ ohm m.
(b) With metals an increase in resistivity as the temperature increases, with semiconductors and insulators a decrease.
(c) Metals, about one per atom; semiconductors, about one per million atoms; insulators hardly any.

**7.2** (a) *Intrinsic* semiconductors are pure materials, *extrinsic* ones have had impurities added to them.
(b) Both electrons and holes. The electrons are those which have been 'shaken free' from their atoms by thermal agitation. A *hole* is a position from which an electron is missing. Such a vacancy behaves as though it had a positive charge. Holes move through the material in the opposite direction to electrons.
(c) *Doping* is the deliberate addition of certain impurities to a semiconductor to change its conductivity.
(d) (i) n-type semiconductors have had atoms added which have donated spare electrons. The main contribution to the electric current then comes from electrons, with the holes being the *minority carrier*.
(ii) p-type semiconductors have had atoms added which have donated holes, i.e. vacancies into which electrons can move. The main contribution to the electric current then comes from holes, with electrons being the minority carrier.

**7.3** (a) Movement of holes from the p-type material into the n-type and electrons from the n-type into the p-type material cause a net migration of charge across the junction. The p-type material in the layer next to the junction suffers a net loss of holes and gain of electrons and so becomes negatively charged. The n-type material in the layer next to the junction suffers a net loss of electrons and gain of holes and so becomes positively charged. This charge distribution continues until it becomes large enough to prevent further migration across the junction. Both materials on either side of the boundary have thus lost some of their charge carriers, hence the term 'depletion' layer.
(b) The charge migration results in a potential difference. The barrier potential for germanium or silicon is a few tenths of a volt.
(c) See Figure A7.1. With reverse bias the applied potential difference widens the depletion layer and nothing other than a very small leakage current occurs, due to the minority charge carriers. With forward bias, if the applied potential difference is greater than the barrier potential, a large current can flow.

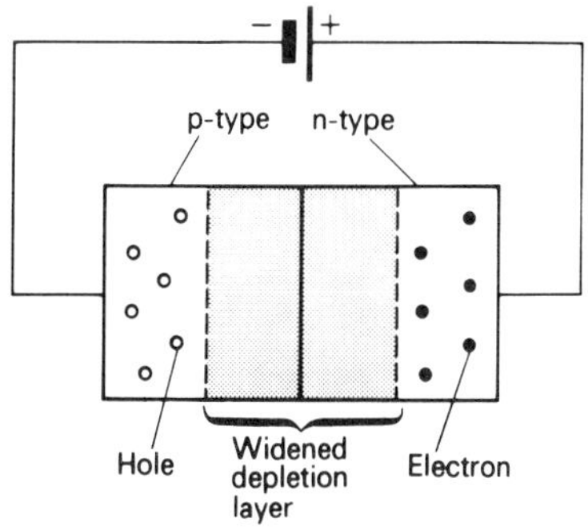


**(a)** Reverse bias

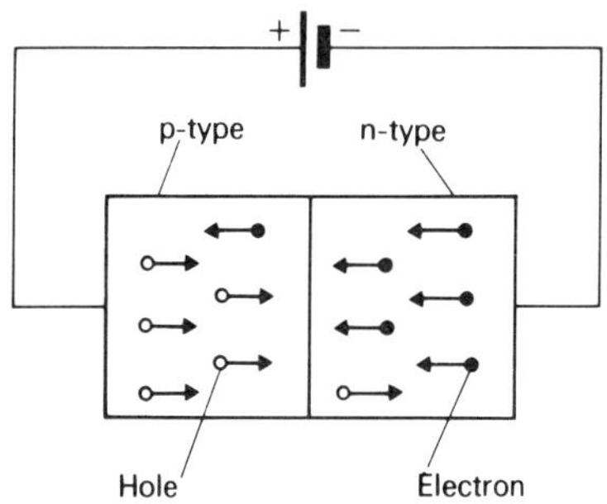


**(b)** Forward bias

**Figure A7.1**

(d) See Figure A7.2. Note that the reverse-current graph scale is different to the forward-current scale. With the forward bias the current becomes large when the barrier potential of about 0.6 V has been overcome. With the reverse bias the current is only leakage until the 'breakdown' potential is reached. At this point the insulation of the barrier layer breaks down.

## Answers II

**7.4** (a) Collector–base junction is reverse-biased, emitter–base is forward-biased.
(b) The base current flows through the low resistance of the forward-biased emitter–base junction. Electrons move across this junction into the very thin base region and about 99% of the electrons go straight through into the collector, even though it is reverse-biased. The result

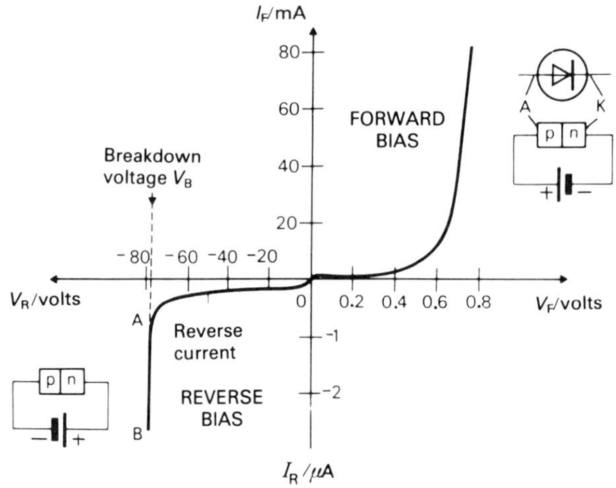


**Figure A7.2**

is that a current occurs in a high resistance, the reverse-biased junction. Hence a current through a low resistance has been transferred to become a current through a large resistance.

(c) $I_C/I_B = 4.5 \times 10^{-3}/40 \times 10^{-6} = 112.5$

**7.5** $C_1$ is to block off any d.c. component of the input and ensure that only an alternating signal from the input is superimposed on the bias and passes on to the base. $C_2$ is to ensure that only an alternating signal is passed through to the output. $R_B$ is to ensure that an appropriate bias is applied to the base to ensure that, for the size of input signal concerned, the collector–base is reverse-biased and the emitter–base is forward-biased. The potential difference $V$ is applied across $R_C$ and the transistor, thus $V = I_C R_C + V_{CE}$. Thus $R_C$ is the load resistance through which the collector current flows and so across which a potential difference is produced which is related to the collector current. Changes in the collector current produce changes in the potential difference. Changes in this potential difference produce changes in $V_{CE}$, since $V$ is constant, and hence in output.

**7.6** (a) When L is connected to M the input is raised to about 6 V and so the output is about 0 V. When M is connected to N the input is about 0 V and so the output is about 6 V.

(b) Changes in the input from below $V_1$ to above $V_2$ cause the output voltage to switch from about 6 V to 0 V. Outside the voltage region $V_1$ to $V_2$ there are no changes in the output for changes in the input.

(c) The input voltage would have to be biased so that all its fluctuations were contained within the voltage range $V_1$ to $V_2$.

**7.7** See Figure A7.3. As an amplifier the transistor is operated well to the right of the knee in the graph, in the region where the collector current varies linearly with $V_{CE}$ for a given base current.

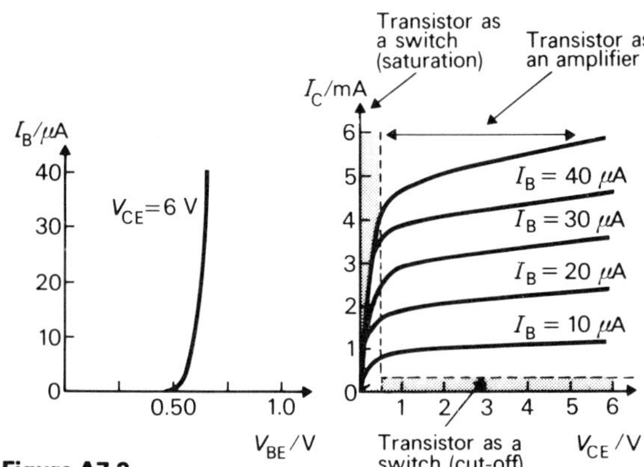


**Figure A7.3**

# Lesson 8 Vacuum electronics

## Aims

After completing this lesson you should be able to

- describe *thermionic emission*
- describe the principle of the thermionic diode valve, using the terms *anode*, *cathode* and *space charge*
- describe the basic components of the cathode ray oscilloscope, i.e. the electron gun, the deflecting system and the display system
- explain the *time base*
- explain how to use a cathode ray oscilloscope to
  - (a) measure d.c. potential differences
  - (b) measure a.c. potential differences
  - (c) display waveforms
  - (d) compare frequencies
  - (e) determine phase differences.

## Prerequisites

Lessons 1 and 4 to give a background of d.c. and a.c. Lesson 8 in Chapter 6 on charged particles in fields is also required.

# STUDY

**I Read** about thermionic emission and the principle of the thermionic diode valve.
References: *B* PP ch. 10, *B* ST5 ch. 4, ST8 ch. 3, *D* ch. 20, *N* ch. 38, *W et al.* ch. 43(33), *Wh&H* ch. 60.

**Self-assessment questions**

**8.1** What is *thermionic emission*?

**8.2** Essentially a diode valve consists of a filament to supply electrons and an electrode, called the *anode*, to collect them, the entire assembly being mounted in a vacuum.
(a) Explain what happens when the filament is heated and thermionic emission occurs, with no external potential difference being applied between the filament and the anode. Use the term 'space charge' in your answer.
(b) Explain what happens when the anode is made positive with respect to the filament (or *cathode*, as it is usually referred to).
(c) For a particular filament temperature, what happens as the potential difference between the anode and the filament is increased?

***8.3** If electrons are emitted from the filament in a diode valve with negligible velocity, with what velocity will they reach the anode when there is a potential difference of 100 V between the anode and the filament? ($e/m = 1.76 \times 10^{11}\,\mathrm{C\,kg^{-1}}$)

**II Read** about the basic components of the cathode ray oscilloscope and the uses of the instrument.
References: *B* ST4 ch. 1, ST5 ch. 4, *D* ch. 21, *N* ch. 38, *W et al.* ch. 43(33), *Wh&H* ch. 60.

**Self-assessment questions**

**8.4** The cathode ray oscilloscope can be considered to be made up of three elements: the electron gun, the deflecting system and the display system. Describe the function of each of these elements.

***8.5** Draw a sketch of the basic construction of the cathode ray oscilloscope, identifying the cathode, the grid, the focusing anode, the accelerating anode, the deflecting plates and the screen.

***8.6** Explain how with a cathode ray oscilloscope (a) the brightness of the image on the screen and (b) the focus are controlled.

***8.7** Explain the action of the deflecting system in a cathode ray oscilloscope.

***8.8** Describe the type of signal used for the time base and explain its form.

***8.9** Explain the function of the TRIG LEVEL control.

***8.10** Explain how a cathode ray oscilloscope can be used to measure (a) a d.c. potential difference, (b) the peak voltage of an alternating potential difference.

***8.11** Figure 8.1 shows a waveform displayed on the screen of a cathode ray oscilloscope. The grid has squares of side 1 cm. If the $Y$ amplification is $2\,\mathrm{V\,cm^{-1}}$ and the time base is $30\,\mathrm{ms\,cm^{-1}}$, what is (a) the peak voltage and (b) the frequency?

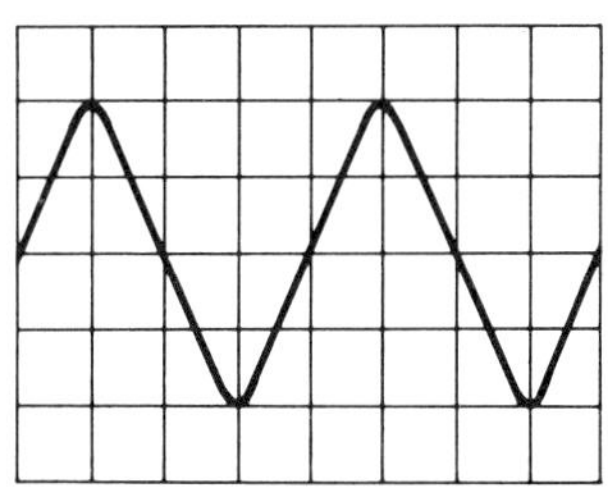

**Figure 8.1**

***8.12** A student has set up an experiment to measure the speed of sound. The experiment involves measuring the time interval between a short burst of sound being sent out and its return after reflection from a distant object. A microphone is used to detect the emitted and the reflected sound, the microphone being connected to the $Y$ input of a cathode ray oscilloscope. Figure 8.2 shows the trace seen on the screen, each square having a side of 1 cm, when the time base is set to 300 ms cm$^{-1}$. What is the time between the emitted and the reflected sounds being detected?

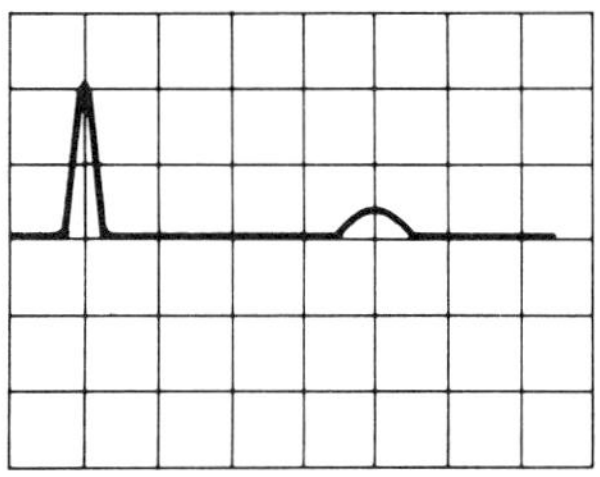

**Figure 8.2**

***8.13** When two potential differences of different frequencies are applied to the $X$- and $Y$- plates of a cathode ray oscilloscope, the trace shown in Figure 8.3 is observed. What is the ratio of the frequencies? If the signal applied to the $X$-plates has a frequency of 50 Hz, what is the frequency used with the $Y$-plates?

**Figure 8.3**

***8.14** Describe or sketch the traces seen on the screen of a cathode ray oscilloscope when two sinusoidal potential differences of the same frequency and amplitude are applied simultaneously to the $X$- and $Y$-plates of a cathode ray oscilloscope but there is a phase difference of (a) 0°, (b) 45°, (c) 90° between them.

# SELF-ASSESSMENT

## Key points

**Thermionic emission** is the emission of electrons from a metal as a result of heating the metal. Dynamic equilibrium occurs when the region around the heated metal returns electrons to the surface as fast as they are emitted. The 'cloud' of electrons above the metal surface is called the **space charge**. In the **thermionic diode** the electrons are attracted away from this cloud by applying a potential difference between an electrode, called the **anode** because it is positive, and the heated metal, called the **cathode** because it is negative.

The **cathode ray oscilloscope** consists of an electron gun which produces a narrowly focused electron beam, a deflecting system by which the beam can be deflected in either the $X$- or the $Y$-direction, and a display system which is a screen coated with a phosphor which absorbs the energy of the incident electron beam and re-emits it as visible light.

## Answers I

**8.1** See the key points.

**8.2** (a) A cloud of electrons is produced in the space immediately adjacent to the filament. This cloud is called the *space charge*. When equilibrium is reached electrons are leaving the filament at the same rate as they enter the filament from the space charge.

(b) When the anode is positive with respect to the filament the electric field causes the electrons to move out of the space charge to the anode. Hence a current flows between the filament and the anode.

(c) As the potential difference is increased, so the current increases as more electrons are attracted out of the space charge. At a high enough potential difference the electrons are being attracted out of the space charge at the same rate as they escape from the filament. This is the maximum current possible. *Saturation* is said to have occurred.

**8.3** $\frac{1}{2}mv^2 = eV$ and so $v^2 = 2(e/m)V = 2 \times 1.76 \times 10^{11} \times 100$ and $v = 5.9 \times 10^6\,\text{m s}^{-1}$.

## Answers II

**8.4** See the key points.

**8.5** See Figure A8.1.

**8.6** (a) By adjusting the potential of the grid which is located immediately in front of the cathode. The grid is negative with respect to the cathode. Making it more negative reduces the number of electrons that get through, making it less negative increases the number.

(b) The electron lens consists of a focusing anode and an accelerating anode. By adjusting the potential, positive with respect to the cathode, of the focusing electrode, the focus of the beam can be changed.

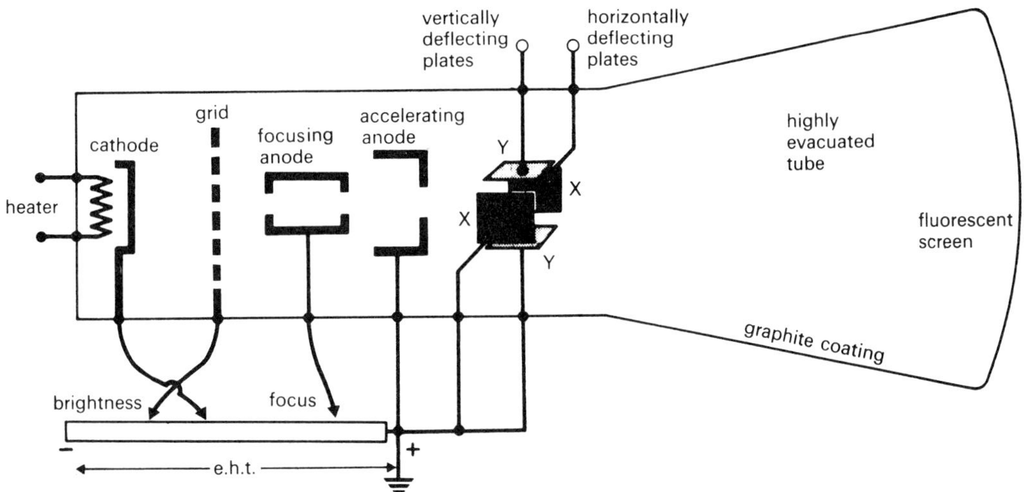


**Figure A8.1 Basic elements of a cathode ray oscilloscope**

**8.7** The deflecting system consists of two pairs of parallel plates, one pair of plates being at right angles to the other. When a potential difference is applied between a pair of plates it produces an electric field which causes the electron beam to become deflected.

**8.8** See Figure A8.2. Along AB the electron beam traverses the screen at a steady rate. Along BC the beam is made to fly back very rapidly to its starting point. It then repeats the motion.

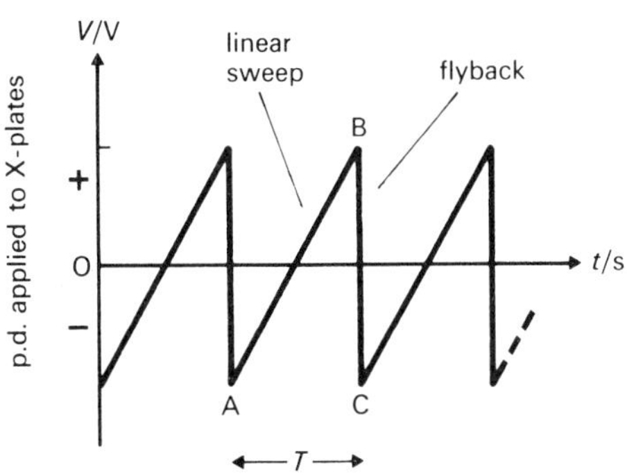


**Figure A8.2 Ideal time base p.d.**

**8.9** To maintain a stable trace on the screen when the input is a regular repetitive signal, part of the input signal is fed to a trigger circuit that gives a pulse which is used to trigger the time base and start the motion of the beam across the screen from left to right. The TRIG LEVEL control chooses the point on the input signal at which the time base will start.

**8.10** (a) Without the time base, the input is connected to the *Y*-input. The spot on the screen is then deflected in the *Y*-direction, the amount of the deflection being proportional to the potential difference. The deflection can be calibrated using a known potential difference.
(b) Without the time base running the alternating input is applied to the *Y*-input. The screen will then show a vertical line, the length of the line representing twice the peak voltage. The deflection can be calibrated using a known d.c. or a.c. potential difference.

**8.11** (a) 2 cm is equivalent to 4 V, the peak value. (b) 4 cm is equivalent to 120 ms. This is the periodic time and hence the frequency $= 1/120 \times 10^{-3} = 8.33$ Hz.

**8.12** 4 cm is equivalent to 1200 ms.

**8.13** $f_Y/f_X = 3{:}1$, hence $f_Y = 150$ Hz. In general for Lissajous' figures, $f_Y/f_X =$ (number of loops touching horizontal line drawn at the bottom of the trace)/(number of loops touching a vertical line drawn at the side of the trace).

**8.14** See Figure A8.3.

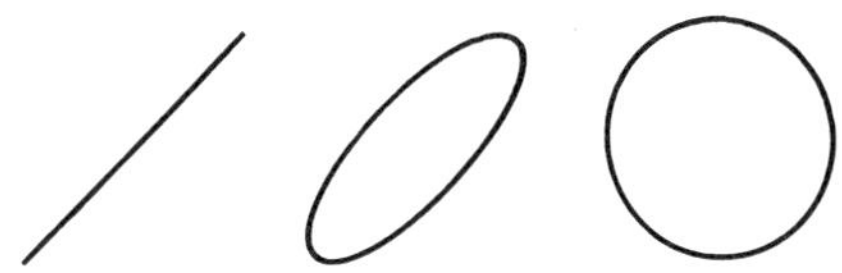

**(a)** 0° **(b)** 45° **(c)** 90°

**Figure A8.3**

# CHAPTER 8

# Atomic and nuclear physics

## Lesson 1 Photoelectricity

### Aims

After completing this lesson you should be able to

- explain what is meant by the *photoelectric effect* and describe the results of experiments on photoelectricity
- explain the results of such experiments in terms of *quanta*
- solve problems involving photoelectric emission by the use of Einstein's photoelectric equation.

### Prerequisites

Lesson 1, Chapter 5, on describing waves, and Lesson 8, Chapter 6, on charged particles in fields.

## STUDY

**I Read** about the photoelectric effect; what it is, and the experiments concerning the charged particles emitted from a photoelectric surface when illuminated. References: *B* PP ch. 7, *B* ST7 ch. 1, *D* ch. 20, *N* ch. 40, *W et al.* ch. 46(44), *Wh&H* ch. 62.

**Self-assessment questions**

**1.1** Ultraviolet light is allowed to fall on a small, freshly cleaned sheet of zinc which is connected to an electroscope. How are the leaves of the electroscope affected if the electroscope is (a) positively charged, (b) negatively charged? Explain your answers.

**1.2** With monochromatic light incident on a photoelectric surface and emission of electrons occurring, what determines (a) the number of electrons emitted per second, (b) the maximum kinetic energy of the electrons?

**1.3** With the photoelectric experiment described in SAQ 1.1, electrons are emitted from the zinc when ultraviolet light is incident on the zinc, but no electrons are emitted if white light is used or a piece of glass is inserted between the ultraviolet light source and the zinc plate. Why?

**II Read** about Einstein's explanation of the photoelectric effect and his photoelectric equation. References: *B* PP ch. 7, *B* ST7 ch. 1, *D* ch. 20, *N* ch. 40, *W et al.* ch. 46(44), *Wh&H* ch. 62.

**Self-assessment questions**

**1.4** How was Einstein able to explain (a) the relation of the number of electrons emitted per second to the intensity of the incident light, (b) the relation of the maximum kinetic energy of emitted electrons to only the frequency of the light and not the intensity, (c) the fact that photoelectric emission from a surface can only occur for light above a certain minimum frequency?

**1.5** State and explain Einstein's photoelectric equation.

***1.6** Yellow light has a wavelength of about $6 \times 10^{-7}$ m. What is the energy of a photon of such light? ($h = 6.6 \times 10^{-34}$ J s, $c = 3.0 \times 10^{8}$ m s$^{-1}$)

***1.7** Zinc has a work function of 4.2 eV. What is (a) the minimum frequency, (b) the maximum wavelength at which photoelectric emission can occur? (1 eV = $1.6 \times 10^{-19}$ J)

***1.8** What is the maximum kinetic energy with which an electron is emitted from a zinc surface when illuminated by light of wavelength (a) $6.0 \times 10^{-7}$ m, (b) $2.0 \times 10^{-7}$ m? ($h = 6.6 \times 10^{-34}$ J s, $c = 3.0 \times 10^{8}$ m s$^{-1}$, 1 eV = $1.6 \times 10^{-19}$ J, work function = 4.2 eV)

***1.9** A surface has a work function of 1.8 eV. Calculate the minimum frequency at which photoelectric emission can occur and the maximum kinetic energy of the emitted electrons when the surface is illuminated with light of wavelength $4.0 \times 10^{-7}$ m. ($h = 6.6 \times 10^{-34}$ J s, $c = 3.0 \times 10^{8}$ m s$^{-1}$, 1 eV = $1.6 \times 10^{-19}$ J)

**III** **Read** about Millikan's photoelectric experiment, or a similar experiment, in which a potential is applied to stop the photoelectrons in order to confirm Einstein's equation.
References: *B* ST7 ch. 1, *D* ch. 20, *N* ch. 40, *Wh&H* ch. 62.

**Self-assessment questions**

**1.10** A potential difference of 1.0 V has to be applied to stop the flow of electrons between an electron-emitting surface and an electrode. What is the maximum kinetic energy of the electrons? ($e = -1.6 \times 10^{-19}$ C)

***1.11** When monochromatic light of wavelength $7.0 \times 10^{-7}$ m is used to illuminate a photoelectric surface, a stopping potential of 0.7 V has to be applied to reduce the photoelectric current to zero. What is the work function of the surface? ($h = 6.6 \times 10^{-34}$ J s, $c = 3.0 \times 10^{8}$ m s$^{-1}$, $e = -1.6 \times 10^{-19}$ C)

***1.12** In a photoelectric experiment, like that of Millikan, a graph is plotted of the stopping potential against the frequency of the light used to illuminate the surface. Sketch the graph and state the value of the slope and the intercept with the frequency axis.

***1.13** Figure 1.1 shows the circuit used with a photocell in a photoelectricity experiment. The current was measured for different potential differences, when monochromatic light was being used to illuminate the cell, and the graph shown was obtained.
(a) What would be the effect on the graph of increasing the intensity of illumination?
(b) What would be the effect on the graph of increasing the frequency of the light used?

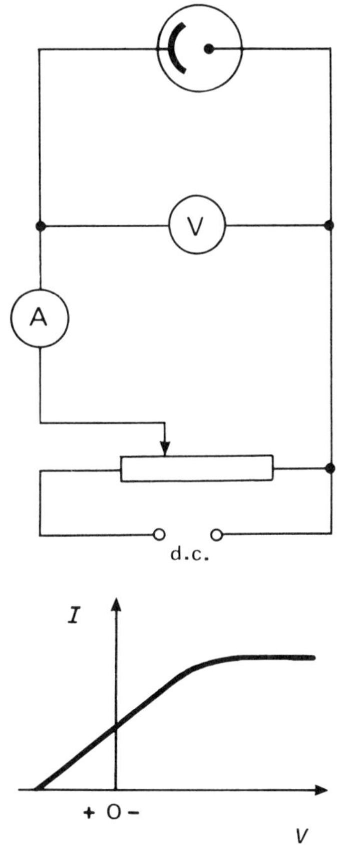

**Figure 1.1**

# SELF-ASSESSMENT

## Key points

In **photoelectric emission** electrons are emitted from a surface when electromagnetic radiation of a sufficiently high frequency is incident on it. The number of photoelectrons emitted per second is proportional to the intensity of the incident radiation. The electrons are emitted with a range of kinetic energies from zero up to a maximum, the maximum value being independent of the intensity but dependent on the frequency of the incident radiation. There is a minimum frequency, called the **threshold frequency**, below which no emission occurs, regardless of the intensity.
Einstein explained the above by considering the incident radiation to be **quantized**, a **quantum** having an energy $E$ given by $E = hf$, where $f$ is the frequency and $h$ the Planck constant. For emission to occur from a surface the energy of the incident quantum must be greater than the **work function** $\phi$ of the surface, this being the energy required to liberate an electron from the surface. **Einstein's photoelectric equation** is

$$hf = \tfrac{1}{2}mv^2_{max} + \phi$$

where $\tfrac{1}{2}mv^2_{max}$ is the maximum kinetic energy of the emitted electron.

## Answers I

**1.1** (a) No effect. (b) The leaves collapse due to a loss of charge. The explanation is that the zinc loses electrons due to the photoelectric emission and these cannot escape from the plate when it is positively charged but can escape when it is negatively charged.

**1.2** (a) The intensity of the light. (b) The frequency, or wavelength, of the light. The intensity does not affect the maximum kinetic energy.

**1.3** The emission of photoelectrons depends on the frequency of the radiation used. Below a certain threshold frequency no emission can occur, no matter how great the intensity. Ultraviolet radiation has a higher frequency, shorter wavelength, than white light and thus in the case of a zinc surface the white light does not have high-enough frequencies for emission to occur. Inserting the sheet of glass eliminates the ultraviolet radiation since it cannot travel through glass. The result is the elimination of the high-frequency radiation and hence no photoelectric emission.

## Answers II

**1.4** (a) The more intense the light the greater the number of photons hitting the surface per second and hence the greater the number of electrons ejected from the surface per second. The term *photon* is used for a quantum of radiation.
(b) The energy of an electron depends on the energy of the incident photon, this depending on the frequency of the radiation. The term 'maximum' is used for the kinetic energy because the energy of the ejected electrons will vary from zero up to a maximum value depending on how much energy is needed to get an electron out and away from the surface.
(c) A certain minimum energy is needed for an electron to escape from a surface, this minimum being known as the *work function*. If the frequency of the incident photon is not high enough it will not have sufficient energy (i.e. more than the work function) to give to an electron for it to escape.

**1.5** It is a statement of conservation of energy for a photon causing an electron to be emitted from a surface. Incident energy = $hf$ = energy of ejected electron ($\frac{1}{2}mv^2_{max}$) + energy to escape ($\phi$).

**1.6** $E = hf = hc/\lambda = 6.6 \times 10^{-34} \times 3.0 \times 10^8/(6.0 \times 10^{-7}) = 3.3 \times 10^{-19}$ J. Another energy unit, the electron volt (eV), is frequently used in such situations, with 1 eV = $1.6 \times 10^{-19}$ J. Thus the energy is 2.1 eV. The electron volt is the energy gained by an electron when passing through a potential difference of 1 V.

**1.7** (a) $hf = \phi$, hence $f = 4.2 \times 1.6 \times 10^{-19}/(6.6 \times 10^{-34}) = 1.0 \times 10^{15}$ Hz. (b) $c = f\lambda$, hence $\lambda = 3.0 \times 10^8/(1.0 \times 10^{15}) = 3.0 \times 10^{-7}$ m. This is in the ultraviolet.

**1.8** (a) Maximum kinetic energy = $hf - \phi = 6.6 \times 10^{-34} \times 3.0 \times 10^8/(6.0 \times 10^{-7}) - 4.2 \times 1.6 \times 10^{-19}$ = a negative answer and so no emission.
(b) Maximum kinetic energy = $6.6 \times 10^{-34} \times 3.0 \times 10^8/(2.0 \times 10^{-7}) - 4.2 \times 1.6 \times 10^{-19} = 3.2 \times 10^{-19}$ J or 2.0 eV.

**1.9** $hf = \phi$, hence $f = \phi/h = 1.8 \times 1.6 \times 10^{-19}/(6.6 \times 10^{-34}) = 4.4. \times 10^{14}$ Hz. Maximum kinetic energy = $hf - \phi = 6.6 \times 10^{-34} \times 3.0 \times 10^8/(4.0 \times 10^{-7}) - 1.8 \times 1.6 \times 10^{-19} = 2.1 \times 10^{-19}$ J.

## Answers III

**1.10** Maximum kinetic energy = $Ve = 1.0 \times 1.6 \times 10^{-19} = 1.6 \times 10^{-19}$ J.

**1.11** $Ve = \frac{1}{2}mv^2_{max} = hf - \phi$, hence $\phi = 6.6 \times 10^{-34} \times 3.0 \times 10^8/(7.0 \times 10^{-7}) - 0.7 \times 1.6 \times 10^{-19} = 1.7 \times 10^{-19}$ J.

**1.12** See Figure A1.1. Slope = $h/e$, intercept = threshold frequency.

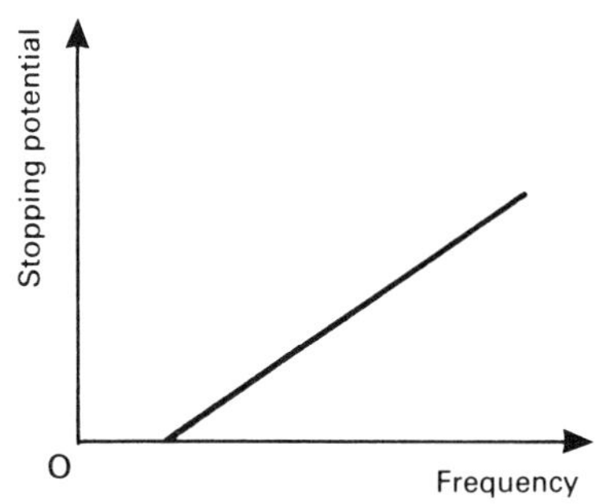


**Figure A1.1**

**1.13** See Figure A1.2.

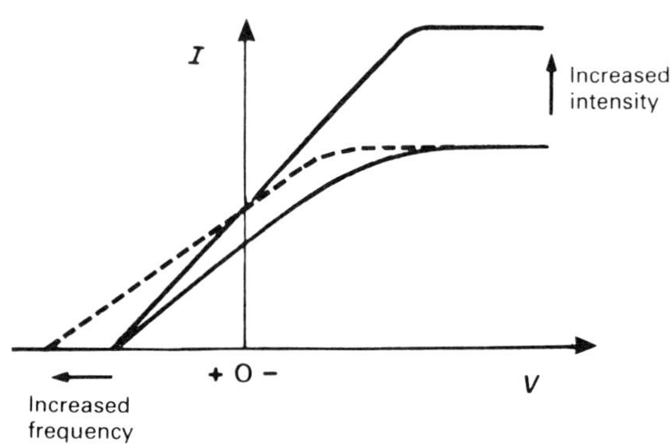


**Figure A1.2**

# Lesson 2 Energy levels

## Aims

After completing this lesson you should be able to

- explain how electron collisions with atoms give evidence for the existence of atomic energy levels
- explain line spectra in terms of transitions between energy levels
- explain the spectral series of hydrogen
- explain electrical conduction in conductors, semiconductors and insulators in terms of energy bands.

## Prerequisites

While Lesson 1 is not essential it could usefully precede this lesson. A basic knowledge of energy (Lesson 5, Chapter 1), waves (Lesson 1, Chapter 5), electric potential (Lesson 2, Chapter 6) and electrical conduction in the solid state (Lesson 7, Chapter 7) is required.

## STUDY

**I Read** about the Franck–Hertz experiment involving electrons colliding with atoms and its evidence for energy levels.
References: *B* PP ch. 7, *B* ST7 ch. 1, *D* ch. 20, *N* ch. 40, *W et al.* ch. 45(43), *Wh&H* ch. 63.

### Self-assessment questions

**2.1** Figure 2.1 shows the basic form of the apparatus used by Franck and Hertz.

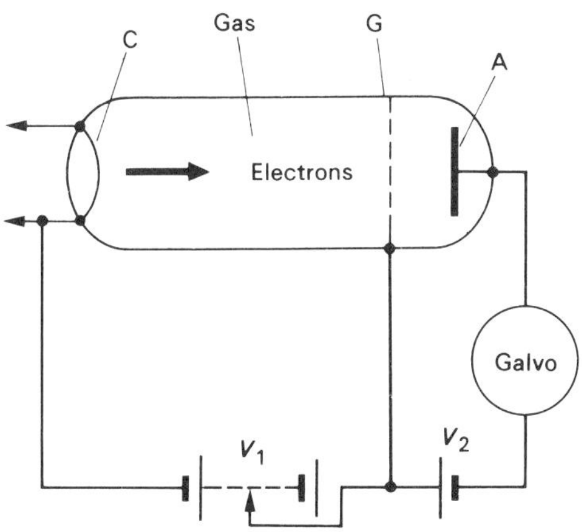


**Figure 2.1**

(a) The electrons emitted at the cathode C are accelerated to the grid by a potential difference of $V_1$, the grid being positive with respect to the cathode. What would you expect to happen to the energy of the electrons passing through the grid as $V_1$ is increased, if the tube did not contain a gas but a vacuum?
(b) The electrode A is made negative with respect to the grid, though $V_2$ is less than $V_1$. Why?
(c) Figure 2.2 shows the results obtained when the tube contains a gas such as helium. Explain what is happening at points P, Q, R and S.

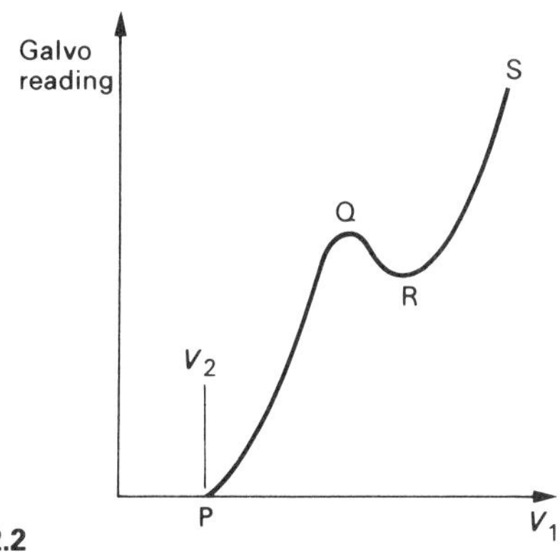


**Figure 2.2**

(d) What happens if the potential difference $V_1$ is increased to higher values than indicated on the graph?

**2.2** Explain the terms *excitation potential* and *ionization potential.*

***2.3** Explain what is meant by *elastic* and *inelastic* collisions between electrons and atoms, and explain what happens to the energy in an inelastic collision.

***2.4** Explain how the Franck–Hertz experiment offers evidence for atomic energy levels.

**II Read** about line spectra and their explanation in terms of transitions between energy levels, also the spectral series of hydrogen and their explanation.
References: *B* PP ch. 7, *B* ST7 ch. 1, ch. 3, *D* ch. 20, *N* ch. 40, *W et al.* ch. 46(44), *Wh&H* ch. 63.

### Self-assessment questions

**2.5** What are line spectra?

***2.6** How can line spectra be explained by transitions between energy levels?

*2.7 What wavelength would the spectral line have that is produced by a transition from an energy level of $-1.5$ eV to one of $-3.4$ eV? ($c = 3.0 \times 10^8$ m s$^{-1}$, $e = 1.6 \times 10^{-19}$ C)

*2.8 Figure 2.3 shows the energy levels for atoms of hydrogen.

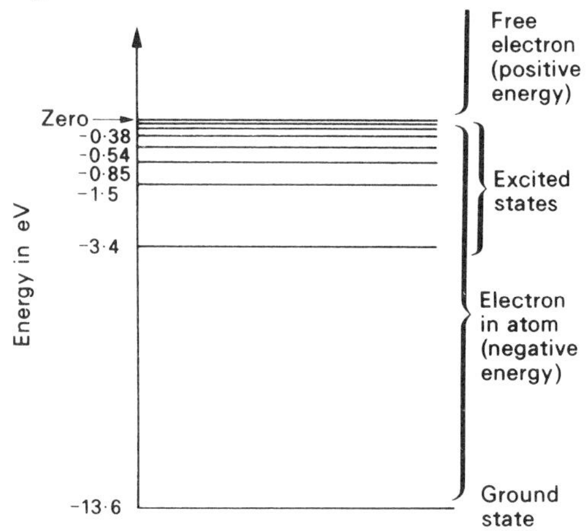


**Figure 2.3**

(a) The ground state of hydrogen is represented by an energy of $-13.6$ eV. What is meant by 'ground state' and what would happen if a hydrogen atom were hit by an electron which gave the atom 13.6 eV energy?
(b) The first and second excited states are at $-3.4$ eV and $-1.5$ eV. What are the wavelengths of the first three lines in the hydrogen spectrum?
(c) The hydrogen spectrum consists of a number of series of lines, the Lyman, the Balmer and the Paschen being three of these series. Explain how these series are produced. ($h = 6.6 \times 10^{-34}$ J s, $e = 1.6 \times 10^{-19}$ C, $c = 3.0 \times 10^{-8}$ m s$^{-1}$)

*2.9 An electron having an energy of $17.0 \times 10^{-19}$ J collides with a hydrogen atom in its ground state. If the result of the collision is that the hydrogen atom emits a photon of wavelength $1.2 \times 10^{-7}$ m, what is the energy of the scattered electron? ($h = 6.6 \times 10^{-34}$ J s, $e = 1.6 \times 10^{-19}$ C, $c = 3.0 \times 10^8$ m s$^{-1}$)

**III** **Read** about electrical conduction in conductors, semiconductors and insulators in terms of energy bands. References: *B* ST8 ch. 4, *D* ch. 21, *N* ch. 39.

**Self-assessment questions**

**2.10** In a solid, instead of the discrete energy levels that individual atoms have, there are energy *bands*. When a potential difference is applied to a solid the electrons can only accept energy if they can move into a vacancy in an energy band. Explain the differences in energy bands that conductors, semiconductors and insulators have and hence account for their different electrical conductivities.

**2.11** Explain how the addition of suitable impurities affects the energy bands and hence electrical conductivity of a semiconductor and produces n-type and p-type materials.

# SELF-ASSESSMENT

## Key points

Electrons in atoms have a series of discrete energy levels which they can move between. They can only accept energy if it is sufficient to enable them to move to a higher energy level. When an electron in an atom has moved to a higher energy level the atom is said to be **excited**. When the electron falls back down the energy level 'ladder' it releases the energy as a photon, or photons if it falls down through a number of energy levels. The frequency of the photon is given by $E_2 - E_1 = hf$, where $E_2$ and $E_1$ are the energies of the levels between which the electron falls.
Conduction in solids can be explained in terms of an **energy band theory**. In a solid the upper energy levels of the discrete atoms have become amalgamated to form energy bands. Electrical conduction can take place if an electron is able to accept a small amount of energy and move into a vacant energy level within a band.

## Answers I

**2.1** (a) The energy would increase since the kinetic energy of the electrons $= \frac{1}{2}mv^2 = Ve$.
(b) This is to stop those electrons which have lost a significant amount of energy in collisions with atoms and only allow through to the galvanometer those electrons which have lost little, if any, energy.
(c) $V_1$ has to be greater than $V_2$ for any electrons to reach A, and hence point P. Up to point Q all the collisions between electrons and atoms have been elastic and no energy has been lost. At Q many of the electrons suffer inelastic collisions with atoms and lose energy, thus failing to reach A. Hence between Q and R the current falls. Further increase in $V_1$ allows even those electrons that have suffered an inelastic collision to gain sufficient energy to overcome $V_2$ and reach A. Hence the current increases to S.
(d) The entire sequence keeps repeating itself since the

electrons that suffered inelastic collisions receive sufficient further acceleration by the potential difference $V_1$ to cause yet a further inelastic collision.

**2.2** The *excitation potential* is the potential that has to be applied to accelerate an electron sufficiently for it to suffer an inelastic collision with an atom. The *ionization potential* is the potential that is necessary for the electron in the atom to be completely ejected from the atom by the collision.

**2.3** *Elastic* collisions are when an electron hits an atom and suffers no loss in kinetic energy as a result of the collision. This means that no energy is given to the atom. *Inelastic* collisions are when the electron loses energy on colliding with the atoms, the energy being absorbed by the atom.

**2.4** The experiment shows that only at certain discrete values of energy will an atom accept energy.

## Answers II

**2.5** Line spectra are emitted by individual atoms and thus occur with low-density vapours and gases when sufficient energy is supplied. The spectra consist of a number of discrete wavelengths (or frequencies).

**2.6** When energy is supplied to an atom, if it is sufficient it can cause the atom to become excited, i.e. an atomic electron to move to a higher energy level. The energy levels can be thought of as rather like rungs on a ladder. The electron can only move up to another rung, not between rungs. When the electron falls back down the rungs of the ladder it releases the energy. Thus packets of energy corresponding to the energy differences between rungs of the ladder can be emitted, the emission taking the form of photons. The frequency of the emitted photon is given by $E_2 - E_1 = hf$, where $E_1$ and $E_2$ are the energies associated with the levels between which the electron falls.

**2.7** $E_2 - E_1 = (-1.5) - (-3.4) = 1.9\,\text{eV} = 1.9 \times 1.6 \times 10^{-19}\,\text{J}$. As this equals $hf = hc/\lambda$, then $\lambda = 6.6 \times 10^{-34} \times 3.0 \times 10^8/(1.9 \times 1.6 \times 10^{-19}) = 6.51 \times 10^{-7}\,\text{m}$.

**2.8** (a) The ground state is the lowest energy state available to an atom. None of its electrons is raised to a higher energy level. If 13.6 eV were supplied to a hydrogen atom in its ground state, this would be just sufficient to take the electron to the point of escaping from the atom, i.e. it is the ionization energy.
(b) The first three lines occur for energy releases of $(-3.4) - (-13.6) = 10.2\,\text{eV}$, $(-1.5) - (-13.6) = 12.1\,\text{eV}$ and $(-1.5) - (-3.4) = 1.9\,\text{eV}$. This last transition gives the wavelength $6.51 \times 10^{-7}\,\text{m}$, as derived in 2.7 above. The other wavelengths, derived in the same way, are $1.21 \times 10^{-7}\,\text{m}$ and $1.02 \times 10^{-7}\,\text{m}$.
(c) See Figure A2.1. Each of the series is produced by transitions from higher energy levels to a particular level. Thus the Lyman series are for the transitions from higher energy levels to the ground state, the Balmer series are for transitions from higher levels to the first excited energy level.

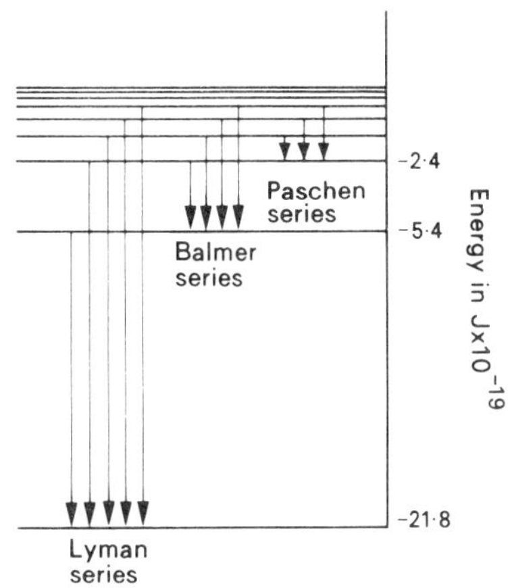


**Figure A2.1**

**2.9** Energy taken by the atom $E = hf = hc/\lambda = 6.6 \times 10^{-34} \times 3.0 \times 10^8/(1.2 \times 10^{-7}) = 1.65 \times 10^{-18}\,\text{J}$. Hence the energy left to the electron is $1.70 \times 10^{-18} - 1.65 \times 10^{-18} = 0.05 \times 10^{-18}\,\text{J}$.

## Answers III

**2.10** See Figure A2.2. Conductors have electrons in partly filled energy bands and thus these electrons are able to accept small amounts of energy. Semiconductors have a small number of electrons in an almost empty band and some vacancies in the almost filled lower band. The result is that only a small number of electrons and holes are available for conduction. Insulators have filled or very nearly filled bands and virtually no electrons are able to accept small amounts of energy, since there is a gap between the filled band and the nearest empty band.

**2.11** See Figure A2.3. With n-type, energy levels containing electrons are donated close to the almost empty conduction energy band and thus electrons are made available for conduction processes. With p-type, empty energy levels are added close to the almost filled energy band and thus holes are made available for conduction purposes.

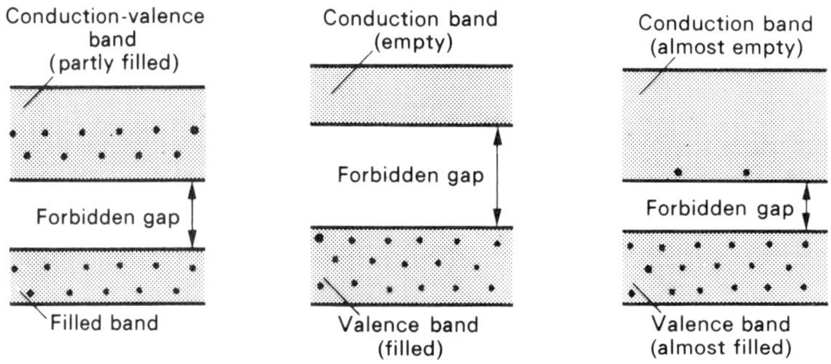


**(a)** Conductor **(b)** Insulator **(c)** Semiconductor at room temperature

**Figure A2.2**

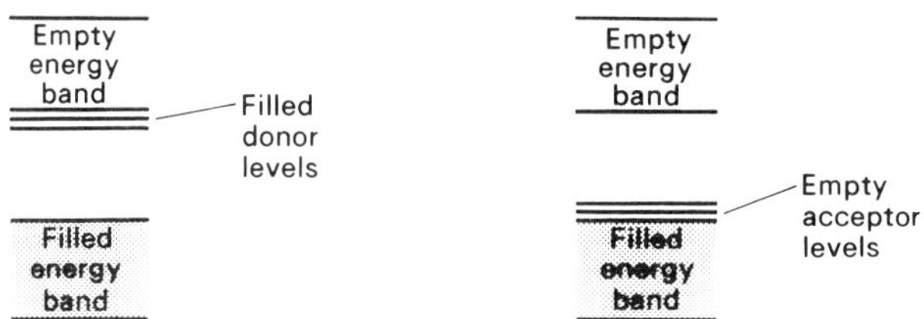


**(a)** n-type **(b)** p-type

**Figure A2.3**

# **Lesson 3** X-rays

## Aims

After completing this lesson you should be able to

- describe how X-rays are produced
- describe the properties of X-rays
- derive Bragg's law and explain the basis of X-ray crystallography
- describe the continuous and the line spectra of X-rays, explaining both in terms of photon emission.

## Prerequisites

Lessons 1 and 2 which gave the basic background of photoelectricity and energy levels, also Lesson 1 of Chapter 7 on d.c. resistive circuits, Lesson 8 of Chapter 6 on charged particles in fields, and the general properties of waves in Lesson 2, Chapter 5.

## STUDY

**I Read** about the production of X-rays and their general properties.

References: *B* PP ch. 12, *B* ST7 ch. 1, *D* ch. 20, *N* ch. 40, *W et al.* ch. 47(33), *Wh&H* ch. 63.

**Self-assessment questions**

***3.1** Describe the basic format of the modern X-ray tube, explaining how the intensity of the X-rays and their penetrating power are controlled.

***3.2** State the basic properties of X-rays.

**3.3** In the X-ray tube, electrons gain kinetic energy by being accelerated through a potential difference. They then strike a target and their kinetic energy is transformed into heat and X-rays.
(a) What is the kinetic energy gained by the electrons if the potential difference across the tube is 40 kV?
(b) If only 0.5% of this kinetic energy is transformed into X-rays and 600 W of heat is produced, what is the electron current in the tube? ($e = 1.6 \times 10^{-19}$ C)

***3.4** An X-ray tube is operated at 20 kV with an electron current in the tube of 16 mA. Estimate the number of electrons hitting the target per second and the rate of production of heat. ($e = 1.6 \times 10^{-19}$ C)

**II Read** about X-ray crystallography and Bragg's equation.
References: *B* PP ch. 9, *B* ST7 ch. 1, *D* ch. 20, *N* ch. 40, *W et al.* ch. 15(14), *Wh&H* ch. 39.

**Self-assessment questions**

**3.5** Describe von Laue's experiment and the type of result he obtained for diffraction with a crystal.

**3.6** Figure 3.1 represents three layers of atoms, $p$, $q$ and $r$, on which a beam of X-rays falls. The X-rays are reflected from each plane of atoms such that the angle of incidence equals the angle of reflection.

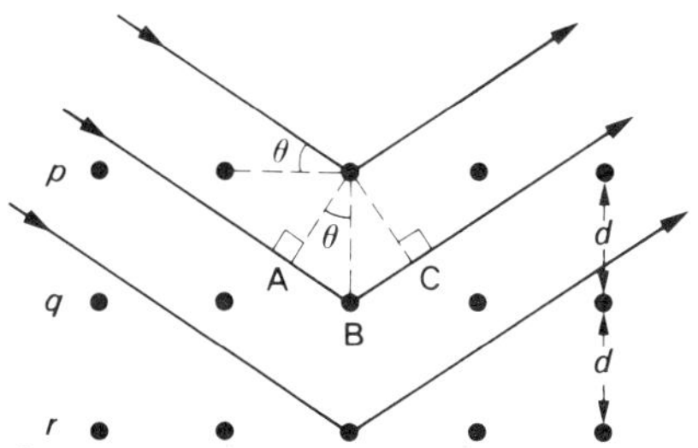


**Figure 3.1**

(a) What is the path difference between X-rays reflected from layers $p$ and $q$?
(b) For constructive interference at an angle $\theta$ the path difference must be an integral number of wavelengths. Hence derive a relationship between $d$, $\theta$ and $\lambda$.

***3.7** A first-order reflection occurs with a copper crystal when the glancing angle is 20° and the wavelength of the X-rays is $1.54 \times 10^{-10}$ m. What is the spacing between the atomic planes which gives this reflection?

***3.8** Describe how the Bragg X-ray spectrometer can be used to measure the wavelengths of X-rays.

****3.9** Explain how X-ray powder photography is used to determine the structure of crystalline materials.

**III Read** about the continuous and line spectra of X-rays and an explanation of their origin.
References: *B* ST7 ch. 1, *D* ch. 20, *N* ch. 40, *Wh&H* ch. 63.

**Self-assessment questions**

***3.10** Sketch a graph showing how the intensity of the X-rays emitted from a typical X-ray tube varies with wavelength.

**3.11** The radiation emitted by an X-ray tube consists of a continuous spectrum with a line spectrum superimposed on it. Which one of these is characteristic of the target material?

***3.12** Explain how the continuous spectrum is produced and derive an equation relating the minimum wavelength of such emissions to the tube voltage.

***3.13** If an X-ray tube is operated at 10 kV what will be the minimum wavelength emitted? ($e = 1.6 \times 10^{-19}$ C, $h = 6.6 \times 10^{-34}$ J s, $c = 3.0 \times 10^{8}$ m s$^{-1}$)

***3.14** Explain how the line spectrum is produced.

****3.15** Moseley in 1913 studied the X-ray line spectra of a large number of elements. Explain how his results indicated that each nucleus has a characteristic number which increases by 1 from one element to the next (the number we now called the *atomic number*).

# SELF-ASSESSMENT

## Key points

**X-rays** are produced by electrons that have been accelerated through a potential difference colliding with a target. Less than 0.5% of the kinetic energy of an electron is transformed into X-rays, the greater part becoming heat. The intensity of the X-rays is controlled by the filament current, the penetrating power by the potential difference.
X-rays travel in straight lines, are not affected by electric and magnetic fields, produce ionization in gases, affect photographic plates, cause fluorescence and penetrate matter.
X-rays are electromagnetic waves with a wavelength of the order of $10^{-10}$ m. The spacing of the atomic planes in crystals is such that they can be used as diffraction gratings with X-rays. **Bragg's equation** $2d \sin \theta = n\lambda$ is obtained by considering planes of atoms to behave like mirrors and constructive interference occurring when the path difference for such reflections is a whole number of wavelengths.
X-rays from an X-ray tube show a line spectrum superimposed on a continuous spectrum. The **continuous spectrum** is produced by the deceleration of the electrons in collisions with the atoms of the target. The minimum wavelength, maximum frequency, of the continuous spectrum is produced when an electron gives up all its kinetic energy to the production of a single X-ray photon, i.e. $hf = \frac{1}{2}mv^2 = Ve$. The **line spectrum** is produced when the electrons cause atomic electrons to move out of the lower energy levels and electrons from higher levels subsequently fall into the vacancies, with the emission of X-ray photons.

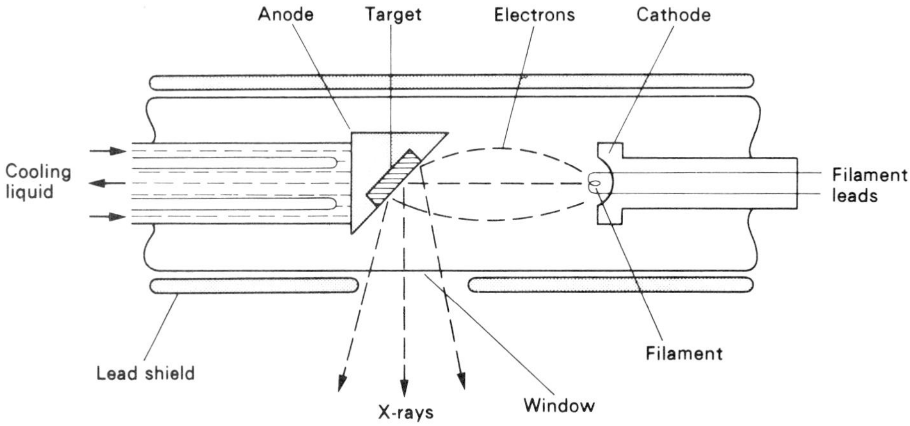


**Figure A3.1 Basic elements of an X-ray tube**

## Answers I

**3.1** See Figure A3.1. The intensity is controlled by the filament current in that this determines the number of electrons being accelerated down the tube and striking the target. The penetrating power is determined by the potential difference through which the electrons are accelerated since this affects the wavelengths produced.

**3.2** See the key points.

**3.3** (a) $\frac{1}{2}mv^2 = Ve = 40 \times 10^3 \times 1.6 \times 10^{-19} = 6.4 \times 10^{-15}$ J (b) $IV = (99.5/100)600$ and so $I = 0.015$ A.

**3.4** $I = ne/t = 16 \times 10^{-3} = n \times 1.6 \times 10^{-19}/t$ and so no. per second $= n/t = 10^{17}$. Power $= IV = 16 \times 10^{-3} \times 20 \times 10^3 = 320$ W. If we assume that of the order of 99.5% of this must be heat then the rate of heat developed will be about 318 W.

## Answers II

**3.5** See Figure A3.2.

**3.6** (a) $AB + BC = 2d \sin\theta$ (b) $2d \sin\theta = n\lambda$

**3.7** $2d \sin\theta = n\lambda$ and so $d = 1 \times 1.54 \times 10^{-10}/(2 \times \sin 20°) = 2.25 \times 10^{-10}$ m.

**3.8** See Figure A3.3. A crystal with known spacing between atomic planes is mounted on the spectrometer table and the angles at which constructive interference gives rise to reflections measured. The Bragg equation can then be used to determine the wavelength.

**3.9** See Figure A3.4. Instead of a single crystal a large number of crystals are used, in powder form. Because of the random orientation of the crystals, many planes of atoms are involved and a large number of reflections occur, resulting in circular arcs on the strip of photographic film. The angles for which reflection occurs can be obtained from measurements made on the film.

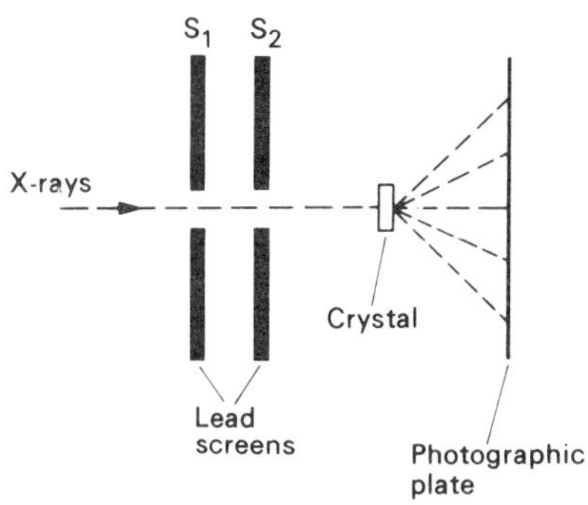


**(a)** Apparatus

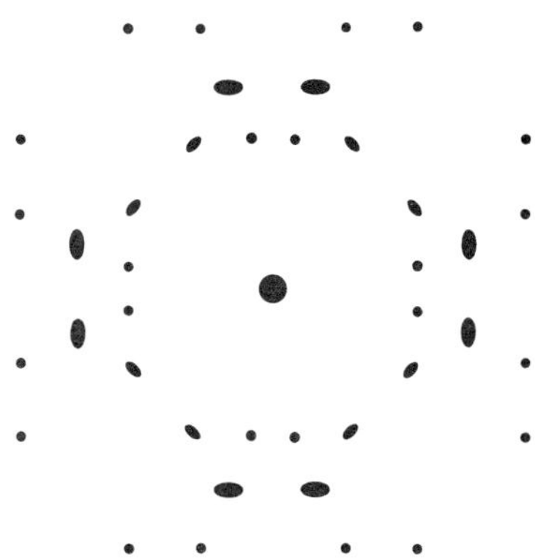

**(b)** Diffraction pattern

**Figure A3.2 Von Laue's experiment**

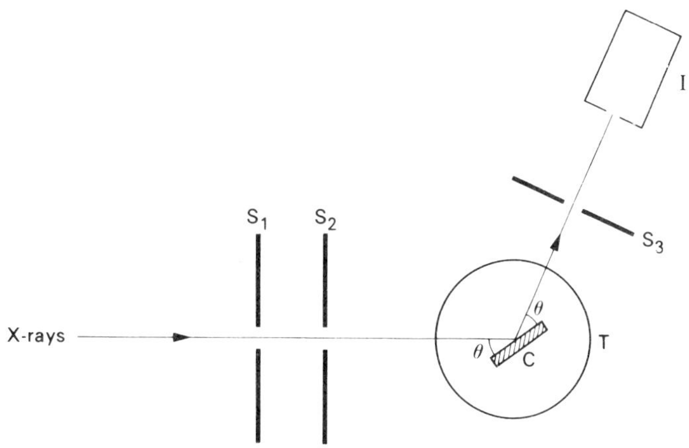


**Figure A3.3 Bragg spectrometer**

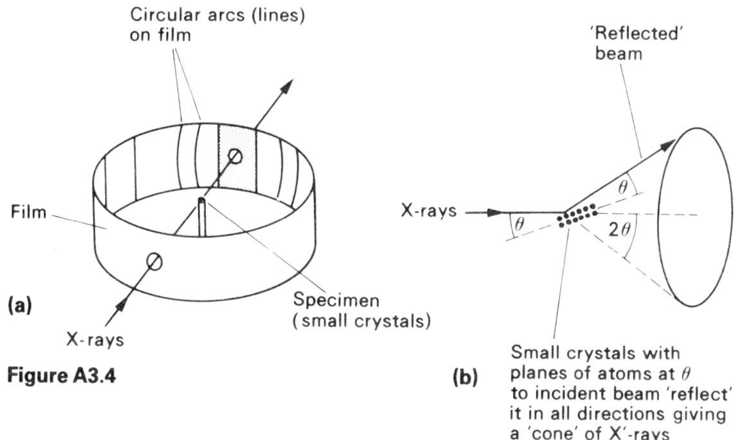


**Figure A3.4**

## Answers III

**3.10** See Figure A3.5.

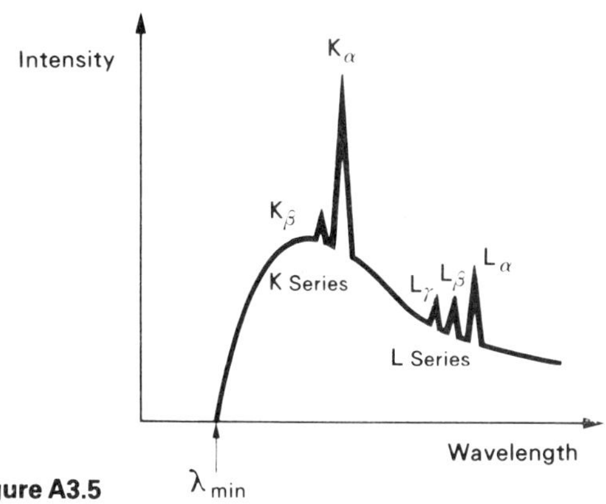


**Figure A3.5**

**3.11** The line spectrum is characteristic of the target material; the positions of the peaks are not affected by the voltage applied to the tube. All targets give a similar continuous spectrum.

**3.12** The continuous spectrum is produced by the electrons being slowed down on encountering the atoms of the target. If an electron gives up all its kinetic energy in a single encounter then $\frac{1}{2}mv^2 = hf = hc/\lambda$. This gives the maximum frequency, the wavelength being the minimum wavelength. As $Ve = \frac{1}{2}mv^2$ then $Ve = hc/\lambda$.

**3.13** $Ve = hc/\lambda$ and so $\lambda = hc/Ve = 6.6 \times 10^{-34} \times 3.0 \times 10^8/(10 \times 10^3 \times 1.6 \times 10^{-19}) = 1.2 \times 10^{-10}$ m.

**3.14** See the key points.

**3.15** Moseley measured the frequency of corresponding line spectra for different elements. He found that the frequency increased from one element to the next, in the order of elements given by the periodic table. On allotting a number $Z$ to each element, which increased by 1 from one element to the next, and plotting $f^{\frac{1}{2}}$ against $Z$, a straight-line graph was produced.

# Lesson 4 Atomic models

## Aims

After completing this lesson you should be able to
- describe the principles of
  (a) the Thomson plum-pudding model
  (b) the Rutherford nuclear model
  (c) the Bohr model
  (d) the wave-mechanics model
- explain the concept of *shells* for atomic electrons
- explain the term *quantum numbers.*

## Prerequisites

Lesson 2 on energy levels and Lesson 2 of Chapter 5 on wave properties.

# STUDY

**I Read** about the Thomson plum-pudding model, the Rutherford model and the Bohr model of the atom.
References: *B* PP ch. 13, *B* ST7 ch. 3, *D* ch. 20, *N* ch. 40, *W et al.* ch. 47(44), ch. 48(47), *Wh&H* ch. 63.

**Self-assessment questions**

**4.1** The Thomson plum-pudding model for the atom had the atom's positive charge and mass spread throughout the entire volume of the atom, with the atomic electrons embedded in it. The Rutherford nuclear model had the atom's positive charge and mass concentrated in a small nucleus, with the electrons in the space surrounding the nucleus. Explain how the scattering of alpha particles by a thin sheet of metal enabled the Rutherford model to be accepted and the Thomson model rejected.

**4.2** According to the Rutherford model, what is the approximate size of (a) the atom and (b) the nucleus?

***4.3** Describe the essential features of the Bohr model of the atom. Explain how it accounts for line spectra.

****4.4** In 1896 J. R. Rydberg arrived at an equation to describe the series of lines that occur in the hydrogen spectrum, the equation being

$$f = cR\left(\frac{1}{m^2} - \frac{1}{n^2}\right)$$

where $c$ is the speed of light, $R$ a constant known as the *Rydberg constant*, $m$ is an integer which can have the value 1, 2, 3, etc. and $n$ is an integer having a value greater than that of $m$.
(a) When $m = 2$ the series of lines arrived at is the Balmer series. Derive the first three frequencies of that series. The value of $cR$ is $32.88 \times 10^{14}$ Hz.
(b) What is the series limit for the Balmer series of lines?
(c) Explain how the Bohr model was able to fit this equation.

**II Read** about the de Broglie equation, which gives the wavelength for 'particles', also the basis of the wave-mechanics model of the atom (not the differential equation).
References: *B* PP ch. 7, ch. 13, *B* ST7 ch. 1, ch. 3, *D* ch. 20, *N* ch. 40, *W et al.* ch. 47(45), *Wh&H* ch. 37.

**Self-assessment questions**

**4.5** State the *de Broglie equation.*

***4.6** What is the wavelength of electrons travelling with a velocity of $1.0 \times 10^6$ m s$^{-1}$? ($h = 6.6 \times 10^{-34}$ J s, $m = 9.1 \times 10^{-31}$ kg)

****4.7** Consider an electron in an atom to be rather like a string fixed at both ends and having a length equal to the diameter of the atom. By considering the modes of vibration of such a string explain why only certain electron energies are possible in an atom. Hint: consider the standing waves possible and hence the possible electron wavelengths.

**III Read** about the concept of shells for atomic electrons and quantum numbers. The reading may best be done in the context of the Bohr atom.
References: *B* ST7 ch. 3, *D* ch. 20.

**Self-assessment questions**

**4.8** The shell model for the arrangement of the atomic electrons postulates, in its simplest form, that the first shell could contain as a maximum just 2 electrons, the second shell could contain a further 8 electrons, the third shell a further 18. In general a shell could contain $2n^2$ electrons, where $n$ is the *principal quantum number* for the shell, i.e. first shell $n = 1$, second shell $n = 2$, etc. The

shells are also labelled in ascending order as K, L, M, N, etc.
(a) Lithium has three electrons. In what shells will its electrons be?
(b) Neon has 10 electrons. In what shells will its electrons be?
(c) The chemical behaviour of an atom is related to the number of electrons in its outer shells. Those atoms with complete outer shells are inert, those with just one electron in the outer shell are very reactive. On this basis state the chemical properties that might be expected of lithium and neon.

****4.9** Explain how quantum numbers arise in the wave-mechanics model of the atom.

# SELF-ASSESSMENT

## Key points

The **Thomson plum-pudding model** of the atom had the atom's positive charge and mass spread throughout the entire volume of the atom, with the atomic electrons embedded in it.
The **Rutherford nuclear model** had the atom's positive charge and mass concentrated in a small nucleus, with the electrons in the space surrounding the nucleus.
The **Bohr model** proposed a nucleus with the electrons in orbits around it. Electrons are only allowed to revolve in certain 'allowed' orbits. When in one of these orbits the electron does not emit any radiation; it has a definite, fixed amount of energy. This means that as only certain orbits are allowed the electrons can only have certain allowed energies. An electron can jump from one orbit to another; when it jumps to a lower-energy orbit the energy released is emitted as a photon.
The **wave-mechanics model** is based on the wave nature of electrons, the wavelength being given by the **de Broglie equation** $\lambda = h/mv$. Within an enclosure, the atom, only certain wavelengths give rise to standing waves. These are the permissible wavelengths, which give the permissible energies for the electrons.
Both the Bohr atom and the wave-mechanics atom consider electrons to exist in **shells**, each shell being given a **principal quantum number** $n$. The number of electrons that can exist in a shell is given by $2n^2$.

## Answers I

**4.1** Some alpha particles were scattered through very large angles. This was not feasible if the alpha particles only encountered diffuse matter but was possible if small centres of high mass were encountered.

**4.2** Atom, about $10^{-10}$ m; nucleus, about $10^{-15}$ m.

**4.3** See the key points. It accounts for line spectra in that emissions only occur between defined energy levels, the energies of the electron orbits. Using $E = hf$ these energies give rise to defined frequencies.

**4.4** (a) $f_1 = cR(1/2^2 - 1/3^2) = 4.567 \times 10^{14}$ Hz, $f_2 = cR(1/2^2 - 1/4^2) = 6.165 \times 10^{14}$ Hz, $f_3 = cR(1/2^2 - 1/5^2) = 6.905 \times 10^{14}$ Hz
(b) $f_{lim} = cR(1/2^2) = 8.220 \times 10^{14}$ Hz
(c) Bohr considered that the two terms in the equation represented different orbits with their different energies: $E_n - E_m = hf = hcR(1/m^2 - 1/n^2) = hcR/m^2 - hcR/n^2$. Thus with each orbit we can assign an energy of $-hcR/n^2$. Bohr was able to derive the Rydberg constant from a consideration of the energy of an electron in an orbit, obtaining a very good fit with $R = me^4/(8\epsilon_0^2h^3c)$.

## Answers II

**4.5** $\lambda = h/mv$

**4.6** $\lambda = 6.6 \times 10^{-34}/(9.1 \times 10^{-31} \times 1.0 \times 10^6) = 7.3 \times 10^{-10}$ m

**4.7** For standing waves to occur between two walls a distance $L$ apart, only certain wavelengths are possible: $\lambda = 2L, \lambda = L, \lambda = \frac{2}{3}L, \lambda = \frac{1}{2}L$, etc. This condition can be written as $n\lambda = 2L$, where $n$ is an integer. Since $\lambda = h/mv$ then $E = \frac{1}{2}mv^2 = h^2/(2m\lambda^2)$ and so $E = h^2n^2/(8mL^2)$. Thus without postulating the need for energy levels they have automatically emerged from a consideration of standing waves. The above model is very simple and a more complex model is needed to arrive at the energy levels of atoms; however, it illustrates the principles.

## Answers III

**4.8** (a) K: 2, L: 1 (b) K: 2, L: 8 (c) Lithium very reactive, neon inert.

**4.9** From a consideration of the possible standing waves, see SAQ 4.7.

# Lesson 5 Nuclear physics

## Aims

After completing this lesson you should be able to

- distinguish between *alpha*, *beta* and *gamma* radiations and describe the evidence which establishes their nature
- describe methods of detecting the radiations
- state the nuclear changes that occur as a result of such emissions
- explain the term *isotope*
- state and use the exponential radioactive decay law, defining such terms as *activity* and *half-life*
- explain the term *binding energy* and solve problems involving energy in nuclear reactions.

## Prerequisites

A basic knowledge of the deflection of charged particles in fields (Lesson 8, Chapter 6), wave properties (Lesson 2, Chapter 5), ionization (Lesson 2 of this chapter) and the exponential function.

## STUDY

**I Read** about alpha, beta and gamma radiations, their properties, the evidence which established their nature, and methods of detecting them.
References: *B* PP ch. 13, *B* ST7 ch. 2, *D* ch. 22, *N* ch. 41, *W et al.* ch. 49(46), *Wh&H* ch. 64.

### Self-assessment questions

***5.1** Compare alpha, beta and gamma emissions with respect to (a) the range they can travel in air, (b) the thickness of material required to stop them, (c) the amount of ionization produced in air, (d) their deviation by electric and magnetic fields, (e) their nature.

***5.2** Explain the evidence which results in an identification of alpha particles as helium nuclei.

***5.3** Describe experiments that could be used to (a) determine the range of alpha particles in air, (b) verify that the intensity of gamma radiation varies inversely as the square of the distance from the source, (c) investigate the absorption of beta particles from a source by different thicknesses of aluminium.

**5.4** Describe the types of tracks seen in a cloud chamber by (a) alpha particles and (b) beta particles.

***5.5** Explain the principle of the Geiger–Müller tube.

***5.6** Alpha particles of energy 5 MeV enter a magnetic field of 0.1 T at right angles to their motion. What will be the radius of curvature of the alpha particles' path in the field? Mass of alpha particle = $6.7 \times 10^{-27}$ kg, $e = 1.6 \times 10^{-19}$ C.

***5.7** Alpha particles of energy 5 MeV have a range in air of 65 mm. How many ion pairs will be produced on average per mm along this track if 30 eV is required to ionize the oxygen or nitrogen in the air?

**II Read** about the nuclear changes that take place as a result of alpha, beta or gamma emissions, explaining the term *isotope*.
References: *B* PP ch. 13, *B* ST7 ch. 2, *D* ch. 22, *N* ch. 41, *W et al.* ch. 49(48), *Wh&H* ch. 64.

### Self-assessment questions

***5.8** What happens to the atomic number and the atomic mass when (a) an alpha particle, (b) a beta particle, (c) a gamma photon is emitted?

**5.9** $^{226}_{88}$Ra represents a particular isotope of radium.
(a) How many protons does the nucleus have?
(b) How many neutrons does the nucleus have?
(c) What is the atomic number?
(d) What is the atomic mass?
(e) Which one of the following is an isotope of the radium, $^{222}_{86}$X or $^{221}_{88}$Y?

***5.10** (a) $^{232}_{90}$Th decays by alpha emission. What is the new nucleus?
(b) $^{228}_{89}$Ac decays by beta emission. What is the new nucleus?

***5.11** The first part of the decay series of $^{232}_{90}$Th involves the sequence alpha,beta,beta,alpha,alpha. Plot a graph showing how the number of neutrons varies with the atomic number during this decay series.

**III Read** about the exponential radioactive decay law, including such terms as *activity* and *half-life*.
References: *B* PP ch. 13, *B* ST7 ch. 2, *D* ch. 22, *N* ch. 41, *W* (ch. 46), *Wh&H* ch. 64.

### Self-assessment questions

**5.12** Define *activity* and state its unit.

**5.13** How many disintegrations are produced per second by a radioactive source having an activity of $5 \times 10^4$ Bq?

**5.14** (a) How is the activity of a radioactive isotope related to the number of atoms of that isotope present?
(b) Define the *decay constant*.
(c) State the relationship for the number of atoms of a radioactive isotope as a function of time.
(d) Define *half-life*.

***5.15** A radioactive isotope has a half-life of 60 days and initially has an activity of 6000 disintegrations per second. What will be the activity after (a) 60 days, (b) 120 days have elapsed?

***5.16** What is the half-life of a radioactive source which has decayed to 1/32nd of its initial activity after 25 days?

***5.17** The count rate was measured at various times for a radioactive source and the following results obtained.

| Time/min | 0 | 1 | 2 | 3 | 4 | 5 |
|---|---|---|---|---|---|---|
| Count rate/$s^{-1}$ | 10100 | 5350 | 3070 | 1650 | 890 | 500 |

Estimate the half-life.

***5.18** The radium isotope $^{224}_{88}Ra$ decays by alpha emission and has a half-life of 3.6 days. What is (a) the initial number of atoms of this isotope in a sample of $1.0 \times 10^{-3}$ g, (b) the number of atoms of the isotope after 14.4 days?
(Avogadro number $= 6.0 \times 10^{23}$)

**IV Read** about binding energy, nuclear stability, fission and fusion.
References: *B* PP ch. 13, *B* ST6 ch. 4, *D* ch. 22, *N* ch. 41, *W et al.* ch. 49(48), *Wh&H* ch. 63, ch. 64.

**Self-assessment questions**

**5.19** Define *binding energy*.

***5.20** Sketch a graph showing how the binding energy per nuclear particle (nucleon) varies with mass number. Use the graph to explain how nuclear fission and nuclear fusion are possible.

***5.21** Calculate the energy released in the following decay.

$$^{238}_{92}U \rightarrow {}^{234}_{90}Th + {}^{4}_{2}He$$

Mass of $^{238}U$ = 238.124 92 u, mass of $^{234}Th$ = 234.116 50 u, mass of $^{4}He$ = 4.003 87 u. The mass is measured in *unified atomic mass units* (u), with 1 u being 1/12th the mass of the carbon-12 atom. This mass is equivalent to an energy of 931 MeV (obtained using $E = mc^2$).

# SELF-ASSESSMENT

## Key points

**Alpha particles** are helium nuclei, **beta particles** are high-speed electrons and **gamma radiation** consists of electromagnetic waves with wavelengths shorter than $10^{-10}$ m.
Alpha radiation is intensely ionizing and penetrates only a few centimetres of air. Beta particles are less ionizing, and while low-energy beta are less penetrating than alpha particles, higher-energy beta can penetrate a few millimetres of aluminium. Gamma radiation is even less ionizing and is able to penetrate significant thicknesses of matter.
Alpha emission causes the atomic number to decrease by 2 and the atomic mass to decrease by 4. Beta emission causes the atomic number to increase by 1 and gives no significant change in atomic mass. Gamma emission has no effect on the atomic number nor any significant effect on the atomic mass.
**Isotopes** are atoms that have the same atomic number but different atomic mass.
**Activity** $= -dN/dt = \lambda N$, where $\lambda$ is the **decay constant** and $N$ is the number of active nuclei. Hence $N = N_0 e^{-\lambda t}$.
The **half-life** is the time taken for the number of active nuclei to be reduced by a half. Half-life $= (\ln 2)/\lambda$.
The energy needed to take all the nucleons apart so that they are completely separated is called the **binding energy**. The total mass of the separated nucleons is greater than the total mass of the nucleus, this difference in mass being a measure of the binding energy.

## Answers I

**5.1** (a) Alpha, a few centimetres; beta, from a few centimetres up to a few metres depending on the beta energy; gamma, very long distances.
(b) Alpha, a thin aluminium foil; beta, from a thin aluminium foil up to a few millimetres; gamma, several centimetres of aluminium or lead.
(c) Alpha, highly ionizing; beta, less ionizing; gamma, least ionizing.
(d) Alpha and beta are deviated, in opposite directions, while gamma is not deviated.
(e) Alpha, helium nuclei; beta, fast electrons; gamma, electromagnetic waves.

**5.2** Deflection in electric and magnetic fields enables the charge-to-mass ratio to be measured; it is the same as that of helium nuclei. Rutherford and Royd carried out an experiment in which they collected alpha particles in an evacuated space. They then found that the helium line spectrum was given.

**5.3** (a) Figure A5.1 shows how an ionization chamber with a d.c. amplifier could be used. The ionization current in a gauze ionization chamber is measured as a function of the distance of the source from the gauze. The *range* is the distance at which the ionization current drops to virtually zero.

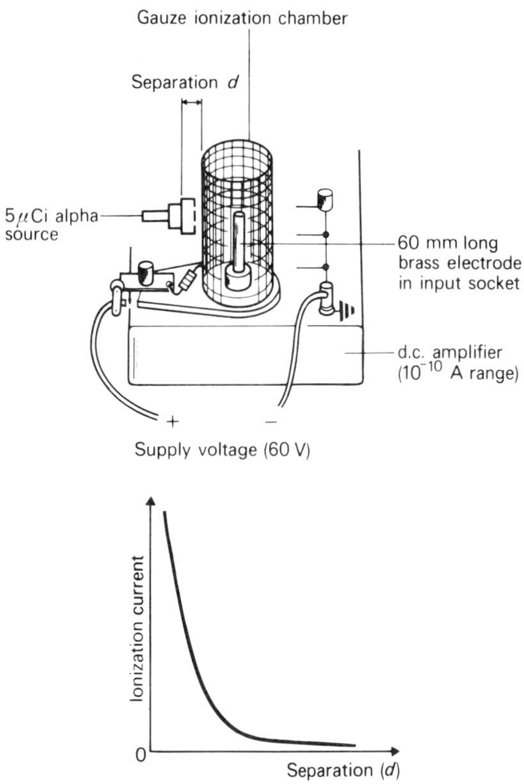


**Figure A5.1 Determination of range of alpha particles**

(b) The count rate obtained with a Geiger–Müller tube placed different distances from a gamma source can be measured. After correcting for the background count, a graph of distance $d$ plotted against $1/C^{\frac{1}{2}}$, where $C$ is the corrected count rate, will yield a straight line. It is preferable to plot the graph in this form rather than $C$ against $1/d^2$, since errors in $d$ can become very significant close to the source and affect this graph.
(c) A Geiger–Müller tube can be positioned a fixed distance from a beta source and the count rate determined when different thicknesses of aluminium are placed between them. After correcting for the background a graph can be plotted of corrected count rate against absorber thickness.

**5.4** (a) Fairly straight, heavy-density tracks.
(b) Short curly tracks of lower density than the alpha tracks.

**5.5** The Geiger–Müller tube consists of a cylindrical metal cathode and a coaxial wire anode. The tube contains argon at a low pressure. A potential difference of about 450 V is maintained between the electrodes. When an ionizing particle passes through the tube, ion pairs are produced in the argon. The electrons accelerate towards the anode and gain sufficient energy to produce further ion pairs. The electrons resulting from this produce yet further ion pairs. The result is an avalanche of electrons reaching the anode and a large pulse of anode current. The positive ions produced in the ion-pair events move more slowly towards the cathode. They are prevented from causing a further avalanche by the presence in the tube of a *quenching agent*, such as bromine: the positive ion energy is used to break apart the molecules of the bromine and so becomes dissipated.

**5.6** $Bqv = mv^2/r$, hence $r = mv/Bq$. As $E = \frac{1}{2}mv^2$ then $r^2 = 2mE/B^2q^2 = 2 \times 6.67 \times 10^{-27} \times 5 \times 10^6 \times 1.6 \times 10^{-19}/(0.1^2 \times 1.6^2 \times 10^{-38}) = 41.9$ m.

**5.7** Total number of ion pairs = $5 \times 10^6/30 = 1.7 \times 10^5$. Average number per mm = $2.6 \times 10^3$.

## Answers II

**5.8** See the key points.

**5.9** (a) 88 (b) 226−88 = 138 (c) 88 (d) 226
(e) The one with the same atomic number, i.e. Y.

**5.10** (a) $^{228}_{88}$Ra (b) $^{228}_{90}$Th

**5.11** See Figure A5.2.

## Answers III

**5.12** *Activity* is the rate of disintegration and the unit is the Becquerel (Bq), with 1 Bq = 1 disintegration per second.

**5.13** $5 \times 10^4$

**5.14** (a) Activity = $dN/dt \propto N$ (b) $dN/dt = -\lambda N$, where $\lambda$ is the decay constant (c) $N = N_0e^{-\lambda t}$ (d) The half-life is the time taken for the number of active atoms, or the activity, to decrease to half its initial value.

**5.15** (a) 3000 disintegrations per second. (b) 1500 disintegrations per second, i.e. half of half the initial activity.

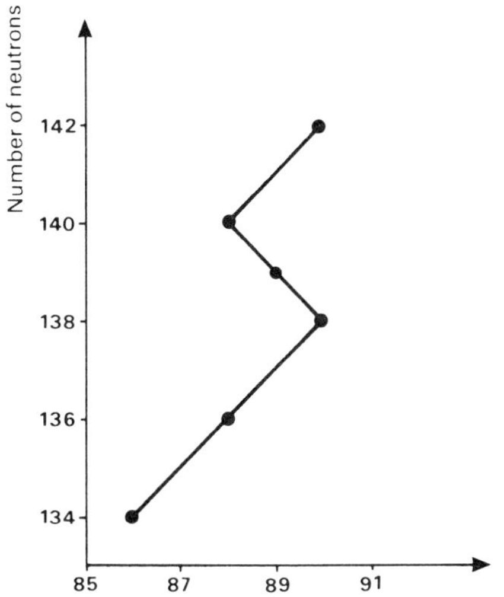

**Figure A5.2**

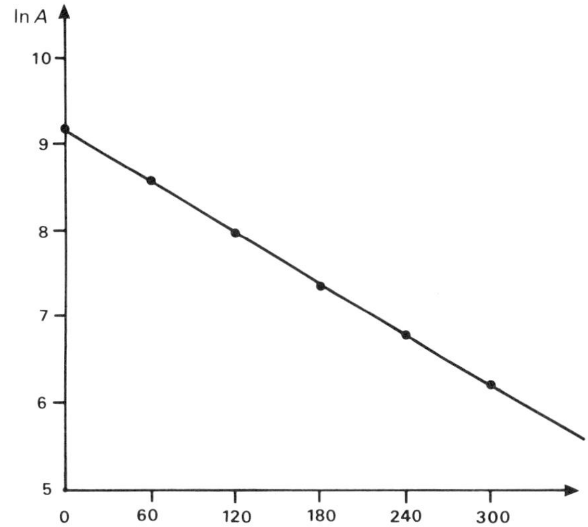

**Figure A5.3**

**5.16** $1/32 = \frac{1}{2} \times \frac{1}{2} \times \frac{1}{2} \times \frac{1}{2} \times \frac{1}{2}$. Thus five half-lives equal 25 days and so the half-life is 5 days.

**5.17** The count rate is the activity. One method would be to plot the activity against the time and by inspection of the exponential graph determine for a number of pairs of points the time taken for the activity to decrease by a half. Another possibility is to plot a log graph. Activity $A = dN/dt = -\lambda N = -\lambda N_0 e^{-\lambda t} = A_0 e^{-\lambda t}$. Thus $\ln A = \ln A_0 - \lambda t$. A graph of $\ln A$ against $t$ will be a straight line with a slope of $-\lambda$. Figure A5.3 shows the graph. $\lambda = 3.0/300 = 1.0 \times 10^{-2}\,s^{-1}$. Half-life $= (\ln 2)/\lambda = 69.3\,s$.

**5.18** (a) 228 g contains $6.0 \times 10^{23}$ atoms, hence the initial number is $2.6 \times 10^{18}$.

(b) 14.4 days is four half-lives, hence number of active nuclei is $1.6 \times 10^{17}$.

## Answers IV

**5.19** See the key points.

**5.20** See Figure A5.4. The nuclei with the highest binding energy and hence the most stable nuclei are those with mass numbers from about 40 to about 120. When heavy nuclei undergo fission the nuclei produced are more stable. Similarly, with fusion of low-mass-number nuclei the result is more stable nuclei.

**5.21** Mass lost = 238.124 92 − 234.116 50 − 4.003 87 = $4.55 \times 10^{-3}$ u. This shows as an energy of $4.55 \times 10^{-3} \times 931 = 4.24$ MeV.

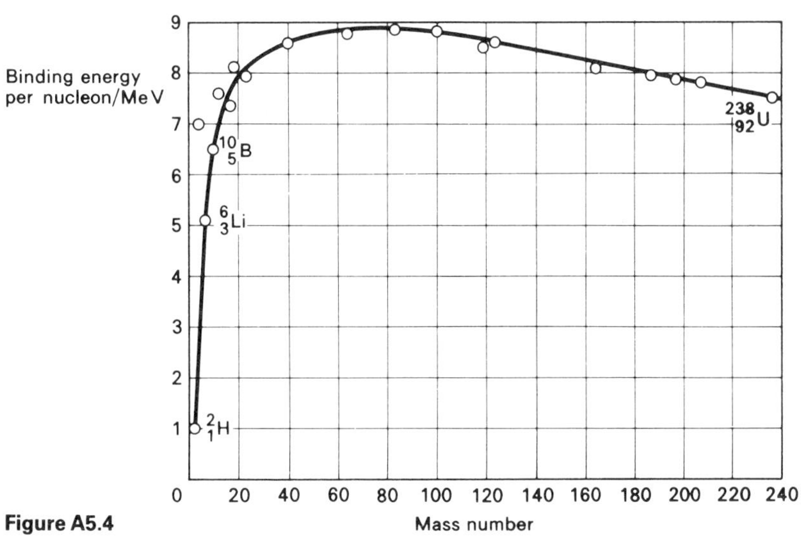

**Figure A5.4**

# Index

# Index